Lecture Notes in Chemistry 62

S. J. Cyvin   J. Brunvoll   R. S. Chen
B. N. Cyvin   F. J. Zhang

Theory of Coronoid Hydrocarbons II

Springer-Verlag Berlin Heidelberg GmbH

Authors

S. J. Cyvin
J. Brunvoll
B. N. Cyvin
Department of Physical Chemistry
The University of Trondheim
N-7034 Trondheim-NTH, Norway

R. S. Chen
College of Finance and Economics
Fuzhou University
Fuzhou, 350002 Fujian
The People's Republic of China

F. J. Zhang
Department of Mathematics
Xinjiang University
Wulumuqi (Urumchi), 830046 Xinjiang
The People's Republic of China

ISBN 978-3-540-58138-3   ISBN 978-3-642-50157-9 (eBook)
DOI 10.1007/978-3-642-50157-9

Cip data applied for

© Springer-Verlag Berlin Heidelberg 1994
Originally published by Springer-Verlag Berlin Heidelberg New York in 1994

Typesetting: Camera ready by author
SPIN: 10128834   51/3140 - 543210 - Printed on acid-free paper

PREFACE

The present monograph is a continuation of Cyvin SJ, Brunvoll J and Cyvin (1991c), a reference to be found in Bibliography. Naturally, the previous volume is cited frequently here. For the sake of brevity, it is referred to as "Volume I". References to different chapters, sections or paragraphs are given like Vol. I–1, I–1.2 or I–1.2.2, respectively. Also tables and equations in "Volume I" are cited; the very last equation therein, for instance, is Vol. I–(9.9).

The present text spans from references to organic syntheses or attempted organic syntheses – – to stringent mathematical theorems proved by graph–theoretical methods. Enumerations of coronoid systems is a substantial part of the work. Algebraic methods involving combinatorics and generating functions are employed on one hand, and computer programming on the other. The whole book is supposed to demonstrate a piece of mathematical chemistry, which can be characterized as lying on the "interfaces between mathematics, chemistry and computer science", a formulation used for the MATH/CHEM/COMP Conferences; cf. Cyvin SJ, Brunvoll and Cyvin (1989d) in Bibliography.

Financial support to BNC from the Norwegian Council for Science and the Humanities is gratefully acknowledged.

March 1994   *Sven Cyvin*

CONTENTS

Chapter 1

INTRODUCTION AND CHEMICAL RELEVANCE

1.1 Reiteration

It is repeated from Vol. I–1.1 that only two coronoid hydrocarbons have been synthesized and belong to a class called cycloarenes (Staab and Diederich 1983). They have also been called coronaphenes (Peter and Jenny 1966). These compounds are cyclic catacondensed benzenoid rings forming a macrocyclic system so that a cavity is present, into which carbon–hydrogen bonds are pointing. The two chemically known polycyclic conjugated hydrocarbons of this category are represented as chemical graphs (Trinajstić 1992b) by C_1 and C_2 in Fig. 1, while a synthesis of C_3 has been attempted.

1.2 Motivation

1.2.1 *General Viewpoints*

Polansky and Rouvray (1976) started their series of pioneering papers, entitled "Graph–Theoretical Treatment of Aromatic Hydrocarbons", at a time when no coronoid hydrocarbon was known chemically. This fact did not prevent the authors to include corona–condensed systems (coronoids) in their first paper in the series (Polansky and Rouvray

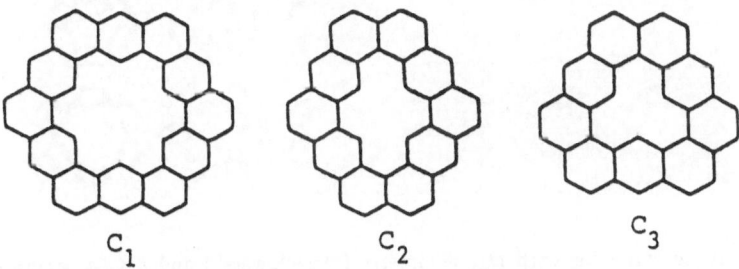

Fig. 1.1. Three cycloarenes (macrocyclic conjugated hydrocarbons):
$C_1 - C_{48}H_{24}$ cyclo[$d.e.d.e.d.e.d.e.d.e.d.e$]dodekakisbenzene, kekulene ([12]coronaphene);
$C_2 - C_{40}H_{20}$ cyclo[$d.e.d.e.e.d.e.d.e.e$]decakisbenzene ([10]coronaphene);
$C_3 - C_{36}H_{18}$ cyclo[$d.e.e.d.e.d.e.e$]nonakisbenzene ([9]coronaphene).

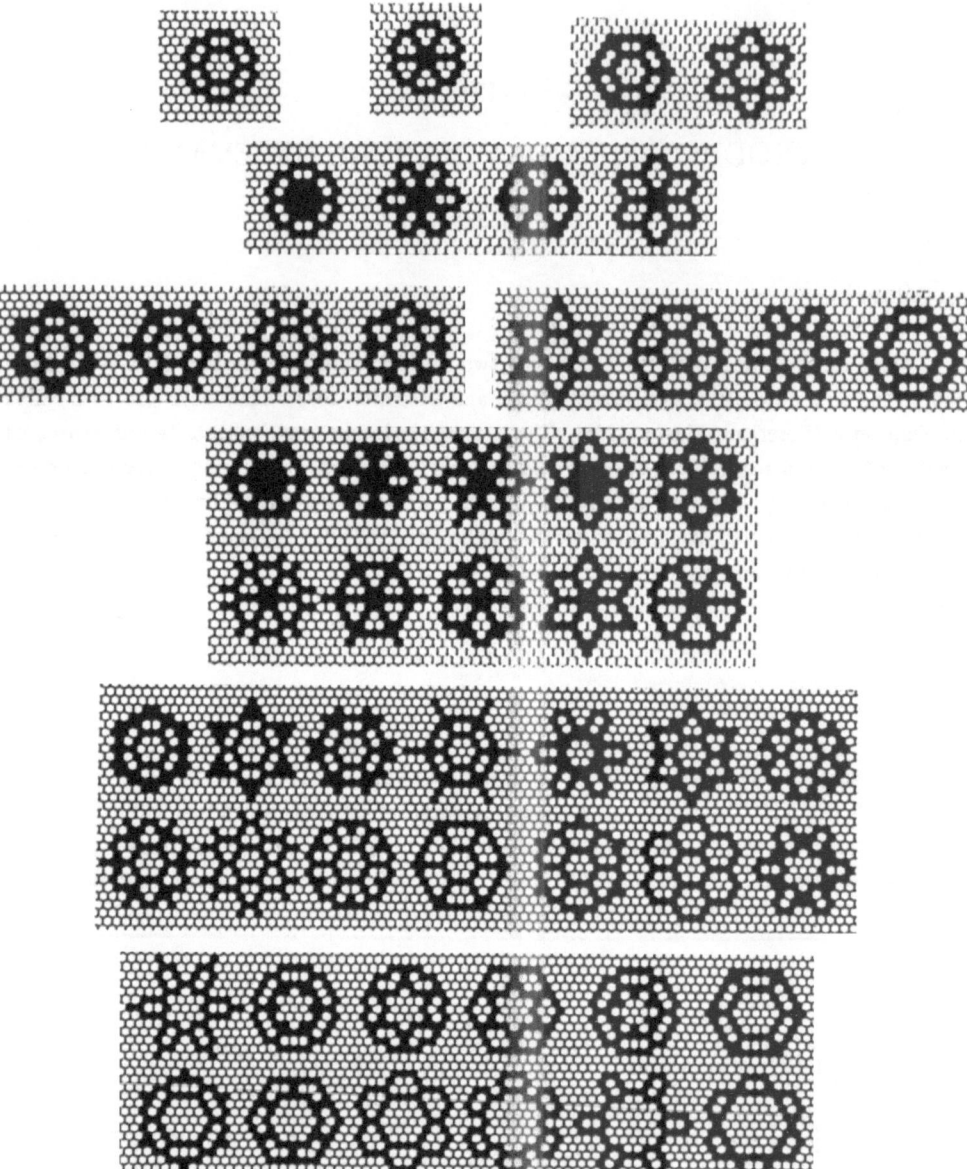

Fig. 1.2. Multiple coronoids with D_{6h} symmetry ("laceflowers") and $h \leq 54$, where h designates the number of hexagons.

1976). Then Balaban and Harary (1968) had already introduced the class of corona—condensed systems. They depicted exactly the systems C_1, C_2 and C_3 of Fig. 1. Polansky and Rouvray (1977) pursued their graph—theoretical studies in a third paper devoted entirely to coronoids. Finally, in the fourth paper of the mentioned series, Polansky and Gutman (1979) could write with reference to a synthesized cycloarene: "Die einzige bis jetzt bekannte Verbindung dieser Art ist der vor einiger Zeiten dargestellte Kekulen." From this description it should be clear that the research in mathematical chemistry does not necessarily presuppose the chemical existence of the relevant compounds in order to be interesting. However, this view is not generally accepted among chemists.

These thoughts about the motivations in mathematical chemistry were elaborated in an introduction by Cyvin SJ, Cyvin and Brunvoll (1989b). The authors start saying: "It seems to be a widespread opinion among chemists that a problem in mathematical chemistry must have chemical relevance in order to be worthwhile considering." Further on they give an example: "a recent work accounts for 1 075 554 systems of primitive coronoids .. [Brunvoll, Cyvin BN, Cyvin, Gutman, Tošić and Kovačević (1989)]. So far *two* [italicized here] of the corresponding hydrocarbons have been synthesized after a great deal of trouble (during 21 years). Here we wish to emphasize that the skill of the preparative chemist is irrelevant for our motivation of choosing the problem, in which we find interest." In summary, we are advocating that a purely mathematical interest should be accepted as a worthy motivation in mathematical chemistry. However, this is not supposed to imply any discrimination of the chemical interest, which also is important.

A third aspect has been drawn to attention (Cyvin SJ, Cyvin and Brunvoll 1989b): "Apart from the mathematical and chemical interest .... we wish to point out the aesthetic motivation." It was perhaps Hosoya (1986a) who first mentioned the aesthetic aspect in the context of mathematical chemistry: "Thus, all the discussions in this paper have been performed from purely mathematical and, if possible, aesthetic points of view."

Multiple coronoids (systems with more than one hole each) of D_{6h} symmetry have been called laceflowers (Cyvin SJ, Brunvoll and Cyvin 1989a; Cyvin BN, Brunvoll, Chen and Cyvin 1993). The two cited references contain depictions of these pretty systems up to 54 hexagons; they are reproduced in Fig. 2. Also the aesthetic qualities of fractal benzenoids (Klein, Cravey and Hite 1991) can hardly be denied. A representative is shown in Fig. 3. The particular set to which it belongs, starts with the two benzenoids C_6H_6 (benzene) and $C_{24}H_{12}$ (coronene). Then comes the single coronoid $C_{132}H_{48}$, while all higher members are multiple coronoids and actually laceflowers. These systems have been called coro[n]enes (Klein, Živković and Balaban 1993); in coro[k]ene, $k = 0$ and 1 correspond to benzene and coronene, respectively. The chemical formulas, C_nH_s, for members of this particular set obey the recurrence relations (Klein, Cravey and Hite 1991)

$$n_k = 6n_{k-1} - 12 \; , \; s_k = 6s_{k-1} - 24 \quad (k > 0) \tag{1.1}$$

with the initial conditions $n_0 = s_0 = 6$ (C_6H_6).

Fig. 1.3. $C_{4668}H_{1560}$ coro$^{[4]}$ene: a member of fractal benzenoids, which is a laceflower.

In general, for $k \geq 0$:

$$n_k = (3/5)(6^{k+1} + 4) \ , \ s_k = (1/5)(6^{k+1} + 24) \tag{1.2}$$

The $C_n H_s$ formula in the legend of Fig. 3 pertains to $k = 4$.

1.2.2 Alkane Isomers

In the present paragraph it is departed from coronoid hydrocarbons. Some historical aspects of the enumeration of $C_N H_{2N+2}$ alkane isomers are exposed. This problem is usually the first enumeration problem encountered in an elementary course of organic chemistry: one methane, one ethane, one propane, two butanes, three pentanes, five hexanes, etc. Yet it is a relevant background to the topics of the present book, especially because more than a whole chapter (the largest one) is devoted to isomer enumeration.

It is not the intention to repeat in details the story of alkane isomer enumeration, since this has been done several times before. One of the papers cited above (Cyvin SJ, Cyvin and Brunvoll 1989b) contains a part of this story; otherwise the reader is referred to a review by Rouvray (1974) and to selected chapters of two recent monographs: Rouvray (1991); Balaban (1991); Trinajstić, Nikolić et al. (1991). Instead, we wish to use the alkane example for illustration of different viewpoints on mathematical chemistry. The historical quotations, which are offered in the following, go back to more than a hundred years ago. Nevertheless, much of the same arguments can be heard even today. It seems that this discussion about motivations in mathematical chemistry will never end.

During the studies of the problem in question perhaps the first substantial interaction between mathematics and chemistry took place. The mathematician Arthur Cayley (1821–1895) published a paper on the number of alkane isomers in a chemical journal (Cayley 1875). This, in a sense, was the start of a special branch of mathematical chemistry (long before this term was coined) and perhaps computational chemistry (without computers). Thus the work of Cayley (1875) had an immense impact on scientific research. Furthermore, it stirred up a hot debate, partly because of two numerical errors in Cayley's analysis, but also on more principal grounds.

Schiff (1875a, 1875b) assumed a positive attitude to a famous mathematician's penetration into chemistry. He envisages at once a new chapter of future chemistry (Schiff 1875a): "Man kann dieses der Zukunftchemie angehörige neue Kapitel der Stöchiometrie als 'c h e m i s c h e S t a t i s t i k' bezeichnen." Furthermore, Schiff (1875a) announces that he must postpone the publication of his enumeration method and gives full tribute to Cayley in the following words:

"Im Augenblicke ist es mir aber kaum möglich, mich so vollständig in jenen Gedankengang zu versetzen, und ich muss mir die Mittheilung der Methode auf eine spätere Zeit versparen, ohne dass ich damit die Mittheilung weiterer ähnlicher Untersuchungen C a y l e y 's auch nur im Geringsten verzögern möchte. Wir Chemiker dürfen es im Gegentheil mit Dank anerkennen, dass ein so ausgezeichneter Mathematiker sich unserer Wissenschaft angenommen hat und wir hoffen, dass er auf dem eingeschlagenen Wege fortfahren und auch künftighin seine Thätigkeit der Behandlung chemischer Fragen zuwenden möge."

In a subsequent paper, Schiff (1875b) makes interesting confessions. In the beginning he considered the enumeration of isomers as scientific playing ("wissenschaftliche Spielerei"):

"Ein reichliches, auf chemische Statistik sich beziehendes Material, hat sich mir während mehrerer Jahre auf Spaziergängen und als J e a n P a u l 'sches Mittel zum Einschlafen angesammelt. Absichtlich mache ich diese ganz persönliche Bemerkung, gerade um einzuge— stehen, dass ich diese Beschäftigung nicht als eine ernsthafte, sondern als eine wissenschaftliche Spielerei betrachtete und dies ist auch der Grund, weshalb ich mich bisher nicht entschliessen konnte, darüber etwas zu veröffentlichen."

But now Schiff (1875b) became convinced through Cayley's engagement, about the justification of this kind of research. The following unique period deserves to be cited in extenso:

"Wenn ich nun heute diesen Anstand überwinde und mir erlaube anzudeuten, wie Probleme chemischer Statistik in einer den Postulaten des Chemikers sich anpassenden Weise gelöst werden können, so mag mir als Entschuldigung dienen, dass ein so hervorragender Mathematiker wie C a y l e y es nicht verschmäht hat, diesen Fragen seine Aufmerksamkeit zuzuwenden; aber man wird es auch entschuldigen, wenn ich mich damit begnüge, diese Methoden nur anzudeuten, und es vorerst nicht für geboten erachte, dem besprochenen Gegenstande eine ausführlichere Darlegung zu widmen."

Below we quote a remark by Flavinskii (1876) on the treatise of Schiff (1875b). The argumentation is not quite clear to us, but it seems that Flavinskii is positive:

"Die genaue Bestimmung der Zahl möglicher Isomere können wir kaum 'als eine wissenschaft— liche Spielerei' betrachten, denn die grosse Zahl derselben macht es kaum möglich, alle Isomere zum Gegenstand der Untersuchung zu machen und man wird sich nur auf solche Fälle, die Stoff zu allgemeinen Folgerungen zu geben versprechen, beschränken."

On the other hand, it is no doubt that Herrmann (1880) is opposed to accepting the mathematical methods in question for publication in a chemical journal. He only published a short note prompted by two errors in Cayley (1875), viz. 357 and 799 for the numbers of $C_{12}H_{26}$ and $C_{13}H_{28}$ isomers, respectively, which he corrected to 355 and 802. He said (Herrmann 1880): "Ich muss es mir versagen, an dieser Stelle die gefundene Berechnungsmethode auch nur dem Principe nach anzudeuten, da ein derartiger Stoff von dem Verhandlungsmaterial dieser Blätter zu weit abliegt." Several years later, Herrmann (1897), apparently provoked by a relatively long paper of Losanitsch (1897a), finds himself compelled to publish a few details on his enumeration method. But he makes clear that he has not changed his mind with regard to the justification of this kind of papers in a chemical journal (Herrmann 1897):

"Von einer Reihe von Jahren habe ich in diesen Berichten eine kurze Notiz .. [Herrmann (1880)] über das fragliche Problem erscheinen lassen un dabei angegeben, dass es mir gelungen sei, eine Methode ausfindig zu machen, welche in etwas concreterer Fassung als die C a y l e y 'sche die sichere Berechnung der isomeren Paraffine von der Formel C_nH_{2n+2} ermöglicht. Ich habe es mir damals versagt, die gefundene Berechnumgsmethode auch nur dem Principe nach anzudeuten, da ein derartiger Stoff von dem Verhandlungsmaterial dieser Blätter zu weit abliegt. Obgleich in dieser Beziehung meine Meinung nicht geändert ist, so möchte ich mir doch erlauben über die damals gefundene Berechnungsmethode einige Andeutungen zu geben, um den Lesern der Abhandlung von L o s a n i t s c h zu zeigen, auf wieviel einfachere und vor allen Dingen übersichtlichere Weise man durch Betrachtungen von gleich elementarer Natur zur Lösung des besprochenen Problems gelangen kann."

Here we cite a contribution from modern times to virtually the same discussion (Pólya and Read 1987):

"It must be admitted that many chemical enumerations that have appeared in the literature are more in the nature of academic exercises than results of practical interest to the chemist. Thus, for example, no chemist *really* needs to know that the number of alkanes having 60 carbon atoms is 22,158,734,535,770,411,074,184 ... [Perry (1932)]. However, the enumeration of compounds with a given frame can produce results of practical importance."

It is interesting that, from the very beginning, the importance of producing exact numbers for the isomers was recognized. We have heard that Herrmann (1880) corrected Cayley's number of 357 $C_{12}H_{26}$ alkane isomers to 355. Later on, the (erroneous) claim of Losanitsch (1897a) that this number should be 354, aroused a further debate (Herrmann 1897; 1898; Losanitsch 1897b).

After a long interim, the two chemists Henze and Blair (1931) published the first satisfactory general solution for the numbers of alkane isomers. It is observed that the pertinent chemical journal apparently had no objections against publishing really large numbers of isomers, far beyond those of pure chemical interest – including the number of twenty–three digits in the above citation from Pólya and Read (1987). Yet the view prevailed that the produced numbers should be exact. Quite adequately, Perry (1932) corrected the number 147284 for $C_{19}H_{40}$ from Henze and Blair (1931) to 148284 (but published himself two large numbers erroneously). In the same spirit, we find it adequate to point out that another wrong number from Henze and Blair (1931), viz. the fourteen–digit number for $C_{40}H_{82}$, has been reproduced in a modern book (Joesten et al. 1991); see below for the correct number.

It may be questioned whether the enumerations of alkane isomers referred to above may be considered as works in computational chemistry. However, this is certainly so for later works in this area (Lederberg et al. 1969; Davis et al. 1971; Knop, Müller, Jeričević and Trinajstić 1981), in which elaborate computer programs were employed. Here the publication of 1981 comes from the Düsseldorf–Zagreb group. A few years later, the same group announced their computations for alkane isomers up to $C_{280}H_{562}$ (Trinajstić, Jeričević et al. 1983). At that time Kornilov and Zamkovii (1981) had already published the pertinent numbers up to $C_{100}H_{202}$ in a Ukrainian journal. These "Ukrainian numbers" are quoted, unfortunately with some misprints, in a recent booklet (Papulov et al. 1990). One year later the "Düsseldorf–Zagreb numbers" up to $C_{100}H_{202}$ appeared in a monograph (Trinajstić, Nikolić et al. 1991); they are reproduced in Chart 1. It is noteworthy that the two sets of numbers under consideration (Kornilov and Zamkovii 1981; Trinajstić, Nikolić et al. 1991) agree digit by digit, except for a misprint in one single digit for $C_{99}H_{200}$ in Kornilov and Zamkovii (1981). It is emphasized that these Ukrainian– and Düsseldorf–Zagreb numbers were produced independently and by entirely different approaches.

Table III.8
The number of alkanes with N sites

N	Number of alkanes	N	Number of alkanes
1	1	51	2994664179967370611
2	1	52	8031081780535296591
3	1	53	21557771913572630901
4	2	54	57919180873148437753
5	3	55	155745431857549699124
6	5	56	419149571193411829372
7	9	57	1128939578361332867936
8	18	58	3043043571906827182530
9	35	59	8208615366863753915949
10	75	60	22158734535770411074184
11	159	61	59858097847706865855186
12	355	62	161805725349297357221898
13	802	63	437671691526158936922623
14	1858	64	1184616185385310843585573
15	4347	65	3208285066181475821271463
16	10359	66	8694130712024868414002815
17	24894	67	23573796134448175745408811
18	60523	68	63955159527348138708694312
19	148284	69	173603007393950249896865875
20	366319	70	471484798515330363034639871
21	910726	71	12811513151764638215613845510
22	2278658	72	3482965749140691245110434511
23	5731580	73	9473447386804490449091871124
24	14490245	74	25779306238954404972323916397
25	36797588	75	70183211512214096492433058105
26	93839412	76	1911156381393249393027319384769
27	240215803	77	520874195248906781713044332539
28	617105614	78	1419908915343952137338409797325
29	1590507121	79	3872282575137005474139119076135
30	4111846763	80	10564476906946675106953415600016
31	10660307791	81	28833609436277333169440806135431
32	27711253769	82	78725585464391037293036629979444
33	72214088660	83	215027809474796675607407513633870
34	188626236139	84	587531723826577193455385789266377
35	493782952902	85	1605913778494711520354663202536756
36	1295297588128	86	4391002908093323425994602631972445
37	3404490780161	87	12010257907756938974208750945664835
38	8964747474595	88	32861295558120887536942123568548502
39	23664748933969	89	8994095902489157699739691928932689
40	62481801147341	90	246245150242821439632304475956113295
41	165351455535782	91	674391606297983432514229725117306224
42	438242894769226	92	1847515048012613337782670842346319120
43	1163169707886427	93	5062818112121161180862827915688625902
44	3091461011836856	94	1387785752958452138432441995o411729295
45	8227162372221203	95	38051836070803837001309074456088423358
46	21921834086683418	96	104363664561059273927704242814298678658
47	58481806621987010	97	286312976836850192359345859166390622180
48	156192366647590639	98	785684759853087702778573182234297830503
49	417612400765382272	99	2156596319845084996862701478402986311496
50	1117743651746953270	100	5921072038125809849884993369103538010139

Chart 1.1. Numbers of alkane $C_N H_{N+2}$ isomers.

The absurdity of this kind of computations from a purely chemical point of view was pinpointed by Davies and Freyd (1989) in the very title of their contribution (accepted in a chemical journal!): "$C_{167}H_{336}$ Is the Smallest Alkane with More Realizable Isomers than the Observed Universe Has 'Particles'."

1.3 Cycloarenes

1.3.1 *The Story of Kekulene Revisited*

In Vol. I–1.2.1 the interesting story on the synthesis of $C_{48}H_{24}$ kekulene (C_1 of Fig. 1) was told very briefly. It seems to be warranted , in the first place, to fill out some details. Next, we shall report some more recent theoretical studies on kekulene.

In the terminology of Peter and Jenny (1966), kekulene is synonymous with [12]coronaphene. In some of the attempts to synthesize this molecule, $C_{48}H_{48}$ [2.2.2.2.2.2]–metacyclophane was employed (Jenny and Burri 1967; Burri and Jenny 1967). The attempts to produce "[12]coronaphene" directly by dehydrogenization of $C_{48}H_{48}$ turned out to be discouraging, but the authors managed to isolate another dehydrogenization product, viz. $C_{48}H_{36}$ [2.2.2.2.2.2]metacyclophane–*all–trans*–hexaene; see I of Fig. 4. The preparation of this compound was characterized as a significant step towards the synthesis of [12]coronaphene. Another isomer of $C_{48}H_{36}$, viz. [2.2.2.2.2.2]metacyclophane–*all–cis*–hexaene, depicted as II of Fig. 4, would seem to be a still more promising product on the route to [12]coronaphene, but Burri and Jenny (1967) find evidence against the formation of this compound. A few years later, Jenny and Paioni (1969), in a fresh approach to the synthesis of "[12]coronaphene", started from $C_{48}H_{36}$ hexahydro–[2.2.2](2,7)phenanthrenophane, from which they obtained $C_{48}H_{30}$ [2.2.2](2,7)–phenanthrenophane–triene; see III of Fig. 4. A spontaneous dehydrogenization of $C_{48}H_{30}$ yielded a product which actually might contain some [12]coronaphene, but a possible separation and

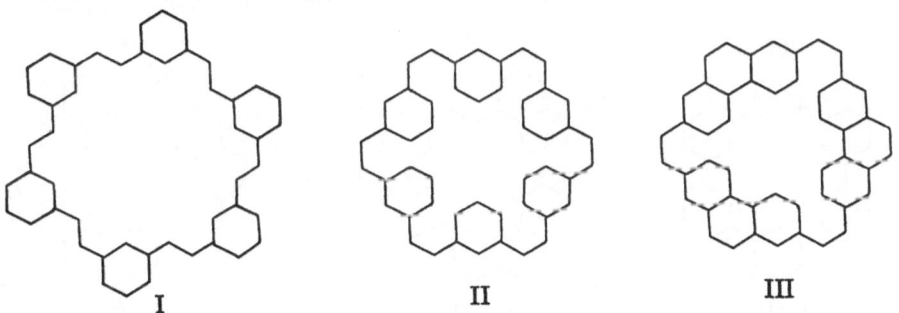

Fig. 1.4. Three macrocyclic conjugated hydrocarbons:

I – $C_{48}H_{36}$ [2.2.2.2.2.2]metacyclophane–*all–trans*–hexaene;

II – $C_{48}H_{36}$ [2.2.2.2.2.2]metacyclophane–*all–cis*–hexaene;

III – $C_{48}H_{30}$ [2.2.2](2,7)phenanthrenophane–*all–cis*–triene.

identification of this compound remained an open problem. In a paper containing a more detailed description of the synthesis of III (Fig. 4), there is no mentioning of [12]coronaphene (Baumgartner et al. 1971).

The activity of H. A. Staab with collaborators, which led to the definitely successful synthesis of cyclo[*d.e.d.e.d.e.d.e.d.e*]dodecakisbenzene (kekulene), went on parallel with the attempts of W. Jenny and collaborators described above. It is noted (Vögtle and Staab 1968) that the unambiguous nomenclature for cycloarenes was chosen according to an advice from "Beilstein–Institut, Frankfurt/Main". The macrocyclic compound $C_{36}H_{24}$ hexa–*m*–phenylene was available at a relatively early stage (Staab and Binnig 1964; 1967); see IV of Fig. 5. It may be referred to as a condensed annulene like the three structures of Fig. 4. Specifically, IV emerges from a condensation to the outside of [18]annulene. Another condensed [18]annulene, viz. V of Fig. 5. ($C_{42}H_{24}$), was prepared shortly after IV (Staab and Bräunling 1965; Staab, Bräunling and Schneider 1968). Notice that the skeleton of [18]annulene forms the boundary of the hole of kekulene, as was pointed out by Staab and Binnig (1967). However, a serious attempt to synthesize kekulene did not start from IV or V (Fig. 5). Vögtle and Staab (1968) reported on the dehydrogenization of a certain macrocyclic compound ($C_{48}H_{32}$), which probably yielded a mixture of $C_{48}H_{28}$ (VI of Fig 5) and $C_{48}H_{24}$ kekulene. However, a preparative separation was deemed to be impossible because of the small yield. So the main problem was yet unsolved. Not before ten years later, after a fresh approach to the problem, a successful synthesis of kekulene as greenish–yellow microcrystals was reported (Diederich and Staab 1978). Somewhat later a more detailed description of this refined synthesis in several steps was offered (Staab and Diederich 1983). Plavšić et al. (1992) have put it this way: "It took a while before the imaginative and clever synthesis design coupled with hard work produced kekulene, ... ". We shall not go into details here, but we wish to make a correction to Vol. I, where S_{12} in I–Fig. 1.9 was said to play a role in the synthesis of kekulene. This misunderstanding stems from an error in Staab and Diederich (1983), where S_{12} in one place was depicted instead of S_{11} (or VI of Fig. 5).

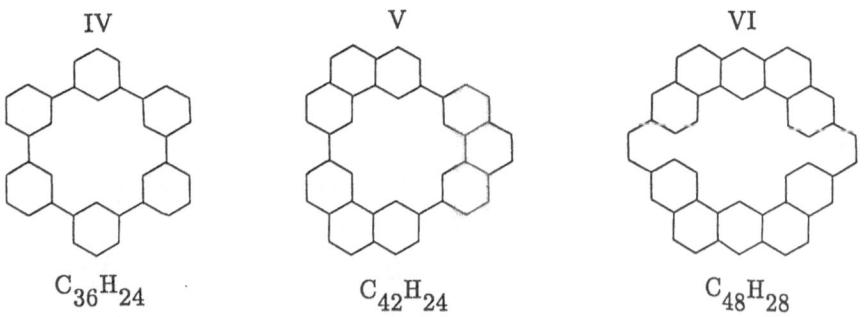

$$IV \qquad\qquad V \qquad\qquad VI$$

$$C_{36}H_{24} \qquad\qquad C_{42}H_{24} \qquad\qquad C_{48}H_{28}$$

Fig. 1.5. Additional three macrocyclic conjugated hydrocarbons.

In a footnote, Krieger et al. (1979) pointed out that ".... [Jenny et al.] claimed, without giving any details, to have prepared kekulene (*1*) — which they named '[12]coronaphene' — as long ago as 1970." Here Krieger et al. (1979) refer to a contribution by W. Jenny, D. Baumgarten and R. Paioni in "Proceedings of the International Symposium on the Nonbenzenoid Aromatic Compounds, Sendai, Japan (1970)"; p 183. This contribution has been cited by others (Aihara 1976; Bergan et al. 1986). In consequence, Krieger et al. (1979) found themselves compelled to repudiate the claim of Jenny, Baumgarten and Paioni by the following statement: "The authors named have until now published neither a synthesis of (*1*) nor any account of its properties in the chemical literature. Although they have not repeated their claim to the synthesis of (*1*) in later topically related publications, they have failed so far to withdraw this claim."

1.3.2 *Other Cycloarenes*

Also the synthesis of $C_{40}H_{20}$ cyclo[*d.e.d.e.e.d.e.d.e.e*]decakisbenzene (C_2 of Fig. 1) has a long story. Probably the first attempts in this direction are mentioned by Peter and Jenny (1965). They deemed a certain macrocyclic compound ($C_{40}H_{28}$) to be promising for the preparation of C_2 ("[10]coronaphene"); see also Peter and Jenny (1966). These authors claimed to have arrived at a dihydroderivative $C_{40}H_{22}$ of [10]coronaphene (Jenny and Peter 1965). Here $C_{40}H_{22}$ was supposed to have the skeleton of [10]coronaphene. At a considerably later time, this claim was denied by Staab, Diederich and Čaplar (1983) in the words: ".... the formation of the carbon skeleton of 2 [$C_{40}H_{20}$] from 14 [$C_{40}H_{28}$] which had been claimed previously by these authors .. [Jenny and Peter (1965)] obviously was not achieved." The citation is from a report on studies towards the synthesis of cyclo[*d.e.d.e.e.d.e.d.e.e*]decakisbenzene (C_2), wherein the formations of the macrocyclic conjugated hydrocarbons $C_{40}H_{24}$ and $C_{40}H_{22}$ (VII and VIII of Fig. 6, respectively) are documented. The finally successful synthesis of C_2 had to wait still a few years (Funhoff and Staab 1986).

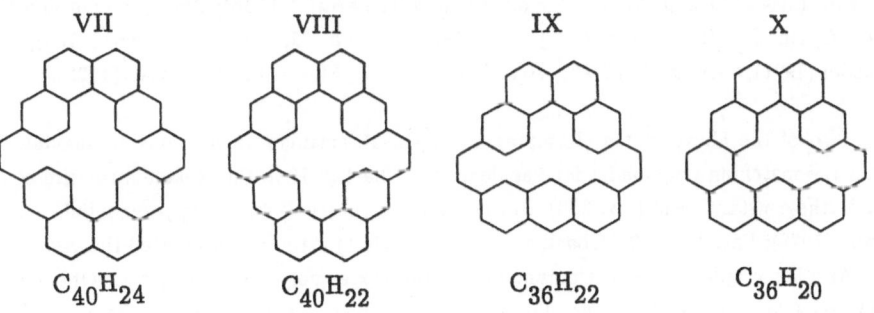

$$\text{VII} \qquad \text{VIII} \qquad \text{IX} \qquad \text{X}$$

$$C_{40}H_{24} \qquad C_{40}H_{22} \qquad C_{36}H_{22} \qquad C_{36}H_{20}$$

Fig. 1.6. Additional four macrocyclic conjugated hydrocarbons.

Errata: In Vol. I the reference Jenny and Peter (1965) has a wrong page number. In Peter and Jenny (1965) the author names are inadvertently switched.

In the reported attempts to synthesize $C_{36}H_{18}$ cyclo[*d. e. e. d. e. e. d. e. e*]nonakisbenzene (C_3 of Fig. 1), Staab and Sauer (1984) arrived at the compounds $C_{36}H_{22}$ and $C_{36}H_{20}$ which are depicted as IX and X, respectively in Fig. 6. Preparations of other compounds with relevance to C_3 are reported by Wilcox, Lahti et al. (1978).

1.3.3 *The Story of Kekulene Continues*

In connection with the synthesis of kekulene and shortly thereafter, the molecule was subjected to experimental investigations of the crystal structure and of a number of physical properties. Also a number of theoretical works are available, both from the time before and after the synthesis. The reader is referred to Vol. I–1.2.1 and I–2.6 for the listing of relevant references.

Another gateway to a part of the existing literature on kekulene is the below list of references to works where the number of Kekulé structures of this molecule ($K = 200$) is derived, rederived or quoted.

Aihara (1976); Krieger et al. (1979); Agranat et al. (1980); Schweitzer et al. (1982); Staab and Diederich (1983); Staab, Diederich, Krieger and Schweitzer (1983); Ramaraj and Balasubramanian (1985); Vogler (1985); Bergan, Cyvin SJ and Cyvin (1986); Randić (1986); Hosoya (1986a); Bergan, Cyvin BN and Cyvin (1987); Cyvin SJ, Bergan and Cyvin (1987); Cyvin SJ, Cyvin, Brunvoll and Bergan (1987); Brunvoll, Cyvin BN and Cyvin (1987); Balasubramanian (1987); Cyvin SJ (1988a; 1988b); Cyvin BN, Cyvin and Brunvoll (1988); Cyvin SJ, Brunvoll, Cyvin and Brendsdal (1988); Cyvin SJ, Brunvoll and Cyvin (1989a); Brunvoll, Cyvin BN, Cyvin, Gutman, Tošić and Kovačević (1989); Cyvin SJ, Brunvoll, Cyvin, Tošić and Kovačević (1989); Cyvin SJ, Brunvoll and Cyvin (1989d); Gutman and Cyvin (1989); Cyvin SJ (1990); Cyvin SJ, Brunvoll and Cyvin (1990a); Knop, Müller, Szymanski and Trinajstić (1990b); Zhang, Cyvin and Cyvin (1990); Cyvin SJ, Brunvoll and Cyvin (1991c); Cyvin SJ, Brunvoll, Cyvin, Bergan and Brendsdal (1991); Trinajstić (1992b); John (1992a; 1993); El–Basil and Hussean (1993).

Two of the above works (Ramaraj and Balasubramanian 1985; Balasubramanian 1987) contain the matching polynomial for kekulene; see Chart 2. Here the Kekulé structure count is found as the constant term (viz. 200). For the general theory of matching polynomials, see, e.g.: Gutman (1979); Farrell (1979); Godsil and Gutman (1981) with references cited therein.

Another graph–theoretical polynomial is the sextet polynomial; see, e.g.: Ohkami et al. (1981) and references cited therein; Ohkami 1990. For kekulene it reads (Ohkami et al. 1981; Zhang, Cyvin and Cyvin 1990):

$$\sigma(x) = 1 + 14x + 48x^2 + 76x^3 + 48x^4 + 12x^5 + x^6 \tag{1.3}$$

Here the sum of the seven coefficients is $200 = K$. The pertinent sextet patterns are depicted in Zhang, Cyvin and Cyvin (1990). Gutman and El-Basil (1984) have derived the uncorrected sextet polynomial for kekulene, in which the coefficient of x is 12 rather than 14; cf. eqn. (3). The reason is that they did not include the enlarged sextet and the super–sextet.

Aromaticity (interpreted or defined in different ways) is a classical property which has been attributed to certain organic compounds. Indeed, the benzenoid hydrocarbons which exist chemically, are frequently referred to as polycyclic aromatic hydrocarbons (PAHs). Also kekulene is a typical PAH. Balaban (1980) asked the rhetorical question: " Is aromaticity outmoded?" In his treatise an extensive historical introduction (going back to 1608) is found. The question is of course answered in the negative. A modern treatment of aromaticity can hardly avoid reference to Clar (1972). The concept is associated with both the matching and the sextet polynomials, and the conjugated circuits (cycles with alternating single and double bonds) are highly relevant. Randić (1977) referred to the [4k+2]–cycles as aromatic conjugated circuits, and to the [4k]–cycles as antiaromatic conjugated circuits. Notice that in kekulene the boundary of the cavity (corresponding to [18]–annulene), the outer boundary (like [30]annulene),

$$
\begin{aligned}
& x^{60} \\
& - 60x^{46} \\
& + 1674x^{44} \\
& - 28850x^{42} \\
& + 344127x^{40} \\
& - 3016998x^{38} \\
& + 20152013x^{36} \\
& - 104913492x^{34} \\
& + 431969433x^{32} \\
& - 1419382254x^{30} \\
& + 3740060904x^{28} \\
& - 7914718788x^{26} \\
& + 13431639205x^{24} \\
& - 18200982024x^{22} \\
& + 19552772649x^{20} \\
& - 16479660654x^{18} \\
& + 10743316299x^{16} \\
& - 5315219724x^{14} \\
& + 1945680262x^{12} \\
& - 509172702x^{10} \\
& + 90806961x^{8} \\
& - 10292946x^{6} \\
& + 665136x^{4} \\
& - 20328x^{2} \\
& + 200
\end{aligned}
$$

Chart 1.2. The matching polynomial of kekulene.

and all the six–membered rings represent [4k+2]–cycles. A special treatment of aromaticity in cycloarenes is due to Randić, Henderson et al (1988); it contains C_2 and C_3 of Fig. 1, but not C_1 (kekulene).

Cioslowski with collaborators performed extensive ab initio calculations for several structures including kekulene (Cioslowski, Mixon and Edwards 1991; Cioslowski, O'Connor and Fleischmann 1991). The following conclusion in the latter reference is of special interest here: ".... the kekulene molecule is stabilized not only by the conjugation within benzene rings (aromaticity) but also by conjugation within the super–ring of benzene rings (super–aromaticity)." Aihara (1992) seems to have clearly disproved this conclusion. He says: "Thus, there is no reason to believe that annulenoid conjugation contributes much to the thermodynamic stability of kekulene. Kekulene is a regular benzenoid in all aspects." Aihara (1993) followed up this work by a general theory of superaromaticity.

Lahti (1988), in his studies by molecular mechanics, argues also against superaromaticity in kekulene. His investigation includes C_1, C_2 and C_3 (Fig. 1), along with two additional cycloarenes. Another, more recent study of kekulene by molecular mechanics is available (Allinger et al. 1990).

The supersextet of kekulene is also mentioned by Vogler (1986) in his semiempirical investigations. In a recent preparative work in organic chemistry (Bell and Jousselin 1991) it is made reference to kekulene. However, these authors were primarily interested in heterocyclic compounds, as is also the topic of other works, which are still more related to kekulene (Katritzky and Marson 1983; Ranshoff and Staab 1985). Detailed treatments of these topics are outside the scope of the present book.

1.4 Annulenes, Annulenoannulenes, and Annulene Derivatives

Although only two cycloarenes (coronoid hydrocarbons) have been synthesized so far, many other macrocyclic conjugated hydrocarbons are known chemically, and they keep attracting constant interest in organic chemistry.

The [n]annulenes (C_nH_n) are reviewed extensively in Vol. I–1.4.3, as also is the case with the annulenoannulenes (I–1.4.4). Here we shall not repeat the pertinent references therefrom, but a supplementary reference to a recent work by Nikolić and Trinajstić (1990) may be useful.

Many annulene derivatives are known among the annelated and condensed annulenes. Again, many examples are found in Volume I (I–1.4.4). Also all the compounds in Figs. 4–6 are condensed annulenes. Here we give some supplementary references to Mitchell et al.: Mitchell, Carruthers et al. (1982); Mitchell, Yan and Dingle (1982); Mitchell, Williams and Dingle (1982); Mitchell, Williams, Mahadevan et al. (1982).

1.5 Antikekulene

The trivial name "anti–kekulene" for a certain macrocyclic conjugated structure ($C_{36}H_{12}$) was coined by Diercks and Vollhardt (1986). This structure has twelve rings which alternate between four–membered and six–membered, and are arranged around a twelve–membered central ring (see Fig. 7). A resemblance with kekulene (C_1 of Fig. 1) is obvious. Several years before the appearance of the above cited work, Hellwinkel (1970) depicted the structure in question and gave it the characterization "rational" species [here translated from German]. Already Diercks and Vollhardt (1986) contemplated a synthesis of antikekulene. A later work (Schmidt–Radde and Vollhardt 1992) led this project substantially nearer to the goal when the authors prepared $C_{30}H_{14}$ angular [5]phenylene, but the last closure into a macrocyclic arrangement has so far not been achieved.

In contrast to kekulene (which is aromatic), antikekulene is antiaromatic (Aihara 1993; Gutman and Cyvin 1993). Notice that the central twelve–membered ring, the twenty–four membered boundary, as well as the six four–membered rings all represent [$4k$]–cycles. The

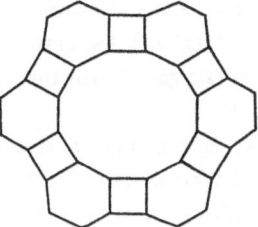

Fig. 1.7. $C_{36}H_{12}$ antikekulene.

prospects of synthesizing antikekulene versus kekulene have been discussed (Trinajstić, Schmalz et al. 1991; Plavšić et al. 1992), and it seems to be a general agreement that the preparation of antikekulene would be still more difficult than the one of kekulene. It does not help that the number of Kekulé structures of antikekulene, viz. 324 (Trinajstić, Schmalz et al. 1991; Gutman and Cyvin 1993) is larger than the corresponding count in kekulene ($K = 200$). The algebraic structure count, ASC (Wilcox 1968; Gutman, Trinajstić and Wilcox 1975; Wilcox, Gutman and Trinajstić 1975) – also called "corrected structure count", CSC (Herndon 1973) – is relevant in this connection. For antikekulene the relatively small number ASC = 16 was found (Gutman and Cyvin 1993).

1.6 Corannulene

In connection with the first synthesis of $C_{20}H_{10}$ corannulene (Barth and Lawton 1966; 1971) and studies of the anion and cation radicals of this molecule (Janata et al. 1967), a theoretical work (not included in Vol. I) is worth mentioning (Gleicher 1967). The crystal and molecular structures of $C_{20}H_{10}$ have been investigated (Hanson and Nordman 1976); the mole—cule has a bowl—shaped structure represented by a planar chemical graph as in Fig. 8. Different theoretical works include corannulene among other molecules (Wynberg et al. 1973; Kao and Allinger 1977; Kao 1987; Dias 1988; Schulman et al. 1989). Calculated electronic spectra of corannulene are available (Kataoka and Nakajima 1986; Feng et al. 1990) and also the infrared— (and Raman—) active frequencies from a normal coordinate analysis (Cyvin SJ, Brendsdal et al.

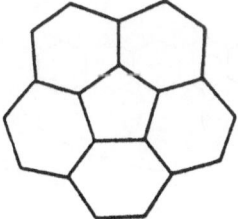

Fig. 1.8. $C_{20}H_{10}$ corannulene.

1991), as well as the vibrational frequencies from an ab initio study (Bakowies and Thiel 1991). Cyclic conjugation in circulenes, including corannulene ([5]circulene), has been studied (Gutman and Lee 1993). Corannulene has eleven Kekulé structures (Bergan, Cyvin BN and Cyvin 1987; Brendsdal and Cyvin 1989; Cyvin SJ, Bréndsdal et al. 1991).

Corannulene has recently attracted new interest among organic chemists. Two new syntheses have been reported (Scott, Hashemi, Meyer and Warren 1991; Borchardt et al. 1992), and the bowl–to–bowl inversion has been discussed (Scott, Hashemi and Bratcher 1992; Borchardt et al. 1992).

1.7 Cyclacenes

Considerable interest is attached to cyclacenes (cyclic acenes, $C_{4h}H_{2h}$) and particularly to [12]cyclacene ($C_{48}H_{24}$), which is a constitutional isomer of kekulene (cf. Fig. 9) and (Ashton et al. 1988): ".... a challenging and intriguing target on account of the mystery surrounding .. [Vögtle (1983); Alder and Sessions (1985); Angus and Johnson (1988)] its molecular and supramolecular properties." Ashton et al. (1988) made significant steps towards a synthesis of $C_{48}H_{24}$ [12]cyclacene or rather $C_{48}H_{36}$ [12]collarene. Collarenes like beltenes (Alder and Sessions 1985), e.g. $C_{48}H_{48}$ [12]beltene, are partial–hydrogenated analogues of cyclacenes. Beltenes have also been called columnenes (Angus and Johnson 1988). Some theoretical works with predictions on cyclacenes are available (Ege and Vogler 1974; Kivelson and Chapman 1983).

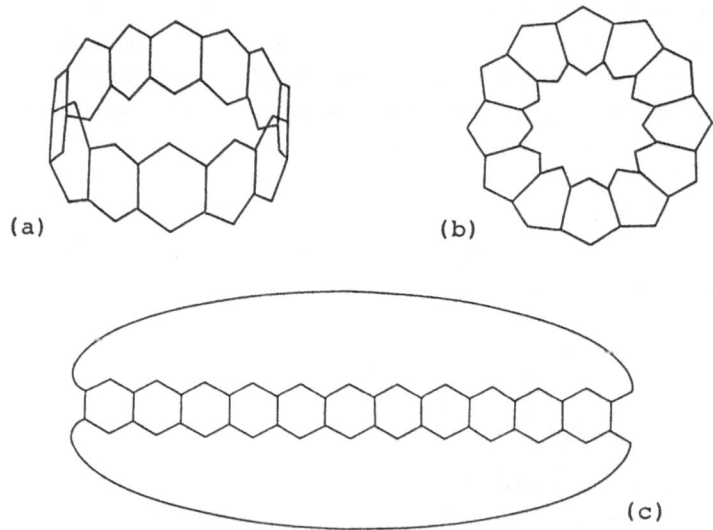

Fig. 1.9. (a) $C_{48}H_{24}$ [12]cyclacene, and (b,c) two representations of [12]cyclacene as planar chemical graphs.

Other theoretical works which include cyclacenes, are more mathematically oriented. Already Heilbronner (1954) represented these structures by planar graphs as in Fig. 9(b), and was followed by others (Agranat et al. 1980; Bergan, Cyvin BN and Cyvin 1987). In the two latter works it was ascertained that [6]cyclacene has four Kekulé structures, a number which applies to all [h]cyclacenes irrespective of the h number of rings (Bergan, Cyvin BN and Cyvin 1987). In spite of this low number ($K = 4$), the fact that both twelve–membered boundaries are [4k]–cycles (antiaromatic conjugated circuits), and the absence of aromatic sextets, Agranat et al. (1980) calculated a surprisingly high resonance energy for [6]cyclacene indicating a high degree of aromaticity. Two additional works are cited: Derflinger and Sofer (1968); Hosoya et al. (1987). In the latter reference (Hosoya et al. 1987) and elsewhere (Cyvin SJ 1990) cyclacenes are represented as in Fig. 9(c).

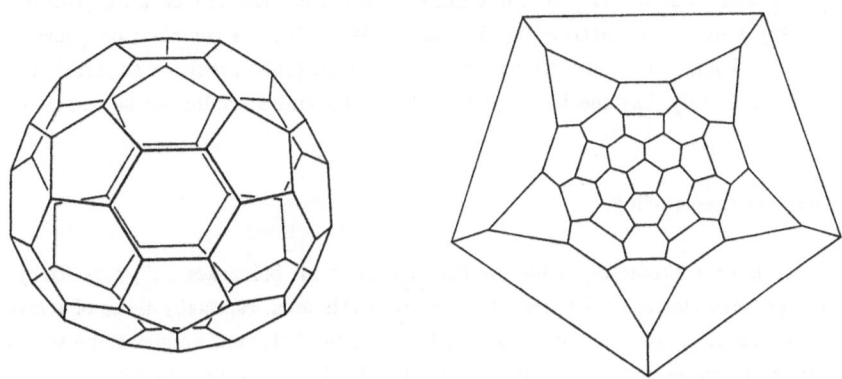

Fig. 1.10. The famous structure of C_{60} buckminsterfullerene, a truncated icosahedron like a football (in America soccerball). It is represented by a planar chemical graph in the right–hand drawing.

1.8 Buckminsterfullerene

In a treatise entitled "New Dimensions in Polynuclear Aromatic Compounds", Herndon (1988) has described a number of polycyclic conjugated compounds, including corannulene, a cyclacene and C_{60} buckminsterfullerene (Osawa 1970; Kroto et al. 1985; Krätschmer et al. 1990); see Fig. 10. The reader understands presumably very well that a full survey of the literature on this interesting molecule (and other fullerenes) is too voluminous (counting thousands of references) to be included here. In the following small extract the number of Kekulé structures ($K = 12500$) for C_{60} is reported (not necessarily as an original finding).

Schmalz et al. (1986); Hosoya (1986a); Trinajstić, Klein and Randić (1986); Klein, Schmalz et al. (1986); Klein, Seitz and Schmalz (1986); Randić, Nikolić and Trinajstić (1987); Aihara and Hosoya (1988); Schmalz et al. (1988); Brendsdal and Cyvin (1989); Klein, Seitz and Schmalz (1990); Elser et al. (1990); Trinajstić (1992b).

Hosoya (1986a) deduced the Z–counting polynomial for C_{60} buckminsterfullerene, a polynomial of degree 60, where K is the coefficient of the highest power (z^{60}). This polynomial is strongly related to the matching polynomial, which has K as its constant term, and has also been given explicitly for C_{60} (Aihara and Hosoya 1988; Elser et al. 1990).

The corannulene skeleton (C_{20}) is recognized as a fragment of C_{60} buckminsterfullerene. This feature has been pointed out several times and is implied, for instance, in a most detailed account of "Elemental Carbon Cages" by Schmalz et al. (1988). More recent works of relevance are: Diederich and Whetten (1992); Fukushima (1992). Here the former work (Diederich and Whetten 1992) is mentioned particularly because it also includes a comment on [9]cyclacene in connection with higher fullerenes. The work of Feng et al. (1990), which was mentioned under corannulene (Sect. 1.6), has the investigation of C_{60} buckminsterfullerene as one of its main topics.

1.9 Nomenclature and Coding

"There is no shortage of coding methods available for polyhexes ....", says Kirby (1990) and follows up with thirteen references. References in this area, especially those of relevance to coronoid hydrocarbons, are also found in Vol. I–5.1. In the following we give some supplements and references to more recent relevant works: Elk (1985); Balaban, Biermann and Schmidt (1985); Jovanovic (1988); Knop, Müller, Szymanski, Nikolić and Trinajstić (1990); Goodson (1991); John (1992b); Elk (1993).

1.10 Conclusion

Several polycyclic conjugated hydrocarbons (Clar 1964) of chemical interest (in addition to the carbon cluster C_{60}) are described above in this chapter. In the next chapter, some aspects of the difficult task of classification of chemical graphs are explained. These classifi–cations/explanations account for all the conjugated hydrocarbons in Sections 1.3–1.7, and also for the carbon cluster (Sect. 1.8).

Chapter 2

CLASSIFICATIONS OF POLYGONAL SYSTEMS,

AND SOME ASPECTS OF KEKULÉ STRUCTURES

2.1 Introduction

A polyhex (system) is a connected geometrical arrangement of congruent regular hexagons so that any two hexagons either share exactly one edge or are disjointed (Balaban 1976; Trinajstić 1992b). A full classification of polyhexes is difficult (Trinajstić 1990; 1992a; Trinajstić, Nikolić et al. 1991; Cyvin BN, Brunvoll and Cyvin 1992b). However, within this broad class of systems the definitions of benzenoids and single coronoids are clear—cut, and various classifications of these systems are well explained. Nevertheless, the terminology in this area is still controversial and subject to discussions (Nikolić et al. 1991; Brunvoll, Cyvin SJ and Cyvin 1991; Trinajstić 1992a).

In the present book we adhere to well—established definitions of benzenoids and single coronoids (Gutman 1982; Cyvin SJ and Gutman 1988; Gutman and Cyvin 1989). A benzenoid is a simply connected, nonhelicenic (geometrically planar) polyhex. A *coronoid* is a similar system, but multiply connected so that it possesses one or more holes, which are called *corona holes*. Each corona hole should have a size of at least two hexagons. A *single coronoid* has exactly one corona hole. Precise definitions of single coronoids are detailed in Vol.I—2.1.1 and elsewhere. The very term "coronoid" appeared first in a paper by Cyvin BN, Brunvoll, Cyvin and Gutman (1986). One year later a detailed account on the enumeration and classification of coronoids was published (Brunvoll, Cyvin BN and Cyvin 1987), along with additional papers consolidating the term "coronoid" (Cyvin SJ, Bergan and Cyvin 1987; Cyvin SJ, Cyvin and Brunvoll 1987; Cyvin SJ, Cyvin, Brunvoll and Bergan 1987; Balaban, Brunvoll et al. 1987). Loosely speaking, a single coronoid may be said to be a benzenoid with a hole. This characterization is formulated as a strict definition in the following way.

Definition 2.1: A single coronoid is obtained from a benzenoid by deleting some internal vertices and/or edge(s) so that a hole with the size of at least two hexagons emerges and is completely surrounded by hexagons. More precisely, this process consists of (a) deleting one internal vertex (together with its incident edges) or (b) deleting one internal edge, whereby it is allowed for a succession of the operations (a) and (b).

The below diagram shows an example for each of the operations (a) and (b). The deleted objects are indicated by a white dot and dotted lines.

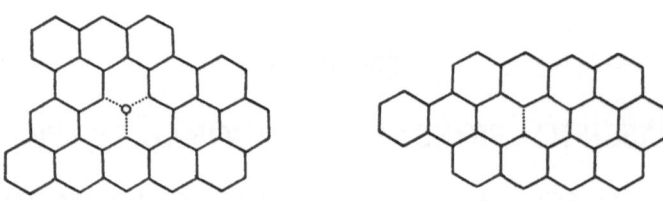

The below scheme of classification applies to benzenoids and coronoids, and to the single coronoids in particular. It is consistent with the classifications adopted previously, which are detailed in Vol. I and elsewhere (Gutman and Cyvin 1989; Brunvoll, Cyvin SJ and Cyvin 1991a).

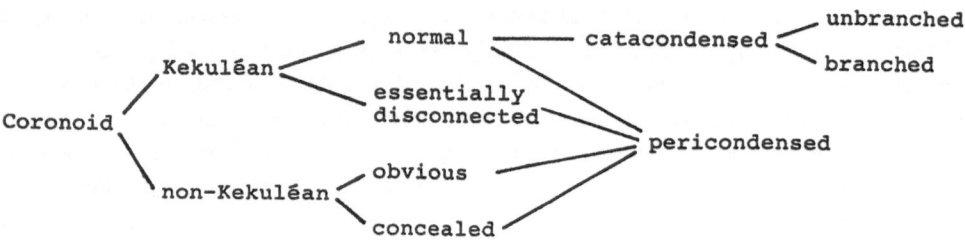

The *Kekulean* and *non–Kekulean* systems are those which possess Kekulé structures $(K > 0)$ or do not possess Kekulé structures $(K = 0)$, respectively. In an *essentially disconnected* system there are fixed bonds, viz. edges which correspond to double and/or single bonds in the same positions of all Kekulé structures. Kekuléan systems without fixed bonds are called *normal. Obvious–* and *concealed non–Kekulean* systems have $\Delta > 0$ and $\Delta = 0$, respectively, where Δ is the color excess, viz. the absolute magnitude of the difference between the numbers of peaks and valleys (I–3.2.4). Finally, the *catacondensed* (unbranched or branched) and the *pericondensed* systems have $n_i = 0$ and $n_i > 0$, respectively, where n_i designates the number of internal vertices (I–3.3.1).

2.2 Classification of Single Coronoids in Relation to Kekulé Structures

Whereas the scheme in Sect. 2.1 applies to both benzenoids and single coronoids, a finer classification for the Kekuléan single coronoids was found to be needed during the studies of Kekulé structure counts for these systems. In this way the concept *half essentially disconnected* (single) coronoids was introduced by Cyvin SJ, Cyvin and Brunvoll (1987). This concept is explained in the book of Gutman and Cyvin (1989) and in Vol. I–3.3.

In the first place, the Kekuléan single coronoids are classified into normal (n) and essentially disconnected (e) systems. Obviously, $n + e$ accounts for all the Kekuléan systems.

Together with the non–Kekuléans (o), the total number of single coronoids is accounted for. This classification, referred to as the *neo classification*, applies to benzenoids as well as (single) coronoids.

Next, among the normal coronoids, some peculiar systems were identified, which exhibited two schemes of Kekulé structures, each being associated with fixed bonds. The Kekulé structures of the two schemes gave the complete set of Kekulé structures. These two schemes showed some characteristic features not easy to define in precise terms. It was decided to call these systems "half essentially disconnected", HED, but the difficulty with a precise definition remained. It was attempted to overcome this difficulty by the definition of *regular* single coronoids, which should form a subclass of the normal single coronoids and exclude the HED systems. After defining the regular systems the term *irregular* was introduced for all the Kekuléan single coronoids which are not regular. Then obviously a classification into regular (r), irregular (i) and non–Kekuléans (o) accounts for all single coronoids. It was referred to as the *rio classification* and is included in the below scheme

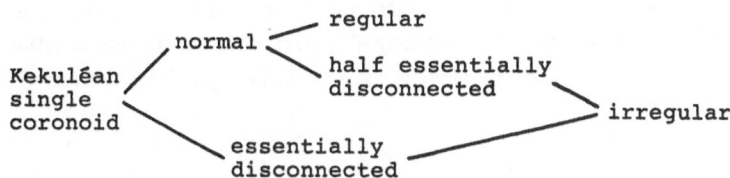

This scheme also suggests a provisional definition of HED single coronoids as the normal single coronoids which are not regular. Then the n systems would be classified into regular (r) and HED (he) single coronoids: $n = r + he$. At the same time one would have $i = he + e$. In summary, the *rheo classification*, which counts the regular (r)–, half essentially disconnected (he)–, essentially disconnected (e)– and non–Kekuléan (o) single coronoids, would account for all single coronoids.

It remains to define the regular single coronoids. In order to give an idea of the approach which was adopted (precise definitions are saved for a subsequent chapter), consider the regular coronoid:

(regular)

The purpose of this diagram is twofold: (1) It should recall the notation of modes of hexagons; cf. Vol. I–3.2.1, especially I–Fig. 3.1, and elsewhere (Cyvin SJ and Gutman 1988; Gutman and Cyvin 1989). (2) It should illustrate the definition of a regular single coronoid, which implies that it should be possible to tear down such a system completely (down to benzene) by deleting only hexagons of the modes L_1, L_3 or L_5, one at a time, in addition to deleting an L_2– or A_2–mode hexagon once during this process in order to open the corona hole.

A similar property for benzenoids is known. Originally it was conjectured (Cyvin and Gutman 1986; 1988; Cyvin SJ, Brunvoll and Cyvin 1989b) that all normal benzenoids with $h + 1$ hexagons are generated by all the possible (i) one–, (iii) three– and (v) five–contact additions (cf. Vol. I–Fig. 2.2) to all normal benzenoids with h hexagons. It is understood that one hexagon at a time should be added, and the added hexagon acquires the mode (i) L_1, (iii) L_3 or (v) L_5, respectively. If the above conjecture is sound, it should also be possible to tear down any normal benzenoid completely (right down to benzene) by deleting successively hexagons of the modes L_1, L_3 or L_5, only. The above conjecture and its corollary were proved by He WC and He (1990b) and thereby raised to the status of theorems.

For the regular coronoid depicted above there are different possibilities. One of the L_2– or A_2–mode hexagons can be deleted so that the system is converted to a normal benzenoid, whereupon the process of tearing down can be continued smoothly. As another possibility, the L_5 hexagon can be deleted, whereupon a smaller, regular coronoid is created; then an L_2– or A_2–mode hexagon can be deleted, etc. One may also start with the deletion of either the top or bottom L_3 hexagon. Let us now try to delete the middle L_3 hexagon. The result is:

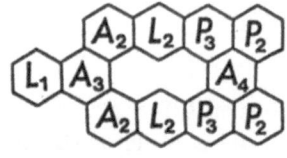

(half essentially disconnected)

Notice that the top and bottom L_3 hexagons have changed their modes to P_2 (and L_5 to A_4). The new system is an HED coronoid and cannot be torn down in the prescribed way. If one of the L_2– or A_2–mode hexagons are deleted, one obtains an essentially disconnected benzenoid, also if the L_1–mode hexagon was deleted first. This example shows that the sequence of hexagons which are deleted, is not arbitrary. Now it is instructive to demonstrate the two characteristic schemes of Kekulé structures for the above HED coronoid. The hatched parts (in the below diagram) indicate hexagons with fixed bonds.

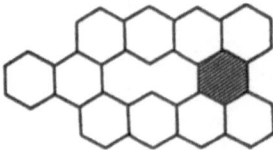 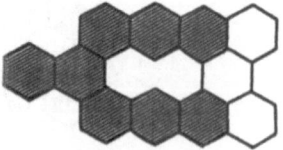

The number of Kekulé structures is $K = 66 + 5 = 71$ (cf. Vol. I–Fig. 7.2).

It was conjectured that the provisional definition of HED single coronoids as the normal systems which are not regular, always would lead to systems reasonable to be classified as HED;

they should exhibit two characteristic schemes of Kekulé structures. On this basis extensive enumerations according to the *rheo* classification (see above) were performed. The main results, with documentation in footnotes, are collected in Table 1. In addition, the corresponding enumerations for the subclass of single coronoids with the phenalene corona hole are available (Cyvin SJ, Brunvoll and Cyvin 1990c). In these computations, all the classes were defined clearly except *he*, for which the numbers, in accordance with the provisional definition, were taken as differences from the totals. We shall presently refer to the *rheo conjecture*, which relates to the above discussion, but needs to be put on a firmer basis. That is accomplished in a subsequent chapter on theorems through a full characterization of the HED single coronoids, which defines these systems strictly. Furthermore, the *rheo* conjecture is proved by demonstrating that the new definition coincides with the provisional definition when the regular single coronoids are defined as previously. In other words, we prove that a normal single coronoid which is not regular, is an HED coronoid according to a strict definition of this class.

All the non–Kekuléan coronoids with $h \leq 14$ (for their numbers, see Table 1) are known to be obvious non–Kekuléans. The smallest concealed non–Kekuléan (single) coronoids have $h = 15$, and there are exactly 23 nonisomorphic systems of this category (Cyvin SJ, Brunvoll and Cyvin 1989c; Vol. I–7.3), each of them possessing the naphthalene hole. For the subclass of single coronoids with the phenalene hole it was found that the smallest concealed non–Kekuléans have $h = 16$, and there are exactly 21 such systems (Cyvin SJ, Brunvoll and Cyvin 1990b).

Table 2.1. Numbers of single coronoids classified according to *rheo*.*

h	r	he	e	o
8	1[a]			
9	3[a]			2[b]
10	18[a]	6[c]		19[a,d,e]
11	90[a]	36[c]	2[d]	155[a,d,e]
12	526[a]	289[c]	39[c,f]	1100[a,d,e]
13	2810[a]	1821[c]	419[c,f]	7313[a]
14	15071[c]	11188[c]	3795[c,f]	46229[c,f,g]

* Abbreviations: *e* essentially disconnected; *he* half essentially disconnected; *o* non–Kekuléan; *r* regular.

[a] Cyvin SJ, Cyvin BN, Brunvoll J (1987). Chem Phys Letters 140: 124

[b] Cyvin BN, Brunvoll J, Cyvin SJ, Gutman I (1986). Match 21: 301

[c] Cyvin SJ, Brunvoll J, Cyvin BN (1991). Lecture Notes in Chemistry 54 (Springer–Verlag, Berlin)

[d] Brunvoll J, Cyvin BN, Cyvin SJ (1987). J Chem Inf Comput Sci 27: 14

[e] Balaban AT, Brunvoll J, Cioslowski J, Cyvin BN, Cyvin SJ, Gutman I, He WC, He WJ, Knop JV, Kovačević M, Müller WR, Szymanski K, Tošić R, Trinajstić N (1987). Z Naturforsch 42a: 863; wrong number therein (for *e* at $h = 12$) is omitted

[f] Brunvoll J, Cyvin SJ, Cyvin BN (1991). J Mol Struct (Theochem) 235: 147

[g] Cyvin SJ, Brunvoll J, Cyvin BN (1991). J Math Chem 8: 63

2.3 Degenerate and Generalized Single Coronoids

Sometimes it is convenient to formulate theorems for a broader class of systems than the single coronoids, but containing the single coronoids as a subclass.

In Vol. I–2.4 a *degenerate single coronoid* system is defined as a connected part of a single coronoid (a single coronoid with vertices and/or edge(s) deleted from it) containing edge(s) not belonging to hexagons. The below definition identifies a subclass of degenerate single coronoids.

Definition 2.2: A *special degenerate single coronoid* is a degenerate single coronoid where the edge(s) not belonging to hexagons are common to the outer boundary and the hole.

In I–Fig. 2.4 the three systems of the upper row are not special degenerate coronoids, but this is the case for the bottom system. Here are four additional examples of degenerate single coronoids:

$C_{36}H_{18}$ $C_{37}H_{19}$ $C_{42}H_{22}$ $C_{26}H_{22}$

Two of them ($C_{42}H_{22}$ and $C_{26}H_{22}$) are not special degenerate single coronoids, while the other two ($C_{36}H_{18}$ and $C_{37}H_{19}$) belong to the category of Definition 2.2. Volume I (I–1.4.4) contains a survey of many special degenerate single coronoids, quite a few of them being chemically known. Also the systems I–X of Figs. 1.4–1.6 are special degenerate single coronoids.

In the case of special degenerate single coronoids, as well as in single (nondegenerate) coronoids, one speaks about the outer and inner perimeter, viz. C' and C", respectively, but now some edge(s) belong to both: C' ∩ C" ≠ ∅.

Definition 2.3: A *generalized single coronoid* is either a single coronoid or a special degenerate single coronoid.

If one wants to avoid the term "special degenerate single coronoid", then one has the synonymous designation "degenerate generalized single coronoid" at hand.

The classifications normal and essentially disconnected are immediately applicable to degenerate single coronoids.

2.4 Examples of Single Coronoids, and Their Kekulé Structure Counts

2.4.1 *Regular Single Coronoids*

When a normal benzenoid B with h hexagons is subjected to a normal addition, then the Kekulé structure count of the new normal benzenoid with $h+1$ hexagons is larger than the Kekulé structure count of B (Cyvin SJ and Gutman 1986; Cyvin SJ, Brunvoll and Cyvin 1989b). In a normal addition, the added hexagon acquires by definition the mode L_1, L_3 or L_5. More about normal (and regular) additions is found in the subsequent chapters. The above property which has been proved rigorously by Cyvin SJ and Gutman (1986), can be adapted to regular single coronoids: when a regular single coronoid C with h hexagons is subjected to a normal addition, then the Kekulé structure count of the new regular single coronoid with $h+1$ hexagons is larger than the Kekulé structure count of C. We shall not give a formal proof of this proposition here, but only illustrate it with a few examples. Starting from the smallest coronoid with the Kekulé structure count $K = 40$, four additional regular single coronoids are built up by successive normal additions of hexagons. The K numbers are inscribed in the last added hexagons:

2.4.2 *Essentially Disconnected Single Coronoids*

The terms "junction" and "effective unit", which were introduced for essentially disconnected benzenoids (Brunvoll, Cyvin BN, Cyvin and Gutman 1988), are transferable to coronoids (cf. Vol. I–3.3.4).

Definitions 2.4: A *junction* of an essentially disconnected coronoid C is the set of hexagons which possess fixed bonds. The *effective units* of C are obtained on deleting all edges which are fixed single bonds, and all fixed double bonds together with their end vertices.

In the below diagram, three essentially disconnected single coronoids and their effective units are depicted. The junctions are painted black.

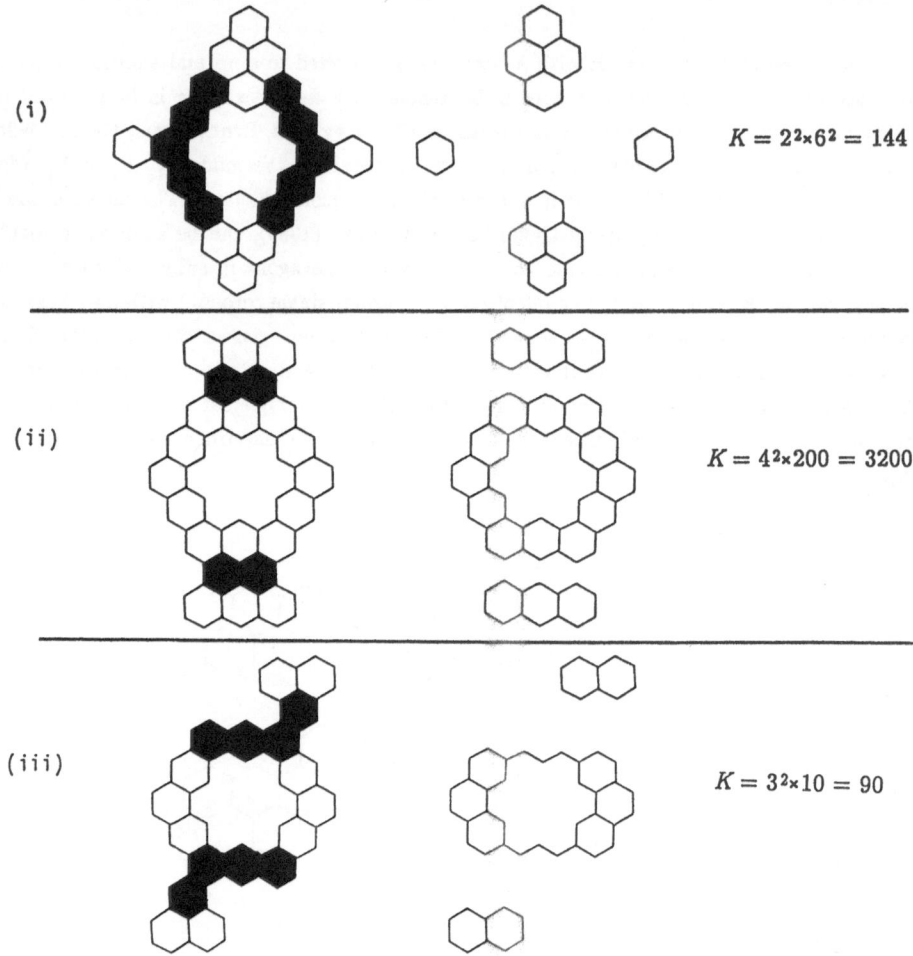

In the first case (i) of the above diagram, the effective units are four benzenoids (two pyrenes and two benzenes); in case (ii) the effective units are two benzenoids (anthracenes) and one coronoid (kekulene); in case (iii) the effective units are two benzenoids (naphthalenes) and one special degenerate single coronoid (condensed [18]annulene, viz. hexabenzo[$bcdef,klmno$][18]–annulene).

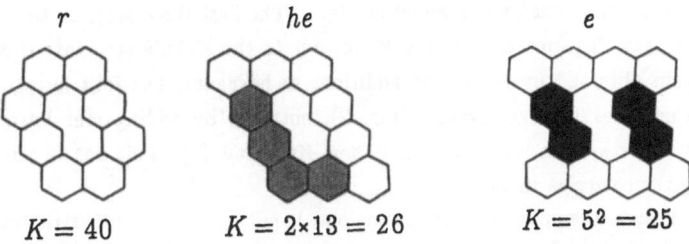

Fig. 2.1. One regular (r), one half essentially disconnected (he), and one essentially disconnected (e) single coronoid. Numbers K of Kekulé structures are indicated.

2.4.3 Additional Instructive Examples

Figure 1 shows three single coronoids of the categories r (regular), he (HED) and e (essentially disconnected), with $h = 8$, 10 and 12, respectively. In the he system one of the two schemes for Kekulé structures is indicated by hatching; the two schemes of this system are symmetrically equivalent. In the e system the junction is painted black. It is noted that the he and e systems are obtained by successive additions of hexagons to the smallest coronoid (r); two hexagons are added each time. However, these additions are not normal, in contrast to those of the examples in Par. 2.4.1. Therefore it is not guaranteed that the Kekulé structure counts increase with increasing numbers of hexagons through this set of coronoids. In fact, the K numbers (which are included in Fig. 1) are found to decrease with increasing h.

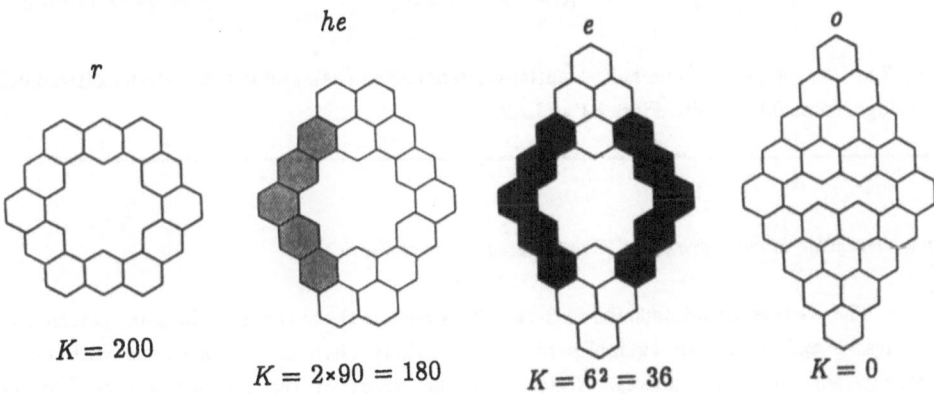

Fig. 2.2. One regular (r), one half essentially disconnected (he), one essentially disconnected (e), and one concealed non–Kekuléan (o) single coronoid; see also Fig. 2.1.

Another instructive example is found in Fig. 2: The first three systems therein ($r + he + e$) illustrate the same features as Fig. 1 with respect to the Kekulé structure counts (K). The systems are again obtained by successive additions of hexagons, but four hexagons are added each time in this case. The last system (Fig. 2), obtained by adding four hexagons into the corona hole of the e system, is a concealed non–Kekuléan (o). This set of single coronoids illustrate nicely the *rheo* classification.

Figures 1 and 2 emphasize the anomalous behaviour of Kekulé structure counts for irregular single coronoids. However, it is not always so that the K numbers decrease through similar sets of coronoids as in these figures. The opposite situation is observed in Fig. 3, where again one he and one e system is produced by successive additions of hexagons to a regular (r) coronoid; six hexagons are added each time. In this case the K numbers happen to increase with increasing h in the "normal" way.

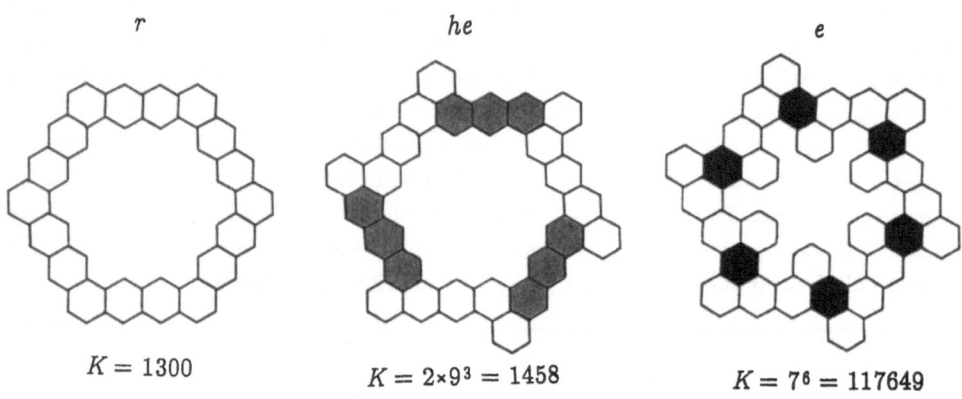

$$r \qquad\qquad he \qquad\qquad e$$

$$K = 1300 \qquad K = 2 \times 9^3 = 1458 \qquad K = 7^6 = 117649$$

Fig. 2.3. One regular (r), one half essentially disconnected (he), and one essentially disconnected (e) single coronoid; see also Figs. 2.1 and 2.2.

2.4.4 *Irregular Single Coronoids With Isolated Internal Vertices*

The HED coronoid depicted in Sect. 2.2, as well as the systems of the same category (he) in Figs. 1 and 3, are all examples of half essentially disconnected coronoids with isolated (nonadjacent) internal vertices. Notice that the colors of the internal vertices alternate successively as white–black–white–black– .... when going around the corona hole; cf. the left–hand column of the below diagram.

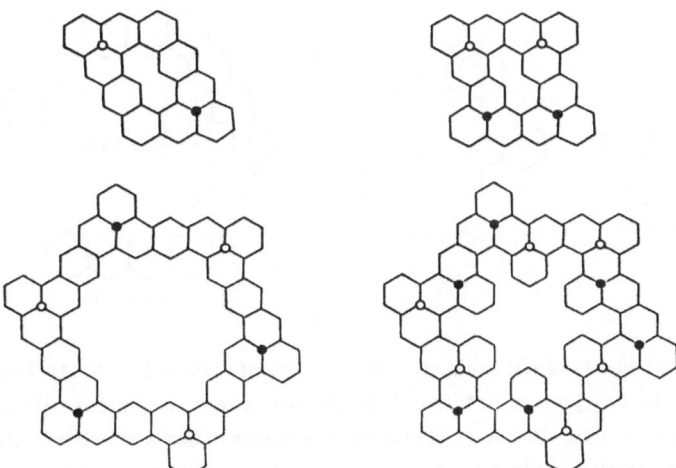

On the other hand, if a Kekuléan single coronoid with isolated internal vertices has two black and two white internal vertices in succession around the corona hole, then it is essentially disconnected. This feature is illustrated for the *e* systems from Figs. 1 and 3 in the right–hand column of the above diagram.

Tomescu and Balaban (1989) investigated the numbers of Kekulé structures for coronoids, both single and multiple, with phenalene fragments. There are clearly overlappings between representatives of their classes and of those considered here. In particular, the systems with two phenalene fragments of Tomescu and Balaban (1989) are clearly HED single coronoids, and the authors have arrived at a formula for Kekulé structure counts which reflects the two schemes which are characteristic for such systems. Klein and Trinajstić (1990) treated the Kekulé structures of two HED single coronoids, one of them having two isolated internal vertices. Finally the work of Randić, Trinajstić et al. (1993) should be mentioned in this connection. These investigators considered the Kekulé structures of exclusively HED systems among the single coronoids, and fourteen out of their fifteen examples deal with systems which each possesses two isolated internal vertices.

2.5 Survey of Kekulé Structure Counts for Single Coronoids

2.5.1 *Combinatorial Formulas*

The Kekulé structure counts (Cyvin SJ and Gutman 1988) for the homologous series coronene, kekulene, ... (see the below diagram) is a manageable problem; the solution appeared the same year in three independent works: Bergan, Cyvin SJ and Cyvin (1986); Hosoya (1986a); Babić and Graovac (1986). See also: Graovac, Babić and Strunje (1986).

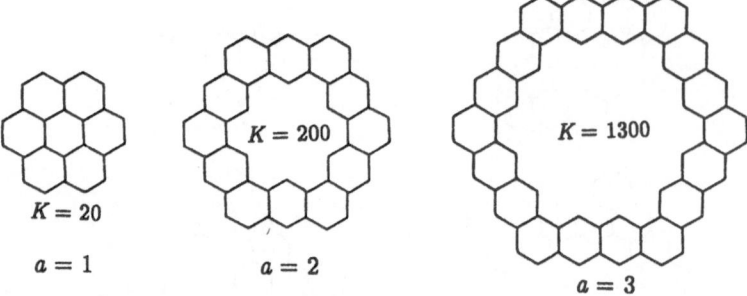

$K = 20$

$a = 1$

$K = 200$

$a = 2$

$K = 1300$

$a = 3$

Here each system consists of six equidistant linear segments (cf. Vol. I–4.2.4). Let $a + 1$ be the length of each of these segments in terms of the number of hexagons, and recall that each corner (A_2 mode hexagon) belongs to two (neighbouring) segments. Then a member of the homologous series under consideration is identified in terms of the sequence of segments as /a+1, a+1, a+1, a+1, a+1, a+1/ or in an abbreviated form: /a+1/^{6}; cf. Vol. I–4.3. The total number of hexagons is $h = 6a$, at least for $a > 1$. For $a = 1$ the system in question is [6]circulene ($h = 6$) or coronene ($h = 7$). The pertinent combinatorial formula for the Kekulé structure counts (K) reads

$$K\{/a+1/^6\} = (a^2 + 1)^2(a^2 + 4) = a^6 + 6a^4 + 9a^2 + 4 \qquad (2.1)$$

Hosoya (1986a) observed a resemblance between the polynomial in (1) and the characteristic polynomial for benzene, viz. $x^6 - 6x^4 + 9x^2 - 4$. He writes: "The reason for this mystic coincidence, however, is not known." On the other hand, a substantial amount of work has been done in the area of combinatorial K formulas in the form of polynomials for primitive coronoids in general, and especially for those of regular hexagonal symmetry (D_{6h}) as in the case above (Cyvin SJ 1988b; 1989; Cyvin SJ, Brunvoll and Cyvin 1989a; Cyvin SJ, Brunvoll, Cyvin, Tošić and Kovačević 1989). For a treatment of primitive coronoids with trigonal symmetry, see Cyvin SJ, Brunvoll and Cyvin (1988). The coronoids /a+1/^{6} represent a special class of hollow hexagons, i.e. primitive single coronoids (Vol. I–4) with exactly six segments each. Let the lengths of these segments be defined so that $h = a + b + c + d + e + f$; i.e. the lengths are $a + 1$, $b + 1$, etc. Then (Cyvin SJ, Brunvoll and Cyvin 1989a; 1989d)

$$K\{/a+1,b+1,c+1,d+1,e+1,f+1/\} = abcdef + abcd + bcde + cdef + defa + efab + fabc$$
$$+ ab + bc + cd + de + ef + fa + ad + bc + cf + 4 \quad (2.2)$$

This formula was derived by Bergan. A determinant form, produced by Hosoya, was communicated by Cyvin SJ, Brunvoll, Cyvin, Bergan and Brendsdal (1991). An interesting general formulation for the Kekulé structure counts of a primitive coronoid with specified lengths of segments reads (Cyvin SJ 1990; Cyvin SJ, Cyvin, Brunvoll, Hosoya et al. 1991)

$$K\{/a_1+1, a_2+1, \ldots, a_S+1/\} = \text{Tr}(\mathbf{a}_1\,\mathbf{a}_2 \ldots \mathbf{a}_S) + 2 \tag{2.3}$$

where

$$\mathbf{a}_i = \begin{bmatrix} a_i & 1 \\ 1 & 0 \end{bmatrix} \tag{2.4}$$

A combinatorial K formula of a different kind, viz. in terms of exponential functions, has been deduced for the class of primitive coronoids with S equidistant segments (Balaban, Brunvoll and Cyvin 1991; Vol. I–6.4). If the length of each segment is $a + 1$, then (Bergan, Cyvin BN and Cyvin 1987)

$$K\{/a+1/^S\} = 2^{-S}[a + (a^2+4)^{1/2}]^S + 2^{-S}[a - (a^2+4)^{1/2}]^S + 2 \tag{2.5}$$

where $S = 12, 14, 16, \ldots$ when $a = 2$, and $S = 6, 10, 12, 14, 16, \ldots$ when $a > 2$. The case $a = 2$ is of particular interest because kekulene belongs to this subclass (for $S = 6$). It is found

$$K\{/3/^S\} = (1 + 2^{1/2})^S + (1 - 2^{1/2})^S + 2 \tag{2.6}$$

The case $a = 1$ is also interesting, especially because Fibonacci numbers and Lucas numbers turn up (Bergan, Cyvin BN and Cyvin 1987; Cyvin SJ 1990):

$$K\{/2/^S\} = 2^{-S}(1 + 5^{1/2})^S + 2^{-S}(1 - 5^{1/2})^S + 2 = F_S + F_{S-2} + 2 = L_{S-1} + 2 \tag{2.7}$$

Here the Fibonacci and Lucas numbers are defined by $F_0 = F_1 = 1$, $F_k = F_{k-1} + F_{k-2}$ ($k > 1$) and $L_0 = 1$, $L_1 = 3$, $L_k = L_{k-1} + L_{k-2}$ ($k > 1$), respectively.

Combinatorial formulas for the Kekulé structure counts of some classes of nonprimitive single coronoids are scattered around in the literature, partly among other topics (Cyvin SJ, Bergan and Cyvin 1987; Cyvin SJ, Cyvin, Brunvoll and Bergan 1987; Cyvin BN, Brunvoll and Cyvin 1988; Cyvin SJ, Brunvoll and Cyvin 1989a; Cyvin SJ, Brunvoll, Cyvin, Tošić and Kovačević 1989; Cyvin SJ 1989; Gutman and Cyvin 1989; Cyvin SJ, Cyvin and Brunvoll 1991). Several of these classes consist of half essentially disconnected coronoids, which often makes the derivation of the pertinent formulas fairly easy. This is for instance the case with the following class of perforated parallellograms (Gutman and Cyvin 1989):

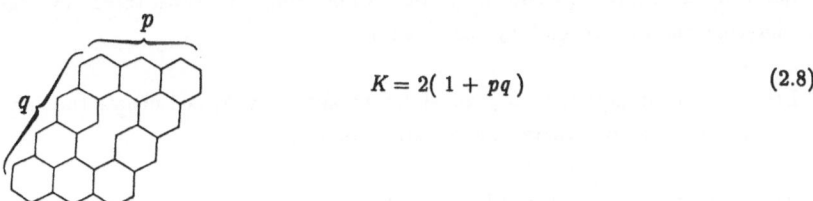

$$K = 2(\,1 + pq\,) \tag{2.8}$$

Each member of this class has exactly two isolated internal vertices.

Probably the most advanced example of a combinatorial K formula for a class of coronoids pertains to the homologous series circumcoronene, circumkekulene, ... :

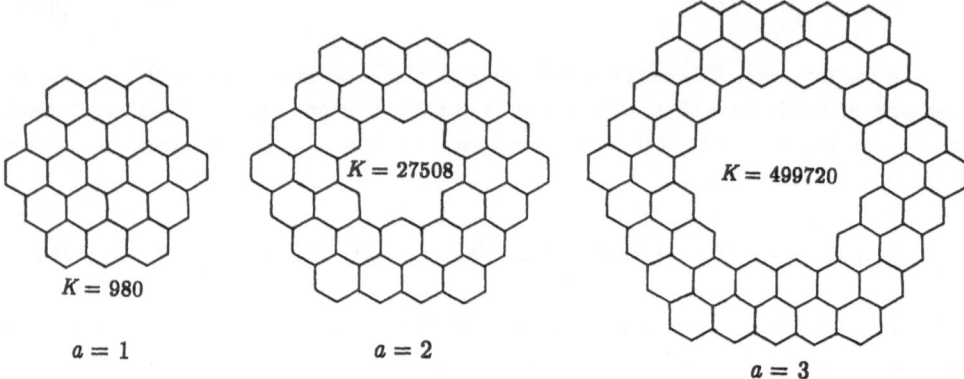

$K = 980$

$a = 1$ $a = 2$

$a = 3$

Here $h = (2a+1)6$ for $a > 1$. Unity must be added for $a = 1$ in order to comply with the benzenoid circumcoronene ($h = 19$). All the other systems ($a > 1$) are regular pericondensed single coronoids. The result for the Kekulé structure counts reads (Cyvin BN, Brunvoll and Cyvin 1988):

$$K(a) = (1/32)(a^2 + 4)(a^2 + 2a + 5)(a^4 + 2a^3 + 9a^2 + 8a + 8)^2$$

$$= (1/32)(a^{12} + 6a^{11} + 39a^{10} + 140a^9 + 483a^8 + 1158a^7 + 2577a^6$$
$$+ 4200a^5 + 6180a^4 + 6464a^3 + 5760a^2 + 3072a + 1280) \qquad (2.9)$$

2.5.2 *Algorithm and Annulenoid Kekulé Structures for Primitive Coronoids*

It has been made diligent use of the method of fragmentation for enumerations of Kekulé structures (Hall 1973; Randić 1976b; Cyvin SJ and Gutman 1988) in the derivations of the combinatorial formulas referred to in the preceding paragraph. This method is also the basis of a useful algorithm for the K numbers of primitive single coronoids (Bergan, Cyvin SJ and Cyvin 1986), which has been quoted before (Gutman and Cyvin 1989; Trinajstić 1992b). In the wording of the original source (where only the word "ring" is substituted by "hexagon", and "corona–condensed benzenoid" by "coronoid"):

(1) Delete an (arbitrary) angularly annelated hexagon, say A, and denote the number of Kekulé structures of the resulting (cata–condensed) chain by k.

(2) Delete A along with the adjacent linear chains (including the two nearest angularly

annelated hexagons). Denote the number of Kekulé structures of the remaining chain by k'.

(3) The total number for the coronoid is $K = k + k' + 2$.

The algorithm is illustrated below; the same example as in Gutman and Cyvin (1989) is used, but a different angularly annelated hexagon (of mode A_2) has been chosen as A.

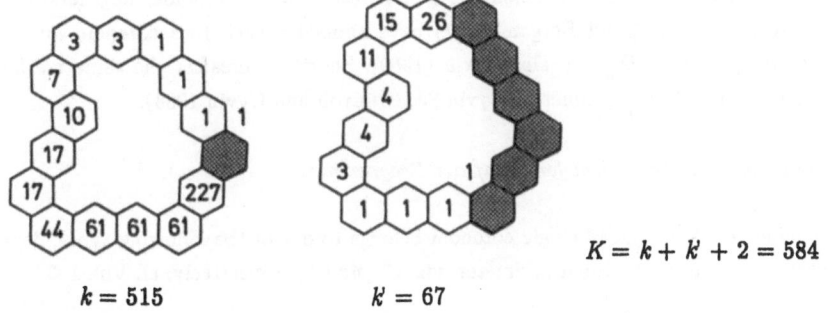

$$K = k + k' + 2 = 584$$

$$k = 515 \qquad\qquad k' = 67$$

The origin to the number 2 under point (3) of the above algorithm are the two annulenoid Kekulé structures, which have been referred to as *proper annulenoid* or in–phase (Cyvin SJ 1990). They are indicated below.

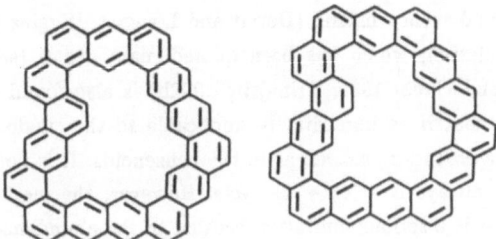

There exist also two *improper annulenoid* or out–of–phase Kekulé structures of a primitive coronoid (Cyvin SJ 1990). In the present example:

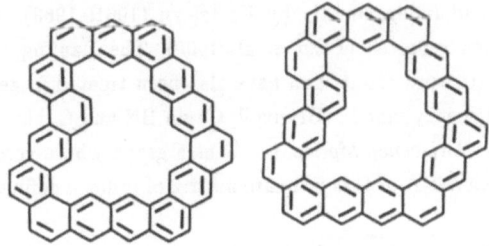

The improper annulenoid Kekulé structures possess aromatic sextets in contrast to the proper ones, and they are included in the sets of k and k' Kekulé structures, one in each set.

The method of fragmentation (see above) has been exploited in an elaborate version referred to as *symmetry–adapted method of fragmentation* (SAMF); cf. Cyvin SJ, Brunvoll and Cyvin (1989a). The SAMF has been applied most frequently to systems of hexagonal symmetry, and perhaps most intensely in the derivation of eqn. (9) by Cyvin BN, Brunvoll and Cyvin (1988). In Cyvin SJ, Bergan and Cyvin (1987) it is observed that the SAMF applied to a benzenoid may lead to fragments which are generalized single coronoids, degenerate or not. Cyvin SJ, Cyvin, Brunvoll and Bergan (1987) have applied SAMF to a coronoid for the first time, to one of symmetry D_{6h}; cf. also Cyvin (1989). Shortly thereafter the same method was applied to a coronoid of D_{3h} symmetry (Cyvin SJ, Brunvoll and Cyvin 1988).

2.5.3 *General Solution Convenient for Computer Programming*

Introduction. A Kekuléan single coronoid belongs to one of the following types, depending on the nature of its outer and inner perimeter, viz. C' and C", respectively; cf. Vol. I–3.2.2.

(a) Both C' and C" are $[4k + 2]$–cycles.
(b) Both C' and C" are $[4k]$–cycles.

If C' is a $[4k + 2]$–cycle and C" a $[4k]$–cycle, or vice versa, then the system is non–Kekuléan. It is well known that for a Kekuléan benzenoid the Kekulé structure count is obtainable by means of the determinant of the adjacency matrix (Dewar and Longuet–Higgins 1952; Cvetković et al. 1974). The precise formulation, which has been quoted many times (see, e.g. Cyvin SJ and Gutman 1988; Gutman and Cyvin 1989; Trinajstić 1992b) is also found in the following. The method of the adjacency matrix is immediately applicable to the single coronoids of type (a) according to the above classification, as well as to the benzenoids. It is noted that the perimeter of a Kekuléan benzenoid always is a $[4k + 2]$–cycle. However, the method fails for the single coronoids of type (b). This is a serious limitation because of the abundance of Kekuléan systems of this category. The smallest representative is cyclo[*d.e.e.d.e.e.d.e.e*]nonakisbenzene (C_3 of Fig. 1.1). Cyvin SJ, Brunvoll and Cyvin (1990c) have adopted a method using the Pfaffian of a skew–symmetric adjacency matrix as a general solution to the enumeration problem of Kekulé structures of single coronoids. The underlying theory was developed by Temperley and Fisher (1961), Fisher (1961), and independently by Kasteleyn (1961; 1963). An application to buck–minsterfullerene has been reported (Elser et al. 1990). The Pauling bond orders are strongly related to the Kekulé structure counts and have also been treated in general terms by means of the skew–symmetric adjacency matrix (Brunvoll, Cyvin BN and Cyvin 1992a).

Application of the Adjacency Matrix. Let G be a graph whose vertices are labeled 1, 2, ..., n. Then the *adjacency matrix* **A** of G is a square matrix of order n defined by:

$$(\mathbf{A})_{uv} = \begin{cases} 1 \text{ if the vertices u and v are adjacent} \\ 0 \text{ otherwise} \end{cases}$$

It is assumed that the vertices u and v have numbers u and v, respectively. The following theorem holds for the K number of Kekulé structures of G.

$$\det \mathbf{A} = (-1)^{n/2}\, K^2 \tag{2.10}$$

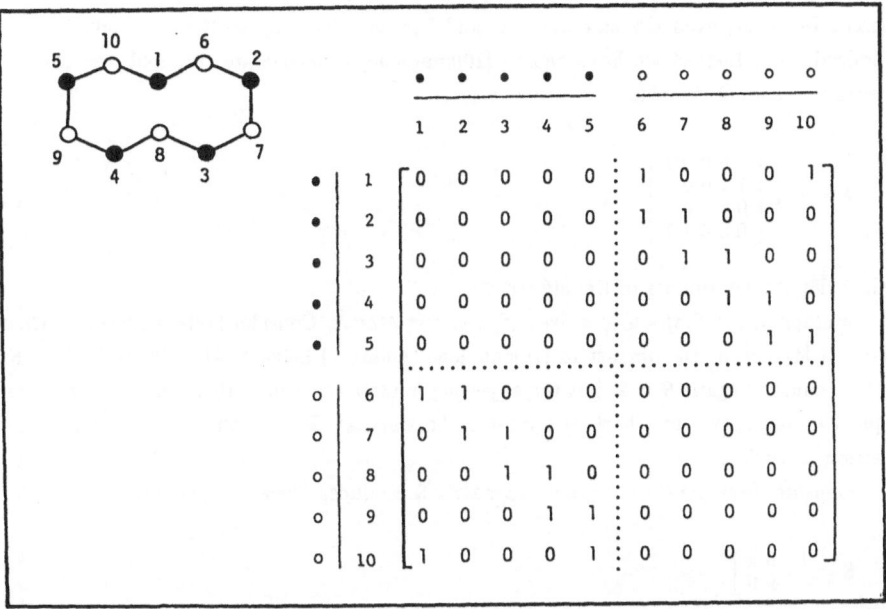

Chart 2.1. The adjacency matrix of [10]annulene and its partitioning; notice the indicated coloring and numbering of vertices.

Assume that G is a benzenoid or a coronoid and that the black and white vertices are numbered consecutively, adopting the convention that the black vertices are numbered first. Then the adjacency matrix will clearly be partitioned into submatrices as:

$$\mathbf{A} = \begin{bmatrix} 0 & a \\ \tilde{a} & 0 \end{bmatrix} \tag{2.11}$$

where each block is a square matrix of the order $n/2$. The **A** matrix is obviously symmetric. For the determinant of (10) one obtains

$$\det \mathbf{A} = (-1)^{n/2}(\det a)^2 \tag{2.12}$$

and therefore

$$K = |\det \mathbf{a}| \tag{2.13}$$

This is the fundamental relation for computing Kekulé structure counts of benzenoids and of coronoids of type (a). The application of (13) is convenient for computer programming (Brown 1983).

An example of the application of eqn. (13) seems to be warranted, and preferably to a coronoid. However, even the smallest coronoid has an adjacency matrix too voluminous to be reproduced here. Instead we have chosen [10]annulene, a degenerate coronoid; see Chart 1. In this example, eqn. (13) yields

$$K = \det \begin{bmatrix} 1\,0\,0\,0\,1 \\ 1\,1\,0\,0\,0 \\ 0\,1\,1\,0\,0 \\ 0\,0\,0\,1\,1 \end{bmatrix} = 2 \tag{2.14}$$

which is the correct number of Kekulé structures.

Method of the Pfaffian to a Skew–Symmetric Matrix. Consider [12]annulene (see Chart 2), which is a [4k]–cycle, in contrast to [10]annulene (Chart 1) being a [4k + 2]–cycle. The Kekulé structure count is again $K = 2$, but the adjacency matrix (as well as its submatrices according to the partitioning shown in Chart 2) appear to be singular. Hence eqn. (13) would give zero. A correction is needed.

Consider first the skew–symmetric matrix $\mathbf{S}$ produced from eqn. (11) as:

$$\mathbf{S} = \begin{bmatrix} 0 & \mathbf{a} \\ -\mathbf{\tilde{a}} & 0 \end{bmatrix} \tag{2.15}$$

By virtue of eqn. (12) one obtains

$$\det \mathbf{S} = K^2 \tag{2.16}$$

and consequently

$$K = (\det \mathbf{S})^{1/2} = \mathrm{Pf}\,\mathbf{S} \tag{2.17}$$

The *skew–symmetric adjacency matrix* $\mathbf{S}$ of a graph G is defined with respect to oriented edges (indicated by arrows): the edge e_{uv} between u and v may be oriented in the sense $u \to v$ or in the opposite sense $v \to u$. Then $\mathbf{S}$ is defined by:

$$(\mathbf{S})_{uv} = \begin{cases} +1 & \text{if } e_{uv} \text{ is oriented as } u \to v \\ -1 & \text{if } e_{uv} \text{ is oriented as } v \to u \\ 0 & \text{otherwise (if } u \text{ and } v \text{ are nonadjacent)} \end{cases}$$

It has been proved that eqn. (17) holds in general for a planar graph if $\mathbf{S}$ is the skew–symmetric

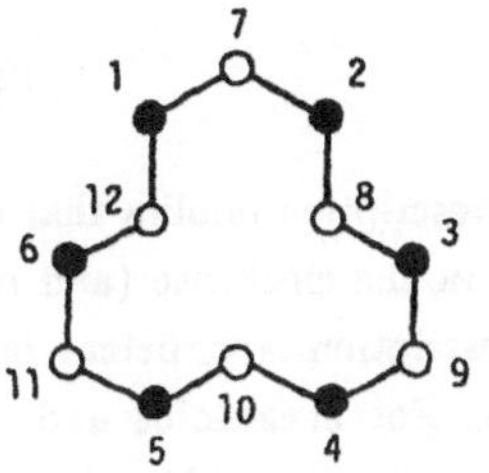

		• 1	• 2	• 3	• 4	• 5	• 6	o 7	o 8	o 9	o 10	o 11	o 12
•	1	0	0	0	0	0	0	1	0	0	0	0	1
•	2	0	0	0	0	0	0	1	1	0	0	0	0
•	3	0	0	0	0	0	0	0	1	1	0	0	0
•	4	0	0	0	0	0	0	0	0	1	1	0	0
•	5	0	0	0	0	0	0	0	0	0	1	1	0
•	6	0	0	0	0	0	0	0	0	0	0	1	1
o	7	1	1	0	0	0	0	0	0	0	0	0	0
o	8	0	1	1	0	0	0	0	0	0	0	0	0
o	9	0	0	1	1	0	0	0	0	0	0	0	0
o	10	0	0	0	1	1	0	0	0	0	0	0	0
o	11	0	0	0	0	1	1	0	0	0	0	0	0
o	12	1	0	0	0	0	1	0	0	0	0	0	0

Chart 2.2. The adjacency matrix of [12]annulene, partitioned in the same way as the similar matrix of Chart 2.1.

adjacency matrix with properly oriented edges. Benzenoids and single coronoids of both types (a) and (b) are planar graphs. Let us make a digression before we give the prescription for a proper orientation of the edges. Assume again a consecutive numbering of the black and white vertices, which resulted in the partitioning of **A** into submatrices as expressed by eqn. (11); cf. also Charts 1 and 2. The corresponding form of the skew–symmetric adjacency matrix is given by

$$\mathbf{S} = \begin{bmatrix} \mathbf{0} & \mathbf{s} \\ -\tilde{\mathbf{s}} & \mathbf{0} \end{bmatrix} \tag{2.18}$$

Hence

$$\det \mathbf{S} = (\det \mathbf{s})^2 \tag{2.19}$$

and, assuming that eqn. (17) is valid, one obtains

$$K = |\det \mathbf{s}| \tag{2.20}$$

It remains to define the proper orientation of the edges in G. The prescription implies that every even–numbered cycle of G should have an odd number of arrows in the clockwise (and in the counterclockwise) direction. For benzenoids and coronoids this prescription is sufficient for the construction of an appropriate skew–symmetric adjacency matrix. For benzenoids and single coronoids of type (a) the prescribed condition is fulfilled if the directions are chosen as (say) consistently going from a black vertex to a white. This rule holds also for [4k + 2]annulenes as in the example of Chart 1:

With this orientation one obtains $\mathbf{s} = \mathbf{a}$ and the skew–symmetric adjacency matrix S as in eqn. (15). Here, in the example of [10]annulene, the "normal" orientation (conventionally black → white) is seen to give five arrows in the clockwise and five in the counterclockwise sense. The important case to consider now is the case of single coronoids of type (b). Here the normal orientation (black → white) would violate the required condition for the two perimeters. A violation occurs also for [4k]annulenes; in the example of [12]annulene (Chart 2) one would get six arrows in the clockwise and six in the counterclockwise sense. We find that a remedy is achieved by introducing one single "defect", namely changing the orientation of one arbitrary edge. This feature is illustrated in Chart 3, where the external arrow points at the edge for which the direction has been changed. Now there are seven arrows in the clockwise and five in the counter– clockwise sense, whereby the requirement about orientations is fulfilled. Chart 3 includes the complete skew–symmetric adjacency matrix S for [12]annulene in consistency with the above discussion. For this matrix eqn. (17) is sound, and eqn. (20) holds for the corresponding submatrix $\mathbf{s}$. It was verified that

$$K = \det \begin{bmatrix} 1 & 0 & 0 & 0 & 0 & -1 \\ 1 & 1 & 0 & 0 & 0 & 0 \\ 0 & 1 & 1 & 0 & 0 & 0 \\ 0 & 0 & 1 & 1 & 0 & 0 \\ 0 & 0 & 0 & 1 & 1 & 0 \\ 0 & 0 & 0 & 0 & 1 & 1 \end{bmatrix} = 2 \tag{2.21}$$

gives the correct Kekulé structure count.

For the primitive coronoid cyclo[$d.e.e.d.e.e.d.e.e$]nonakisbenzene, which is of type (b), we show a proper orientation of the edges in the following diagram. The directions of two edges, viz. one on each of the perimeters (indicated by external arrows), have been changed from the normal (black → white).

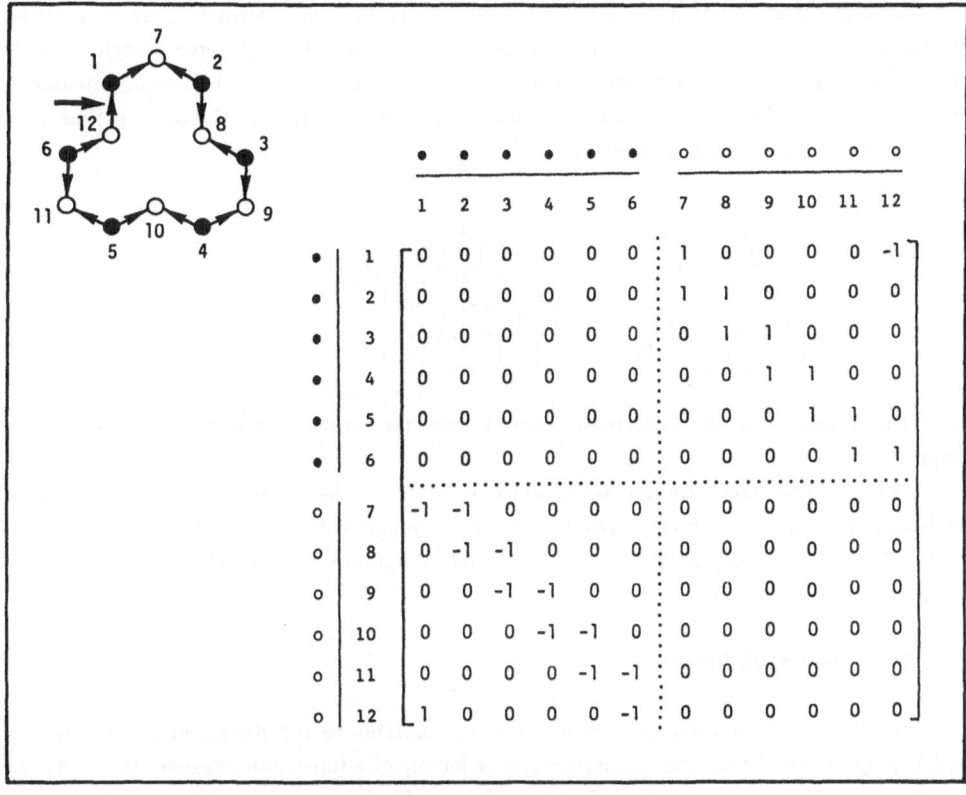

Chart 2.3. The skew–symmetric adjacency matrix of [12]annulene and its partitioning; notice the indicated coloring, numbering of vertices, and orientation of edges (in terms of arrows).

$K = 54$

Here it is adhered to the conventions (I–3.2): (i) All coronoid systems are drawn so that two edges of every hexagon are vertical. (ii) It is agreed that peaks should be colored white, whereby the valleys become black. Under these conventions the normal orientation of edges implies that all vertical edges are supplied with downward arrows, while all oblique edges should have upward arrows, either to the left or to the right. For a single coronoid of type (b) in general, a proper orientation of edges implies that all vertical edges are supplied with downward arrows, while all oblique edges should have upward arrows, either to the left or to the right. For a single coronoid of type (b) in general, a proper orientation is achieved by starting with the normal orientation and changing the directions of a set of parallell edges from the inner to the outer perimeter as illustrated below by two alternatives.

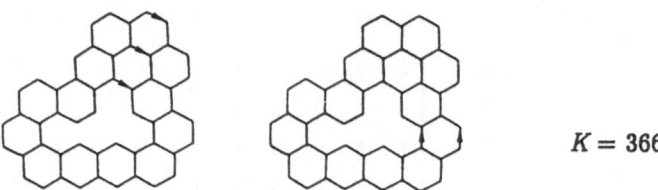

$$K = 366$$

Only the directions which have been reversed from the normal are indicated on the above diagram.

The method under consideration, which is based on the Pfaffian of a skew–symmetric adjacency matrix, has been employed in a computer program (Cyvin SJ, Brunvoll and Cyvin 1990c). Thus it became possible to execute routine computations of Kekulé number counts for single coronoids of type (b).

2.5.4 *Supplementary References*

In Par. 1.3.3 an extensive list of references pertains to the Kekulé structure count of kekulene ($K = 200$). The same paragraph gives a survey of a topic with relevance to the Kekulé structures: the sextet polynomial and sextet patterns of kekulene. Here we shall supplement the above surveys (Par. 1.3.3 and preceding sections of the present chapter) by a few more references.

Some works on Kekulé structure counts of certain classes of single coronoids: primitive coronoids (Sheng 1991) and all–coronoids (Cyvin SJ, Cyvin, Brunvoll and Gutman 1991). The sextet patterns, occasionally formulated explicitly in terms of Clar structures (cf., e.g.: Gutman 1982; Gutman and Cyvin 1989) have been studied by several investigators, who included coronoids in their works, e.g. El–Basil (1988; 1992). The latter work (El–Basil 1992) is closely related to Bergan, Cyvin SJ and Cyvin (1986), a communication referred to several times above. Other works within this area are: Hansen and Zheng (1992a); Guo and Zhang (1992). Other theoretical works of relevance to Kekulé structures: Zhang and Guo (1991); Chen and Guo (1993); Guo and Zhang (1993).

For the sake of completeness we wish to mention (somewhat out of context) that Zhang and Lin (1993), in a note on enumeration of certain equivalence classes, included circumkekulene (which after all is treated within Par. 2.5.1) as an example.

2.6 Isospectral Single Coronoids

Two graphs are said to be isospectral when they have the same characteristic polynomial, viz. $\det|x\, I - A|$, where I and A are the identity and adjacency matrices, respectively; see, e.g. Graovac, Gutman and Trinajstić (1977); Trinajstić (1992b). The characteristic polynomial is a polynomial in x of degree n.

Babić (1993) reported his sensational detection of pairs of isospectral benzenoid systems (cf. also Gutman, Marković and Grbović 1991; Babić and Gutman 1992), although Cioslowski (1991) conjectured: "There are no isospectral benzenoid graphs." However, this conjecture depends heavily on the definition of "benzenoid"; some authors insist that a benzenoid should be Kekuléan (Trinajstić 1990; 1992a; Nikolić et al. 1991). All the systems of Babić (1993) are non–Kekuléan, and it is still an open question whether there exists a pair of Kekuléan isospectral benzenoids. On the other hand, Babić (1993) produced a pair of isospectral coronoids, which are

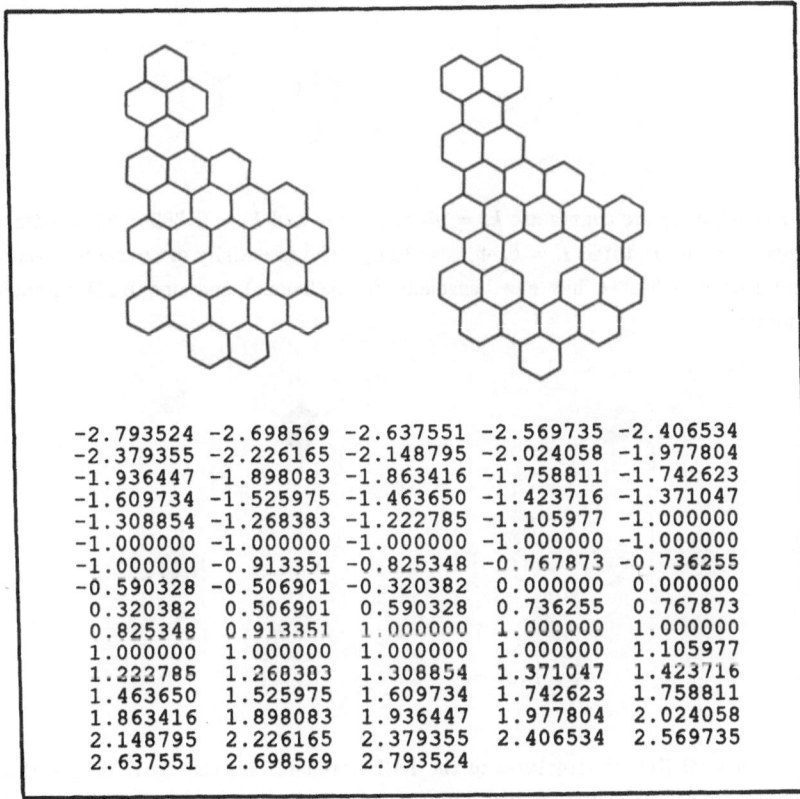

```
-2.793524  -2.698569  -2.637551  -2.569735  -2.406534
-2.379355  -2.226165  -2.148795  -2.024058  -1.977804
-1.936447  -1.898083  -1.863416  -1.758811  -1.742623
-1.609734  -1.525975  -1.463650  -1.423716  -1.371047
-1.308854  -1.268383  -1.222785  -1.105977  -1.000000
-1.000000  -1.000000  -1.000000  -1.000000  -1.000000
-1.000000  -0.913351  -0.825348  -0.767873  -0.736255
-0.590328  -0.506901  -0.320382   0.000000   0.000000
 0.320382   0.506901   0.590328   0.736255   0.767873
 0.825348   0.913351   1.000000   1.000000   1.000000
 1.000000   1.000000   1.000000   1.000000   1.105977
 1.222785   1.268383   1.308854   1.371047   1.423716
 1.463650   1.525975   1.609734   1.742623   1.758811
 1.863416   1.898083   1.936447   1.977804   2.024058
 2.148795   2.226165   2.379355   2.406534   2.569735
 2.637551   2.698569   2.793524
```

Chart 2.4. A pair of isospectral single coronoids and their 78 roots of the characteristic polynomial (x values).

Kekuléan. These two $C_{78}H_{30}$ systems have 24 hexagons each and are reproduced in Chart 4. The 78 roots of the corresponding characteristic polynomials were computed, and it was verified that they indeed are identical; the numerical values are included in Chart 4.

It was also confirmed that the Kekulé structure count is the same for the two isospectral systems under consideration: $K = 12636$. It is interesting that the pair of systems consists of one half essentially disconnected (HED) and one essentially disconnected single coronoid. The two schemes of Kekulé structures for the HED system (left–hand drawing in Chart 4) are indicated below.

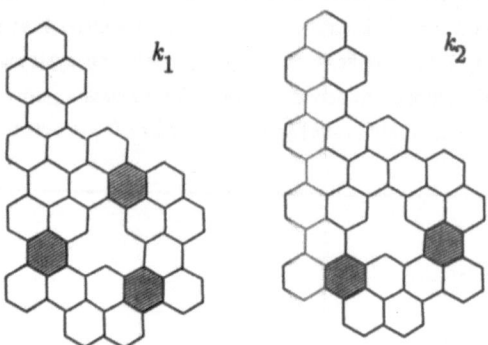

Here the Kekulé structure counts are $k_1 = 9^2 \times 78 = 6318$ and $k_2 = 9 \times 702 = 6318$ when referring to the above diagram. In total, $K = k_1 + k_2 = 12636$. The essentially disconnected system at the right–hand side of Chart 4 has one benzenoid (naphthalene) and one HED coronoid as its effective units:

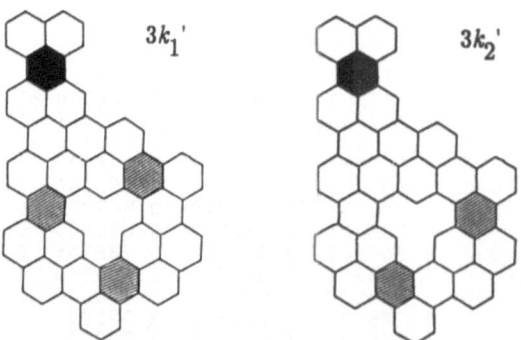

The two schemes of Kekulé structures of the HED coronoid (see the above diagram) are $k_1' = 9^2 \times 26 = 2106$ and $k_2' = 9 \times 234 = 2106$. Hence $k_1' + k_2' = 4212$, which yields for the whole system: $K = 3 \times 4212 = 12636$.

Is the pair of single coronoids in Chart 4 the smallest pair of isospectral coronoids?

2.7 Some Main Classes of Polyhexes

2.7.1 *Introduction*

Benzenoids and single coronoids represent two main classes of polyhexes (Sect. 2.1). The inclusion of multiple coronoids represents an extension in *one* direction. Classes of multiple coronoids are defined and discussed in the next chapter. Here we shall concentrate upon polyhexes without holes (as benzenoids) and those with one hole each (as single coronoids) and discuss an extension which involves geometrically nonplanar (helicenic) systems. We shall find it expedient to define a class where the members are restricted to helicenic systems which can be represented by planar graphs. In the next section, polyhex systems which cannot be represented by planar graphs, are exemplified.

2.7.2 *Hexagonal and Trigonal Lattices, and the Dualist*

A useful restriction on some classes of polyhexes is imposed by means of the hexagonal lattice (Gutman 1982; see also, e.g., Cyvin and Gutman 1988; Gutman and Cyvin 1989; Vol. I–2.1.1), a geometrically planar lattice of congruent hexagons. For benzenoids and coronoids, for instance, it is required that the system can be embedded in the hexagonal lattice. In fact, a benzenoid is defined as a cycle on the hexagonal lattice (the perimeter) and its interior. Similarly, a single coronoid is defined by two cycles on the hexagonal lattice: the inner perimeter, say C'', which defines the corona hole, and the outer perimeter C'. Here C' should completely embrace C''. The coronoid consists of the vertices and edges on C' and C'', as well as those in the interior of C', but outside C'' (cf., e.g., Vol. 2.1.1). An illustration is given below.

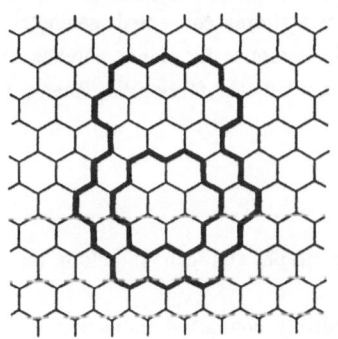 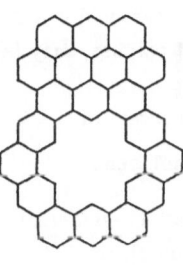

A polyhex which can be embedded in the hexagonal lattice is said to be *embeddable* (in the hexagonal lattice).

It may be convenient to represent an embeddable polyhex system as a dualist (Smith 1961; Balaban and Harary 1968; Bonchev and Balaban 1981; Nikolić, Trinajstić, Knop et al. 1990; Trinajstić 1993b; see also, e.g., Vol. I–2.1.1), in which each vertex corresponds to one hexagon, and two vertices are connected by a dualist edge when the two corresponding hexagons share an edge. A dualist can be embedded in a trigonal lattice as shown below: an example which pertains to the same coronoid as in the above diagram.

The presence of a cycle larger than a triangle in the dualist (Balaban 1969; 1980; Bonchev and Balaban 1981) is characteristic for a coronoid.

2.7.3 Helicenes and Corohelicenes

A helicenic polyhex is embeddable in the hexagonal lattice, but it acquires overlapping edge(s) when drawn in a plane. Collisions of edges are avoided when the systems are thought of as nonplanar so that they can pass from one layer to another on a multilayer lattice. Therefore it is usual to refer to helicenic polyhexes as geometrically nonplanar (cf., e.g., Cyvin BN, Brunvoll and Cyvin 1992b). A helicene has been defined as a simply connected, geometrically nonplanar polyhex, while a benzenoid is a simply connected, geometrically planar polyhex. A fusene, viz. a simply connected polyhex, may be a benzenoid or a helicene (Cyvin SJ, Zhang, Cyvin and Guo 1993):

fusene $\Big\langle$ benzenoid
helicene

Any helicene may be represented by a planar graph and can therefore be referred to as graph–theoretically planar.

The following definition of a class of multiply connected polyhexes is supposed to be analogous to the definition of helicenes in a rational way.

Definition 2.5: A *corohelicene* (system) is a multiply connected, geometrically nonplanar (helicenic) polyhex, which is graph–theoretically planar.

The requirement about graph–theoretical planarity is new. Notice that corohelicenes are not coronoids, but they are included among the "helicenic quasi–coronoids" (Vol. I–2.5.2; Cyvin BN, Brunvoll and Cyvin 1992b). This last term shall be avoided here because of the new and precise definition which is proposed (Definition 2.5). We shall also avoid the term "helicirculene", which has been used previously (Knop, Szymanski, Jeričević and Trinajstić 1984; Knop, Müller, Szymanski and Trinajstić 1985; Cyvin BN, Brunvoll and Cyvin 1992b), and likewise "heli–coronoid" (Trinajstić 1990). Some depictions of corohelicenes are found in the literature (Randić, Nikolić and Trinajstić 1988; Vol. I–Fig. 2.6; Trinajstić 1992a; Cyvin BN, Brunvoll and Cyvin 1992b).

It is reasonable to introduce the term *corofusene* in analogy to fusene according to:

$$\text{corofusene} \Big\langle \begin{matrix} \text{coronoid} \\ \text{corohelicene} \end{matrix}$$

Our term corofusene is not directly interchangeable with "coronafusene" of Balaban (1982). We shall also speak about *single corofusene* and *single corohelicene* when referring to the respective systems with one corona hole each.

Single corofusenes can be produced in copious amounts by additions of hexagons to primitive single coronoids; all the five addition modes (viz. L_1, P_2, L_3, P_4, L_5; cf. Vol. I–Fig. 2.2 and I–Fig. 3.1) are allowed, and hexagons can be added to the inner– as well as to the outer perimeter without limitation. One example:

 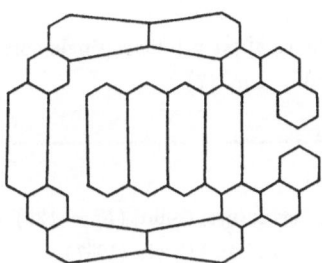

Here the left–hand drawing shows a dualist, while the right–hand drawing demonstrates that the corresponding corohelicene can be represented by a planar graph. The above procedure of generating single corofusenes is not complete for the single corohelicenes. There are unbranched catacondensed corohelicenes which would be missed, as the one in Fig. 4. It is reasonable to refer to the unbranched catacondensed single corohelicenes as *primitive single corohelicenes*, and also the meaning of the designation *primitive single corofusenes* should be obvious (viz. primitive single coronoids + primitive single corohelicenes). Now it is inferred that all the non–primitive single corofusenes are generated by additions of hexagons (as described above) to all the primitive single corofusenes. The smallest primitive single corohelicene has been depicted several times (Bergan, Cyvin BN and Cyvin 1987; Balaban, Brunvoll and Cyvin 1991; Vol. I–Fig. 2.6; Cyvin BN, Brunvoll and Cyvin 1992b) and is also reproduced in Fig. 5.

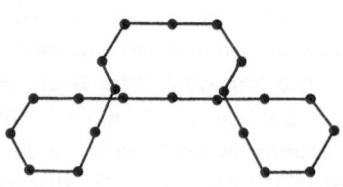

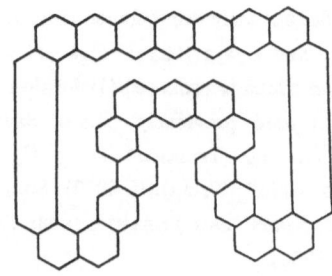

Fig. 2.4. A primitive single corohelicene, $C_{88}H_{44}$ ($h = 22$). It has $K = 8284$, as was found by the algorithm of Sect. 2.5.2.

Fig. 2.5. The smallest primitive single corohelicene, $C_{40}H_{20}$ ($h = 10$). It has $K = L_9 + 2 = 125$, as was found by eqn. (7).

The Kekulé structure count ($K = 125$) for this system was reported previously at least by Bergan, Cyvin BN and Cyvin (1987).

Having the graph–theoretical planarity in mind, it is clear that the outer and inner perimeters of a single corohelicene always can be identified. However, an identification of the corona hole with a benzenoid may be obscured. Therefore we must rephrase the requirement that the corona hole should have a size of at least two hexagons (cf. Sect. 2.1). The following formulation is valid for corofusenes: any inner perimeter should consist of at least ten edges (and vertices).

2.7.4 Planarity and Nonplanarity

It should be made clear that the geometrical nonplanarity of helicenic polyhexes is not taken into account when the isomorphic systems of this category are enumerated and when symmetry groups are attributed to these systems. It has been ascertained, for instance, that all the helicenes (helicenic systems) which correspond to the normal helicenes (polycyclic hydrocarbons) possess the symmetry C_{2v} (Cyvin SJ, Zhang, Cyvin, Guo and Brunvoll 1992; Cyvin SJ, Zhang, Cyvin and Guo 1993). Nevertheless, the $C_{4h+2}H_{2h+4}$ normal helicenes or [h]helicenes which are known chemically, have coiled (helical) structures. Corresponding features apply also to corohelicenes. The below diagram shows an obvious example, where [10]helicene is condensed to a primitive single coronoid.

(a)   (b)   (c)

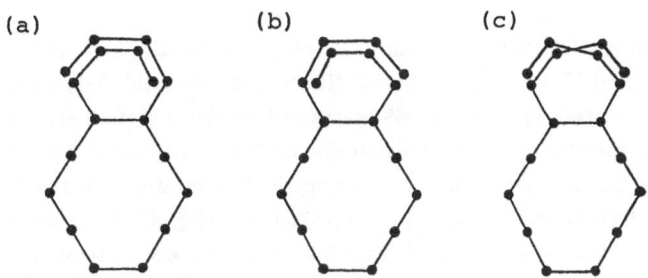

In (a) and (b) the coiled structure of [10]helicene is suggested; it gives rise to two enantiomorphic forms. However, these structures represent one system of C_{2v} symmetry. In (c) this symmetry is accentuated. Considerations of this kind apply also to primitive corohelicenes (and additions to such systems). The above principles have been followed in practically all works on enumerations of polyhexes; cf., e.g. Cyvin BN, Brunvoll and Cyvin (1992b) with references cited therein. However, there is a notable exception described in the following.

Herndon (1990), in his chemically–oriented enumeration of simply connected (Kekuléan) polyhexes, has not only counted enantiomer forms of helicenes. He has got a step further, taking into account nonplanarity and enantiomerism in systems like benzo[c]phenanthrene and dibenzo–[c,g]phenanthrene (pentahelicene), which actually correspond to geometrically planar (nonheli–cenic) systems according to the definitions adopted here. In general, Herndon (1990) has taken into account nonplanarity of structures with coves and fjords in accord with chemical evidence. In order to avoid confusion, we shall speak about *structural planarity* and *structural nonplanarity* when referring to the actual chemical structures of the hydrocarbons in question, either determined experimentally or expected, occasionally with support by theoretical studies. In consequence, these concepts are not defined rigorously in mathematical terms.

Cycloacenes (see Sect. 1.7) are polyhexes which can be represented by planar graphs, as is demonstrated in Fig. 1.9 (a) and (b). The structures are undoubtedly structurally nonplanar; this was apparently the kind of planarity Herndon (1988) had in mind when writing the treatise which was mentioned in Sect. 1.8. A cycloacene polyhex is not embeddable in the hexagonal lattice. Hence it goes beyond the present definitions to specify whether it is helicenic or nonhelicenic (although the latter alternative is the more reasonable one), and whether it is geometrically planar or nonplanar. Graph–theoretically planar polyhexes which are not embeddable, have also been presented elsewhere (e.g. in: Bergan, Cyvin BN and Cyvin 1987; Cyvin SJ, Cyvin and Brunvoll 1991).

It remains to treat graph—theoretically nonplanar polyhexes. We shall not attempt to give any systematic classification of the variety of these systems, but present some examples in the next section.

2.8 Examples of Graph—Theoretically Nonplanar Polyhexes

2.8.1 *Cyclohelicenes*

Already Balaban (1969) mentioned (in a footnote) an interesting polyhex structure, which later has been designated "figure—eight" (Randić, Nikolić and Trinajstić 1988; Balaban 1988); see Fig. 6. Randić, Nikolić and Trinajstić (1988) characterized this structure as: "A twisted loop overlapping without intersection." They called it cyclohelicene, and indeed it can be interpreted as two hexahelicenes fused to each other in a macrocycle. The mentioned authors reported $K = 260$ as the Kekulé structure count for $C_{48}H_{24}$ cyclohelicene (Fig. 6). One year later, Randić, Gimarc et al. (1989) offered a more detailed theoretical study of the same structure.

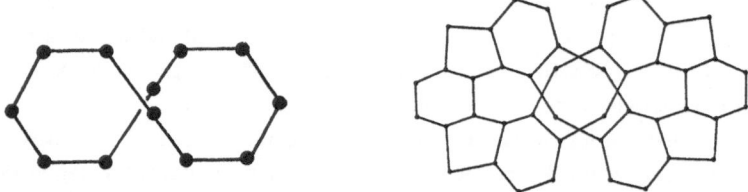

Fig. 2.6. An isomer of $C_{48}H_{24}$ [12]cyclohelicene: the dualist representation (left) and a nonplanar graph (right).

The polyhex under consideration (Fig. 6) is embeddable in the hexagonal lattice, and it has overlapping edges or, more precisely, two overlapping hexagons. Hence it is reasonable to consider this system as helicenic, and that has actually been done previously (Randić, Nikolić and Trinajstić 1988; Randić, Gimarc et al. 1989; Cyvin BN, Brunvoll and Cyvin 1992b). However, cyclohelicene is not a corohelicene, and of course not a coronoid. It is true that two perimeters are present, but they are symmetrically equivalent; none of them qualifies to be identified especially as the outer— or inner perimeter. There are two enantiomers of the structure in question (Fig. 6).

The designation "[12]cyclohelicene" (Randić, Gimarc et al. 1989) is ambiguous. Another "figure—eight" isomer of $C_{48}H_{24}$ is depicted in Fig. 7. It was put forward by Balaban (1988) and is perhaps even more interesting than the structure of Fig. 6. The two $C_{48}H_{24}$ isomers under

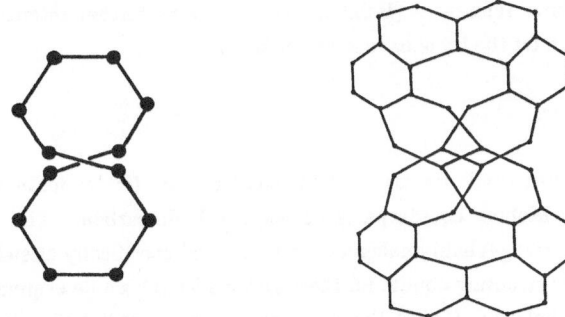

Fig. 2.7. Another isomer of $C_{48}H_{24}$ [12]cyclohelicene; cf. also Fig. 2.6.

consideration display basically the same properties, but some differences are pointed out in the following. All the hexagons in the system of Fig. 7 are angularly annelated, while the system of Fig. 6 has two linearly annelated hexagons. Hence eqn. (7) is applicable to the system of Fig. 7 and yields $K = L_{11} + 2 = 324$. This system, when embedded in the hexagonal lattice, exhibits two pairs of overlapping hexagons.

Balaban (1969; 1988) depicted actually the two enantiomorphic forms of a [14]cyclo-helicene (rather than [12]cyclohelicene) and augmented it to a system with 22 hexagons and two loops:

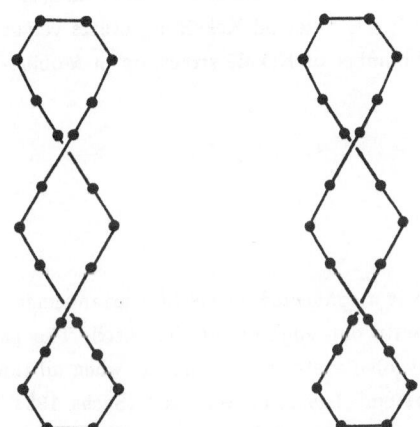

Here the left–hand system is graph–theoretically nonplanar like the systems of Figs. 6 and 7. However, the right–hand isomer is a graph–theoretically planar system and actually a corohelicene similar to the system of Fig. 4.

One of the above references (Balaban 1988) contains further interesting suggestions for polyhexes, constructed on the basis of topological knots.

2.8.2 *Möbius–Polyhexes*

Already Balaban and Harary (1968) pointed out (in a footnote) the hypothetic possibilities to form Möbius counterparts of polyhex hydrocarbons. The Möbius versions of cyclacenes and their zigzag–chain analogues are mentioned specifically elsewhere (Balaban 1980; 1988), and the Kekulé structure counts for these systems have been determined (Cyvin 1990).

Take $C_{48}H_{24}$ kekulene (C_1 of Fig. 1.1), cut it along a radial edge (not on a perimeter), flip around, and put the system together again. Then the Möbius counterpart of kekulene emerges, viz. $C_{48}H_{24}$ Möbius–kekulene (Cyvin 1990):

Notice that the two perimeters of kekulene have merged into one. In consequence, Möbius–kekulene has only two annulenoid Kekulé structures versus the number four in kekulene (cf. Par. 2.5.2). The total number of Kekulé structures in Möbius–kekulene is $K = 198$ (Cyvin 1990).

2.9 Polygonal Systems

2.9.1 *Introduction*

A polygonal system is a connected geometrical arrangement of polygons so that any two hexagons either share exactly one edge or are disjointed. The polyhexes (cf. Sect. 2.1) form clearly a subset of the polygonal systems; they emerge when all the polygons are hexagons. The designation "hexagonal system" has also been used (Sachs 1984, and others). Polypentagons (Cyvin SJ, Cyvin, Brunvoll, Brendsdal et al. 1993) are polygonal systems consisting of exclusively pentagons. The polygonal systems as chemical graphs represent polycyclic conjugated hydrocarbons with arbitrary ring sizes. A useful compilation of chemically known compounds of this category is available (Dias 1988). A recent enumeration work on simply connected polygonal systems (isomers of completely condensed polycyclic conjugated hydrocarbons with arbitrary ring sizes) has been carried out through systems with five polygons each (Cyvin BN, Brunvoll and Cyvin 1993; Cyvin SJ, Brunvoll and Cyvin 1993; Brunvoll, Cyvin BN and Cyvin 1993a; 1993c; Cyvin SJ, Cyvin and Brunvoll 1993).

Many of the definitions for polyhexes can be generalized straightforwardly to the poly–gonal systems, such as: simply/multiply connected; catacondensed/pericondensed; Kekuléan/ non–Kekuléan; graph–theoretically planar/nonplanar. The definition of helicenic systems, coupled with geometrical planarity/nonplanarity on the other hand, is not at all straightforward. Namely, we have no tool in analogy with the hexagonal lattice for polygonal systems in general. This is, however, the case for a subclass of polygonal systems described in the next paragraph.

2.9.2 *Mono–q–Polyhexes*

A mono–q–polyhex (Cyvin 1992a) is a polygonal system with exactly one q–gon and otherwise hexagons (if any). For $q = 6$ these systems reduce to polyhexes. The fluoranthenoids (hydrocarbons) contain $C_{16}H_{10}$ fluoranthene as a key molecule and have in general even–carbon C_nH_s formulas (i.e. with even numbers n and s), while the fluorenoids have odd–carbon C_nH_s formulas (n and s odd); cf.: Dias (1991b; 1992a); Cyvin SJ (1992b); Cyvin SJ, Cyvin and Brunvoll (1993b); Cyvin SJ, Cyvin, Brunvoll, Zhang, Guo and Tošić (1993). Both of these classes (viz. fluoranthenoids/fluorenoids) are represented by monopentapolyhex ($q = 5$) systems. Another important fluoranthenoid (monopentapolyhex) is $C_{20}H_{10}$ corannulene (Sect. 1.6). Similarly, biphenylenoids (Cyvin SJ, Cyvin and Brunvoll 1993a) contain $C_{12}H_8$ biphenylene among both even–carbon and odd–carbon C_nH_s compounds; they are represented by monotetrapolyhexes ($q = 4$).

The mono–q–hexagonal lattices (Cyvin 1992a) are analogous with the hexagonal lattice. Examples for $q = 5$ and $q = 7$ are shown below.

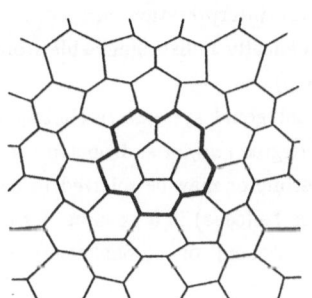 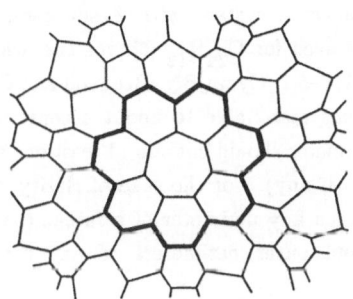

In both cases of the above diagram, the perimeter of [q]circulene viz. [5]circulene and [7]circulene, are drawn by heavy lines (cf. Vol. I–1.3, and notice in particular that corannulene is identical with [5]circulene).

The introduction of mono–q–hexagonal lattices makes it feasible to adapt the notions of geometrically planar/nonplanar and nonhelicenic/helicenic systems to mono–q–polyhexes in general.

2.9.3 *Holes and Polygons*

It was said that kekulene (C_1 of Fig. 1.1) and antikekulene (Fig. 1.7) exhibit an obvious resemblance, but this is a somewhat superficial viewpoint. The resemblance with $C_{48}H_{24}$ kekulene is still more striking for another isomer of $C_{48}H_{24}$, viz. [12]circulene:

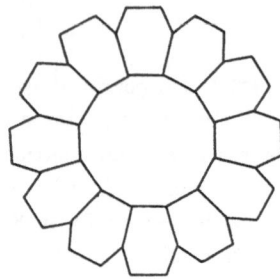

Both kekulene and [12]circulene, as well as antikekulene, can be interpreted as twelve polygons in a macrocycle. As such they are multiply connected polygonal systems with twelve polygons each. That is at least the current interpretation for kekulene. Antikekulene and [12]circulene on the other hand, are simply connected polygonal systems with thirteen hexagons each.

When is a cycle the circumference of a polygon, and when does it surround a hole? This is basically a matter of definition. A [q]circulene, for instance, may either be interpreted as a simply connected polygonal system with $q + 1$ polygons as in the above discussion, but also as a multiply connected system with q hexagons. The latter interpretation has been adhered to several times even for $C_{24}H_{12}$ [6]circulene, which is chemically indistinguishable from coronene (cf., e.g., Vol. I–5.1; Cyvin BN, Brunvoll and Cyvin 1992b).

It seems reasonable to adopt a convention for polygonal systems (including polyhexes) that any perimeter should contain at least one vertex of degree two, in addition to the vertices of degree three (if any). For the sake of clarity, such a perimeter may be referred to as a *genuine perimeter*. Then any perimeter of a fusene (benzenoid or helicene) is a genuine perimeter. Also the inner and outer perimeters of any corofusene (coronoid or corohelicene) are genuine perimeters.

The above convention excludes the interpretation of [q]circulenes as multiply connected systems. Notice that the above definition of a genuine perimeter implies, in the chemical context, that hydrogen atoms are attached to it.

2.9.4 *Cluster Systems*

In a detailed treatment of polypentagons (Cyvin SJ, Cyvin, Brunvoll, Brendsdal et al. 1993) a polygonal lattice consisting of exclusively pentagons was introduced. This lattice is finite and can be realized by congruent regular pentagons, which close into a regular dodecahedron. The complete lattice consists of eleven pentagons, while the "missing" pentagon appears as the boundary of the dodecahedron graph. By means of this lattice the proper polypentagons were defined as simply connected systems without overlapping edges or, in other words, nonhelicenic. It was found that there are exactly 39 proper polypentagons. One of them is the system of eleven pentagons, which can be represented by a planar graph or the complete pentagonal lattice. It corresponds to the (structurally nonplanar) C_{20} dodecahedron cluster and is therefore referred to as a cluster system. Notice that the pertinent graph has no genuine perimeter! The boundary is a pentagon.

The above description explains much of the properties of the cluster system buckminsterfullerene (Sect. 1.8). In the planar graph of Fig. 1.10 a pentagon was chosen as the boundary; one could equally well choose a hexagon.

Chapter 3

BENZENOIDS, SINGLE CORONOIDS

AND MULTIPLE CORONOIDS

3.1 General Considerations, Basic Definitions, and Terminology

This chapter deals with certain nonhelicenic (geometrically planar) polyhexes, which may be simply connected (without holes) or multiply connected with a varying number of corona holes. It is recalled that a corona hole by definition should have a size of at least two hexagons.

The members of the class under consideration shall be referred to as g–polyhexes, where g is the number of corona holes: $g = 0$ for benzenoids, $g = 1$ for single coronoids, and $g > 1$ for multiple coronoids. Following Hall (1988) we refer to the number of corona holes (g) as the *genus* of the system. Multiple coronoids are classified into double –, triple –, quadruple coronoids, etc., which pertain to $g = 2, 3, 4, \ldots$, respectively. In general the term g–*tuple* ($g \geq 1$) coronoid is used about single and multiple coronoids. Within the framework of the present terminology we can also use the designations 0–polyhexes for benzenoids, 1–polyhexes for single coronoids, 2–polyhexes for double coronoids, etc.

It is natural to consider g–polyhexes as a class since many general properties of these systems can be formulated and can, in the best cases, be expressed by general functions of g. Here we are especially interested in such general properties, while specific treatment of single coronoids is the main subject of the remaining chapters. The present chapter is only an introduction to the theory of g–polyhexes.

3.2 Invariants and Relations Between Them

3.2.1 *Specifications*

A basic set of invariants for a g–polyhex is listed in the following (# means "number of"). Some relations, which support the specifications of these invariants, are included in parentheses.

Listing

$h = $ # hexagons;
$n = $ # vertices;
$m = $ # edges;
$n_i = $ # internal vertices;

n_b = # boundary vertices (= $n - n_i$);

s = # vertices of degree two ($\equiv n_2$);

t = # boundary vertices of degree three (= $n_b - s$);

n_3 = # vertices of degree three (= $n - s$);

m_i = # internal edges (= $m - n_b$);

d_s = difference between # disconnections and # connections for the internal edges.

The hydrocarbon which corresponds to a polyhex under consideration, has the chemical formula $C_n H_s$. Here the number of carbon atoms (n) corresponds to the number of vertices (see the above listing). The number of hydrogens (s) is equivalent to the number of secondary carbon atoms and corresponds to the number of vertices of degree two. These vertices are exclusively on the perimeters (inner and outer) of the polyhex. The total number of tertiary carbon atoms on the perimeters is t (see above). The total number of the boundary vertices (on the perimeter), viz. n_b, is also the total number of edges on the perimeters. This number (n_b) is the combined perimeter length (i.e. the sum of the lengths for the outer and all the inner perimeters).

The invariant d_s (Dias 1982a; 1982b; 1987) has been referred to as the Dias parameter (Brunvoll and Cyvin SJ 1990). The interpretation given in the above listing does not apply to benzene (C_6H_6; $g = 0$, $h = 1$), which has $d_s = -1$. However, the functions in terms of other invariants (see below) may be taken as a general definition of d_s.

3.2.2 *Relations*

Most of the pairs among the invariants considered here (Par. 3.2.1) are independent for a given g. Only h and n_3 are not independent, and the three invariants n_b, s and t are mutually not independent. Anyone of the invariants can be expressed by two arbitrary independent invariants through linear combinations, where g in most cases appears as a parameter.

Relations of this kind have been derived independently by several researchers. Gutman and Cyvin (1989) argued that it is easy to deduce such formulas, which therefore should not be associated with anybody's name. Nevertheless these authors give some references to works in this area for benzenoids ($g = 0$); a more extensive list follows: Gutman (1974); Polansky and Rouvray (1976); Dias (1982a; 1982b; 1987); Cyvin SJ (1983); Sachs (1984); Gutman (1985). In Vol. I the relations in question are treated extensively for single coronoids ($g = 1$) with proper documentations of previous works. For g–polyhexes in general the invariant relationships are treated most extensively in Polansky and Rouvray (1976) and in Hall (1988), and in a more rudimentary way elsewhere (Dias 1982a; He WJ and He 1986; Cyvin SJ and Brunvoll 1989; 1990).

In the derivation of the relations between invariants of polyhexes the Euler relation is frequently invoked (se, e.g. Hall 1988). In our notation it reads

$$h + g = m - n + 1 \tag{3.1}$$

Table 3.1. Invariants of g–polyhexes: benzenoids ($g = 0$), single coronoids ($g = 1$) and multiple coronoids ($g > 1$).[*]

Invariant	Function of (h, n_i)	Function of (n, s)
h	h	$(1/2)(n - s) + 1 - g$
n	$4h - n_i + 2 - 2g$	n
m	$5h - n_i + 1 - g$	$(1/2)(3n - s)$
n_i	n_i	$n - 2s + 6 - 6g$
n_b	$4h - 2n_i + 2 - 2g$	$2s - 6 + 6g$
$s \equiv n_2$	$2h - n_i + 4 - 4g$	s
t	$2h - n_i - 2 + 2g$	$s - 6 + 6g$
n_3	$2h - 2 + 2g$	$n - s$
m_i	$h + n_i - 1 + g$	$(1/2)(3n - 5s) + 6 - 6g$
d_s	$h - n_i - 2 + g$	$(1/2)(3s - n) - 7 + 6g$

[*] In Polansky and Rouvray (1976) the relation which corresponds to $m(n, s)$ is given erroneously in their table, but correctly in the text. In Hall (1988) the function $d_s(n, s)$ is wrong, probably due to a misprint.

when adapted to the g–polyhexes. Here we shall not go into further details before presenting complete sets of formulas for the invariants considered above. They are found in Table 1 as functions of the pairs (h, n_i) and of (n, s). The expressions in terms of (n, s) are especially important because of their relevance to the $C_n H_s$ formulas (see above).

It is noteworthy that the function $m(n, s)$ is independent of g (cf. Table 1). The same feature for $n_3(n, s)$ is more obvious.

For the dualist of a g–polyhex we write the Euler relation:

$$t + g = m - n + 1 \tag{3.2}$$

Here t is the number of triangles, while m and n are used to denote the numbers of dualist edges and dualist vertices, respectively. The genus (g) reveals itself as the number of cycles larger than a triangle. It is easily perceived that

$$n = h, \quad t = n_i \tag{3.3}$$

when the definition of a dualist is recalled. On combining eqns. (2) and (3) one obtains

$$m = h + n_i - 1 + g \tag{3.4}$$

Hence (cf. Table 1), as could have been said immediately,

$$m = m_i \qquad (3.5)$$

It is mentioned that Polansky and Rouvray (1976) had chosen the following pairs of independent invariants (in our notation) in a tabulation similar to our Table 1: (n, m), (n_2, n_3), (t, m), (n, s). The corresponding table for benzenoids, which also applies to helicenes (Cyvin SJ, Zhang, Cyvin and Guo 1993) is a special case of Table 1 for $g = 0$.

Example: Figure 1 shows a double coronoid, for which the invariants considered here are specified in the legend. In the representation by hexagons (left–hand drawing) the internal edges are given as heavy lines in order to illustrate the Dias parameter (d_s), which is consistent with the interpretation of the listing in Par. 3.2.1. The system of internal edges is separated into ten disjoint parts; hence the number of disconnections is nine. However, there is also one connection represented by the cycle around a hexagon. The net result is $d_s = 9 - 1 = 8$. The value fits into the formulas of Table 1, which all can be verified for this particular example.

3.2.3 *Outer and Inner Perimeters*

In a g–tuple coronoid, the perimeter length n_b can be split into n_b' for the outer perimeter and $(n_b'')_1$, $(n_b'')_2$, $(n_b'')_3$, ..., $(n_b'')_g$ for the inner perimeters of the g corona holes so that

$$n_b = n_b' + \sum_{k=1}^{g} (n_b'')_k \qquad (3.6)$$

Correspondingly we may split the invariants s and t, which also pertain to the perimeters.

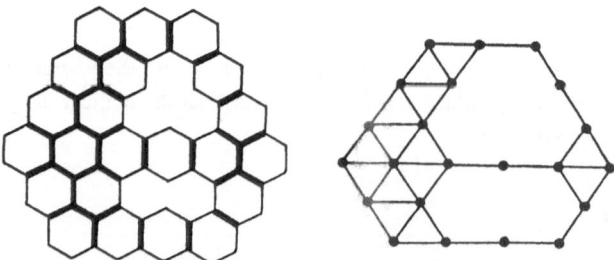

Fig. 3.1. A $C_{72}H_{26}$ 2–polyhex (double coronoid), represented in terms of hexagons (left) and as a dualist (right). It has: $g = 2$, $h = 22$, $n = 72$, $m = 95$, $n_i = 14$, $n_b = 58$, $s = 26$, $t = 32$, $n_3 = 46$, $m_i = 37$, $d_s = 8$; for the dualist representation: $t = 14$, $n = 22$, $m = 37$.

We write:

$$s = s' + \sum_{k=1}^{g} s_k''$$ (3.7)

$$t = t' + \sum_{k=1}^{g} t_k''$$ (3.8)

The invariants of the outer perimeter obey the relations for benzenoids, which may be obtained from the appropriate functions of Table 1 as special cases for $g = 0$. (It is mentioned by passing that n_b' in that case corresponds to the invariant often denoted by n_e in benzenoids and referred to as the number of external vertices, equal to the number of external edges or the unique perimeter length.) Now we have

$$n_b' = s' + t' = 2s' - 6$$ (3.9)

and

$$s' - t' = 6$$ (3.10)

When it comes to the inner perimeters it is expedient to invoke the interpretation of the corona holes as benzenoids. The benzenoid which corresponds to a corona hole is just defined by the inner perimeter of the hole as its unique perimeter. Let the numbers of external vertices of degree two and of degree three on the perimeters of these benzenoids be identified by the symbols $s_1°, s_2°, s_3°, ..., s_g°$ and $t_1°, t_2°, t_3°, ..., t_g°$, respectively. Now we have

$$(n_b'')_k = s_k'' + t_k'' ; \quad k = 1, 2, 3, ..., g$$ (3.11)

as in the first part of eqn. (9). However, for the individual terms on the right—hand side of (11) we observe a reciprocity expressed by:

$$s_k'' = t_k° , \quad t_k'' = s_k° ; \quad k = 1, 2, 3, ..., g$$ (3.12)

Consequently, by virtue of the connection

$$s_k° - t_k° = 6 ; \quad k = 1, 2, 3, ..., g$$ (3.13)

obtained from the relations for benzenoids, we attain at

$$t_k'' - s_k'' = 6 ; \quad k = 1, 2, 3, ..., g$$ (3.14)

It follows

$$(n_b'')_k = 2s_k'' + 6 ; \quad k = 1, 2, 3, ..., g$$ (3.15)

It is implied that the above equations (11) — (15) are valid individually for every corona hole. Let us now take the summations over all k and introduce the notations

$$n_b{}'' = \sum_{k=1}^{g} (n_b{}'')_k , \quad s'' = \sum_{k=1}^{g} s_k{}'' , \quad t'' = \sum_{k=1}^{g} t_k{}'' \tag{3.16}$$

Then eqns. (6) — (8) are immediately converted to

$$n_b = n_b{}' + n_b{}'' , \quad s = s' + s'' , \quad t = t' + t'' \tag{3.17}$$

From eqns. (11), (14) and (15), in combination with (16), it is obtained

$$n_b{}'' = s'' + t'' = 2s'' + 6g \tag{3.18}$$

and

$$t'' - s'' = 6g \tag{3.19}$$

It is especially interesting to compare these two equations, viz. (18) and (19), or alternatively (15) and (14), with eqns. (9) and (10), respectively. These relations emphasize the different properties of an outer perimeter from those of an inner perimeter. This feature has been described in detail by Hall (1988). Also Polansky and Rouvray (1977) have offered an elaborate treatment of the perimeter lengths of coronoids.

Example. Let the naphthalene and phenalene corona holes in the double coronoid of Fig. 1 be identified by the subscript 1 and 2, respectively. Then

$$n_b{}' = 36 , \qquad (n_b{}'')_1 = 10 , \qquad (n_b{}'')_2 = 12 , \qquad n_b{}'' = 22 ;$$
$$s' = 21 , \qquad s_1{}'' = 2 , \qquad s_2{}'' = 3 , \qquad s'' = 5 ;$$
$$t' = 15 , \qquad t_1{}'' = 8 , \qquad t_2{}'' = 9 , \qquad t'' = 17 .$$

3.3 Additional Definitions, Terminology and Relations

3.3.1 Corona Holes

Let the g corona holes of a g-tuple coronoid define a set of benzenoids $B_1{}^o$, $B_2{}^o$, $B_3{}^o$, ..., $B_g{}^o$ with the numbers of hexagons equal to $h_1{}^o$, $h_2{}^o$, $h_3{}^o$, ... , $h_g{}^o$, respectively. Use the corresponding notation for the other invariants. In the following we shall need $(n_i{}^o)_k$, $n_k{}^o$, $s_k{}^o$, $t_k{}^o$ and $(n_e{}^o)_k$ $(k = 1, 2, 3, ... , g)$, where $s_k{}^o$ and $t_k{}^o$ already were encountered in eqn. (12), while $(n_e{}^o)_k$ is used to designate the perimeter length of $B_k{}^o$ (in consistency with a parenthesized remark in Par. 3.2.3). Introduce also the relevant summations, viz.

$$h^o = \sum_{k=1}^{g} h_k^o , \quad n_i^o = \sum_{k=1}^{g} (n_i^o)_k \tag{3.20}$$

$$n^o = \sum_{k=1}^{g} n_k^o , \quad s^o = \sum_{k=1}^{g} s_k^o \tag{3.21}$$

$$t^o = \sum_{k=1}^{g} t_k^o \tag{3.22}$$

$$n_e^o = \sum_{k=1}^{g} (n_e^o)_k \tag{3.23}$$

The individual invariants for each B_k^{O} ($k = 1, 2, 3, \ldots g$) obey, of course, the relations between benzenoid invariants. Thus, for instance (cf. Table 1),

$$n_k^o = 4h_k^o - (n_i^o)_k + 2 , \quad s_k^o = 2h_k^o - (n_i^o)_k + 4 ; \quad k = 1, 2, 3, \ldots , g \tag{3.24}$$

and conversely

$$h_k^o = (1/2)(n_k^o - s_k^o) + 1 , \quad (n_i^o)_k = n_k^o - 2s_k^o + 6 ; \quad k = 1, 2, 3, \ldots , g \tag{3.25}$$

Furthermore,

$$t_k^o = 2h_k^o - (n_i^o)_k - 2 = s_k^o - 6 ; \quad k = 1, 2, 3, \ldots , g \tag{3.26}$$

and

$$(n_e^o)_k = 4h_k^o - 2(n_i^o)_k + 2 = 2s_k^o - 6 = (n_b'')_k ; \quad k = 1, 2, 3, \ldots , g \tag{3.27}$$

The last part of eqn. (27) expresses the fact that the inner perimeter for a corona hole coincides with the unique perimeter of the corresponding benzenoid (B_k^{O}). By taking the summations over all k one attains at relations between the corresponding sums of invariants, viz.:

$$n^o = 4h^o - n_i^o + 2g , \quad s^o = 2h^o - n_i^o + 4g \tag{3.28}$$

$$h^o = (1/2)(n^o - s^o) + g , \quad n_i^o = n^o - 2s^o + 6g \tag{3.29}$$

$$t^0 = 2h^0 - n_i^0 - 2g = s^0 - 6g \tag{3.30}$$

$$n_e^0 = 4h^0 - 2n_i^0 + 2g = 2s^0 - 6g \tag{3.31}$$

Example: For the double coronoid of Fig. 1 one has

$$
\begin{array}{lll}
h_1^0 = 2, & h_2^0 = 3, & h^0 = 5; \\
(n_i^0)_1 = 0, & (n_i^0)_2 = 1, & n_i^0 = 1; \\
n_1^0 = 10, & n_2^0 = 13, & n^0 = 23; \\
s_1^0 = 8, & s_2^0 = 9, & s^0 = 17; \\
t_1^0 = 2, & t_2^0 = 3, & t^0 = 5; \\
(n_e^0)_1 = 10, & (n_e^0)_2 = 12, & n_e^0 = 22.
\end{array}
$$

Here again (as in the example of Par. 3.2.3) the holes corresponding to naphthalene ($C_{10}H_8$) and phenalene ($C_{13}H_9$) have been identified by the subscripts 1 and 2, respectively. In other words $B_1^0 = C_{10}H_8$ and $B_2^0 = C_{13}H_9$, which incidently is an unambiguous designation for these two benzenoids.

3.3.2 Associated Benzenoid and Perforated Benzenoid

Let C_g denote a g–tuple coronoid.

Definition 3.1: The *associated benzenoid* to C_g is defined by the benzenoid which emerges when the g corona holes of C_g are completely filled by hexagons.

It is noted that the associated benzenoid is determined unambiguously when C_g is given.

Let **B** be the benzenoid which is associated with C_g. Assume that **B** has H hexagons, N_i internal vertices, and the formula $C_N H_S$. Then, of course (cf. Table 1),

$$N = 4H - N_i + 2, \quad S = 2H - N_i + 4 \tag{3.32}$$

and conversely

$$H = (1/2)(N - S) + 1, \quad N_i = N - 2S + 6 \tag{3.33}$$

The outer perimeter of C_g coincides with the unique perimeter of **B**. Let its length be N_e when relating it to **B**. Then, in addition to $S = s'$ one has

$$N_e = 4H - 2N_i + 2 = 2S - 6 = n_b' \tag{3.34}$$

where the quantities s' and n_b', as related to the outer perimeter of C_g, are defined in Par. 3.2.3.

Let (h, n_i) be the appropriate pair of invariants of C_g, while its formula is $C_n H_s$. When using the notation of Par. 3.3.1 and some of the relations therefrom it is attained at

$$H = h + h^o , \quad N_i = n_i + n^o = 4h^o + n_i - n_i^o + 2g \tag{3.35}$$

and

$$N = n + n_i^o = n + n^o - 2s^o + 6g , \quad S = s - t^o = s - s^o + 6g \tag{3.36}$$

Furthermore,

$$N_e = 4(h - h^o) - 2(n_i - n_i^o) + 2 - 4g = 2(s - s^o) - 6 + 12g \tag{3.37}$$

Example: The benzenoid which is associated with the double coronoid of Fig. 1 (cf. also the example of Par. 3.3.1), is $C_{73}H_{21}$ dicircumphenalene ($H = 27$, $N_i = 37$). It has the perimeter length $N_e = 36$.

Assume again that C_g is a g–tuple coronoid.

Definition 3.2: C_g is a *perforated benzenoid* B if B is the associated benzenoid to C_g.

We can also say that B is perforated by (specified or unspecified) hole(s).

Usually a benzenoid B is compatible with different corona holes in different constellations. In other words, two or more nonisomorphic coronoids may be associated with the same benzenoid B. In fact the term "perforated B" may or may not characterize a coronoid unambiguously. The ambiguity depends highly on different restrictions, which may be imposed on C_g. Firstly, g may be fixed or not. Moreover, definite shapes of the holes may be assumed. If g is not fixed and "perforated B" is unambiguous, then C_g must clearly be a single coronoid (C_1). Namely, if C_g with $g > 1$ is associated with B, then we can always construct another coronoid C_{g-1} which also is associated with B.

The designation "perforated rectangles" for certain single and multiple coronoids, of which the Kekulé structures have been studied (Cyvin SJ, Cyvin and Brunvoll 1989a; Chen and Cyvin 1989; Cyvin SJ, Cyvin, Brunvoll and Chen 1989; Chen, Cyvin et al. 1990), is compatible with the above definition of perforated benzenoids.

3.3.3 Naphthalenic Coronoid

Coronoids with exclusively naphthalene holes play important roles in different contexts, as we shall see in the following. This is true for both single and multiple coronoids.

Definition 3.3: A *naphthalenic coronoid* is a coronoid (single or multiple) with only naphthalene hole(s).

Assume that C_g is a naphthalenic g–tuple coronoid with (h, n_i) and the formula $C_n H_s$. Furthermore, let the associated benzenoid to C_g be denoted by B, of which some selected invariants are given by (H, N_i), $C_N H_S$ and N_e. For the case of g naphthalene holes, which is considered here, one has:

$$h^o = 2g, \quad n_i^o = 0, \quad n^o = 10g, \quad s^o = t'' = 8g, \quad t^o = s'' = 2g, \quad n_e^o = n_b'' = 10g \qquad (3.38)$$

By means of these expressions one obtains the following special cases of eqns. (35) − (37), which apply to C_g and B.

$$H = h + 2g, \quad N_i = n_i + 10g \qquad (3.39)$$

$$N = n, \quad S = s - 2g \qquad (3.40)$$

$$N_e = 4h - 2n_i + 2 - 12g = 2s - 6 - 4g \qquad (3.41)$$

3.3.4 Extremal Coronoid

An *extremal coronoid* (single or multiple) is defined in the same way as an extremal benzenoid (Harary and Harborth 1976; Cyvin SJ 1992c; Brunvoll, Cyvin BN and Cyvin 1992b).

Definition 3.4: An extremal g–polyhex, where g is fixed, is defined by having the maximum number of internal vertices (n_i) for a given number of hexagons (h): $n_i = (n_i)_{max}(h, g)$.

An extremal g–tuple coronoid shall frequently be identified by the symbol A_g in the following. A coronoid which is not extremal, is sometimes referred to as a *nonextremal coronoid*.

A formula for the upper bound of n_i in g–polyhexes has been launched by Cyvin SJ and Brunvoll (1990) as

$$n_i \leq 2h + 1 - 6g - \lceil (12h - 3 + 24g)^{1/2} \rceil \qquad (3.42)$$

The special case of this equation for $g = 1$ is given in Vol. I − 3.2.3 and elsewhere (Cyvin SJ and Brunvoll 1989). The general form of (42) was found very simply by inserting H and N_i from eqn. (39) into the well known inequality for benzenoids ($g = 0$), viz. (Harary and Harborth 1976; Gutman 1982)

$$N_i \leq 2H + 1 - \lceil (12H - 3)^{1/2} \rceil \qquad (3.43)$$

This derivation is based on the assumption that the number of internal vertices in a g–polyhex with h hexagons, viz. $n_i(h, g)$, cannot be larger than it would be in an extremal benzenoid perforated by g naphthalene holes. At this stage, however, it is an open question whether such an extremal benzenoid always exists for a given pair of values g and h, which are possible in g-polyhexes.

Already in the original work on extremal g-tuple coronoids (Cyvin SJ and Brunvoll 1990) it was pointed out that eqn. (42) is not "perfect" in the sense that the upper bound as given therein is not realized for all possible pairs of (h, g) values. It was verified, however, that this upper bound is realized for all the possible values of h when $g = 1$ and $g = 2$, in addition to $g = 0$. But in general one has to write

$$(n_i)_{\max} \leq 2h + 1 - 6g - \lceil (12h - 3 + 24g)^{1/2} \rceil \tag{3.44}$$

where the sign of equality is valid for any possible h when $g = 0, 1, 2$. From the analysis of the subsequent sections it is inferred that the equality holds in general only for these three g values. For $g > 2$ it always fails for one or more of the smallest ones of the possible h values.

Below we give a useful definition in connection with the above discussion. Consider a g-tuple extremal coronoid A_g, and denote its associated benzenoid by **A**.

Definition 3.5: A_g is a *perfect extremal coronoid* if and only if **A** is an extremal benzenoid.

An extremal coronoid which is not perfect, shall be referred to as an *imperfect extremal coronoid*.

3.4 First Enumeration Results for Benzenoids and Coronoids

In Table 2 the total numbers of nonisomorphic g-polyhexes according to their h values are collected, as far as they are known from the literature. For a voluminous list of references to the relevant original works of enumeration, the reader is referred to the footnotes of the table.

Forms of the members of different classes of g-polyhexes have been depicted many times. For the benzenoids, see a recent review with supplements (Cyvin BN, Brunvoll and Cyvin 1992b) and references cited therein, especially Knop, Müller, Szymanski and Trinajstić (1985). Many forms of single coronoids are displayed in Vol. I with proper documentations to previous works. The smallest ($h = 13$) double coronoid was depicted, perhaps for the first time by Dias (1984b), and later by Brunvoll, Cyvin BN and Cyvin (1987). In the paper of Dias (1984b) also one of the smallest ($h = 18$) triple coronoids is shown. More extensive depictions from systematic generations of the systems are available for double coronoids (Brunvoll, Cyvin BN and Cyvin 1990; Cyvin SJ, Brunvoll and Cyvin 1990a; Knop, Müller, Szymanski and Trinajstić 1990b; Brunvoll, Cyvin BN, Cyvin, Knop et al. 1990) and triple coronoids (Brunvoll, Cyvin BN and Cyvin 1990; Cyvin SJ, Brunvoll and Cyvin 1990a; 1991e).

Table 3.2. Numbers of g–polyhexes: benzenoids ($g = 0$), single coronoids ($g = 1$) and multiple coronoids ($g > 1$).

h	g = 0	1	2	3
1	1^a			
2	1^a			
3	3^a			
4	7^a			
5	22^a			
6	$81^{b,c}$			
7	$331^{b,c}$			
8	$1435^{b,c}$	1^d		
9	$6505^{b,c}$	5^d		
10	$30086^{b,c}$	43^e		
11	141229^f	283^g		
12	669584^h	1954^g		
13	3198256^i	$12363^{j,k}$	$1^{l,m}$	
14	15367577^{n-q}	$76283^{m,n}$	$11^{l,m}$	
15	$74207910^{n,p,q}$	$453946^{n,r}$	$149^{l,r}$	
16	359863778^n	2641506^n	$1618^{l,s}$	
17	$1751594643^{t,u}$	†	15123^v	
18	†	†	125760^v	$4^{l,s,v}$
19	†	†	†	$71^{l,v}$
20	†	†	†	1214^l
21	†	†	†	15907^w

[a] Klarner DA (1965). Fibonacci Quarterly 3: 9

[b] Knop JV, Szymanski K, Jeričević Ž, Trinajstić N (1983). J Comput Chem 4: 23

[c] Trinajstić N, Jeričević Ž, Knop JV, Müller WR, Szymanski K (1983). Pure & Appl Chem 55: 379

[d] Knop JV, Szymanski K, Jeričević Ž, Trinajstić N (1984). Match 16: 119

[e] Knop JV, Müller WR, Szymanski K, Trinajstić N (1986). Match 20: 197

[f] Stojmenović I, Tošić R, Doroslovački R (1986). Proceedings of the Sixth Yugoslav Seminar on Graph Theory, Dubrovnik 1985, Novi Sad: 189

[g] Brunvoll J, Cyvin BN, Cyvin SJ (1987). J Chem Inf Comput Sci 27: 14

[h] He WJ, He WC, Wang QX, Brunvoll J, Cyvin SJ (1988). Z Naturforsch 43a: 693

[i] Müller WR, Szymanski K, Knop JV, Nikolić S, Trinajstić N (1989). Croat Chem Acta 62: 481

[j] Cyvin SJ, Cyvin BN, Brunvoll J (1987). Chem Phys Letters 140: 124

[k] Cyvin SJ, Cyvin BN, Brunvoll J, Bergan JL (1987). Coll Sci Papers Fac Sci Kragujevac 8: 137

[l] Cyvin SJ, Brunvoll J (1990). Chem Phys Letters 170: 364

[m] Knop JV, Müller WR, Szymanski K, Trinajstić N (1990). J Mol Struct (Theochem) 205: 361

[n] Knop JV, Müller WR, Szymanski K, Trinajstić N (1990). J Chem Inf Comput Sci 30: 159

[o] Müller WR, Szymanski K, Knop JV, Nikolić S, Trinajstić N (1990). J Comput Chem 11: 223

[p] Nikolić S, Trinajstić N, Knop JV, Müller WR, Szymanski K (1990). J Math Chem 4: 357

[q] Knop JV, Müller WR, Szymanski K, Trinajstić N (1990). Reports in Molecular Theory 1: 95

[r] Brunvoll J, Cyvin BN, Cyvin SJ, Knop JV, Müller WR, Szymanski K, Trinajstić N (1990). J Mol Struct (Theochem) 207: 131

[s] Cyvin SJ, Brunvoll J, Cyvin BN (1990). J Chem Inf Comput Sci 30: 210

[t] Tošić R, Mašulović D, Brunvoll J, Cyvin BN, Cyvin SJ (1993). J Mol Struct (Theochem) 282: 295

[u] Mašulović D, Tošić R, Cyvin BN, Cyvin SJ (1993). Match 29: 165

[v] Brunvoll J, Cyvin BN, Cyvin SJ (1990). Croat Chem Acta 63: 585

[w] Cyvin SJ, Brunvoll J, Cyvin BN (1991). Rev Roumaine Chim 36: 313

[†] Unknown.

3.5 Smallest Multiple Coronoids

3.5.1 *Introduction*

The interesting and important question about the smallest multiple coronoids is still an open problem. The tentative solution presented below is not supported by rigorous proofs. A precise formulation of the problem is: Find the minimum number of hexagons for a g–polyhex when g is given. In other words, determine the function $h_{min}(g)$. From the systematic enumerations (cf. Table 2) one has the results for the smallest g values: $h_{min}(0) = 1$ for benzenoids; $h_{min}(1) = 8$ for single coronoids; $h_{min}(2) = 13$ for double coronoids; $h_{min}(3) = 18$ for triple coronoids.

3.5.2 *Basic Assumptions*

The following assumptions are made, partly based on the experience from systematic generations and enumerations of the coronoid systems, and partly from more or less systematic trials and errors.

Assumption: A smallest multiple coronoid can be found as a catacondensed naphthalenic system, where the naphthalene holes are parallel.

The assumption about naphthalene holes is very plausible since, loosely speaking, a larger corona hole would imply a waste of hexagons. It is also clear that the holes should be packed together as closely as possible, and then it seems to be an advantage to have no internal vertices. Finally it is assumed that this is achieved by taking parallel holes in a certain grid as described in the below algorithm.

The following condition is obviously a necessary one for the smallest coronoids.

Condition: If any hexagon from a g–tuple coronoid is deleted, then the system is no longer a g–tuple coronoid.

This is clearly not a sufficient condition for a g–tuple coronoid to be a smallest coronoid for the g value in question. Two counterexamples are found below

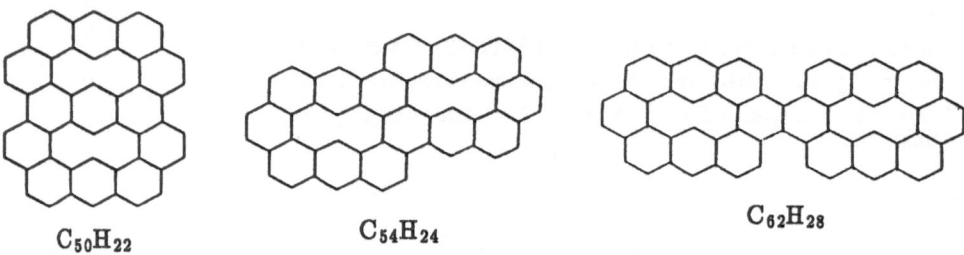

$C_{50}H_{22}$ $C_{54}H_{24}$ $C_{62}H_{28}$

Here the left–hand system ($C_{50}H_{22}$) is the unique smallest double coronoid and has thirteen hexagons ($h = 13$). The middle ($C_{54}H_{24}$) and right–hand ($C_{62}H_{28}$) systems are clearly not smallest double coronoids since they have $h = 14$ and $h = 16$, respectively. All these three systems fulfil the above condition. Notice in particular that a hexagon can be deleted from the right–hand system so that it falls apart into two separated polyhexes: one single coronoid and one benzenoid.

The three systems depicted above are also consistent with the assumptions formulated in this paragraph. Herefrom it is clear that these assumptions are not sufficient for a multiple coronoid to be a smallest system. It is even not implied that the formulated assumptions are necessary.

3.5.3 *Algorithm for Construction of Smallest Multiple Coronoids*

The following algorithm represents a tentative method to construct one or more nonisomorphic smallest coronoids with a given g, viz. systems with $h = h_{\min}(g)$.

Algorithm

1. Consider the extremal benzenoid(s) with g hexagons. By definition they have the maximum number of internal vertices for the given g, say $N_i = (N_i)_{\max}(g)$. Here (Harary and Harborth 1976; Gutman 1982):

$$(N_i)_{max} = 2g + 1 - \lceil (12g - 3)^{1/2} \rceil \tag{3.45}$$

cf. also eqn (43). Extremal benzenoids have simultaneously the minimum total number of vertices for a given number of hexagons; in our case $N = N_{min}(g)$. Here (Harary and Harborth 1976):

$$N_{min} = 2g + 1 + \lceil (12g - 3)^{1/2} \rceil \tag{3.46}$$

2. Inspect the extremal benzenoid(s) with a given g for the numbers of parallel edges, say P. In general there are three directions to be inspected for every system, but one or two of the directions may be equivalent by virtue of symmetry. Detect the minimum value, viz. $P_{min}(g)$.

3. Then, for a smallest coronoid with g holes it is conjectured that

$$h_{min}(g) = N_{min}(g) + P_{min}(g) \tag{3.47}$$

4. In order to depict (one of) the smallest coronoid(s) with g holes: (a) select an extremal benzenoid with $P = P_{min}$ and draw it so that the P_{min} parallel edges are horizontal; (b) convert each hexagon of this benzenoid into a naphthalene corona hole according to:

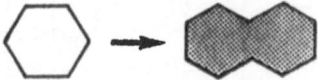

(c) surround the holes with hexagons into a grid without internal vertices.

Examples

1. Consider the case of $g = 6$. There are three extremal benzenoids with six hexagons each, viz. (I) anthanthrene, (II) benzo[ghi]perylene and (III) triangulene. For these systems one has $P = 9$ in most cases as illustrated below, where the horizontal edges (marked by heavy lines) are counted.

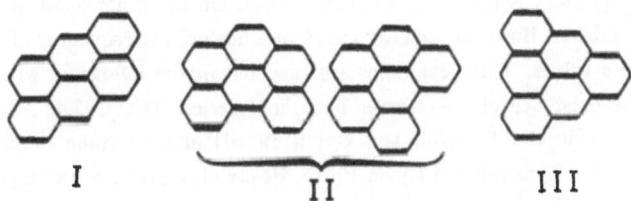

I   II   III

In the remaining nonequivalent orientation for anthanthrene (I) one has $P = P_{min} = 8$. Hence the construction of a presumably smallest 6–tuple coronoid goes like this:

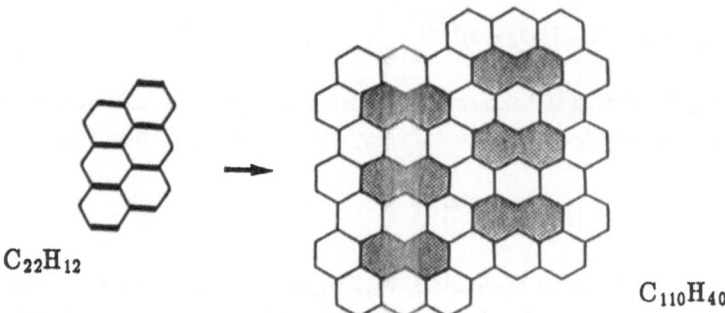

$C_{22}H_{12}$

$C_{110}H_{40}$

From this example it is inferred that $h_{min}(6) = N_{min}(6) + P_{min}(6) = 22 + 8 = 30$.

2. Consider the case of $g = 7$. Coronene is the unique extremal benzenoid with seven hexagons. Furthermore, the three orientations for counting parallel edges are all equivalent. Hence the following construction.

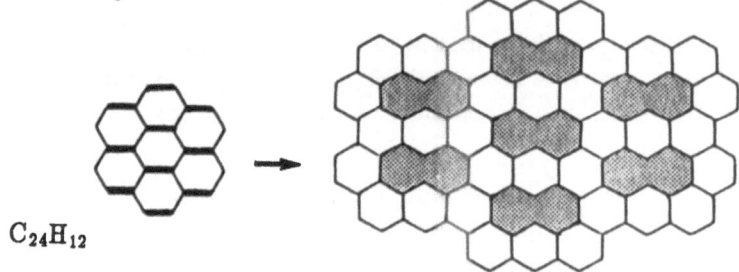

$C_{24}H_{12}$

$C_{124}H_{44}$

In this case, $h_{min}(7) = N_{min}(7) + P_{min}(7) = 24 + 10 = 34$.

The values of $N_{min}(g)$ and $P_{min}(g)$ were derived systematically for $g \leq 15$. The results are collected in Table 3, where also the extremal benzenoids with $P = P_{min}$ are depicted; the P_{min} edges are drawn horizontally.

For $N_{min}(g)$ one has the explicit equation (46). On the other hand, no explicit equation for $P_{min}(g)$ is available. However, one can easily produce infinite sequences of values of P_{min} for non–consequtive g values. Three examples are reported in the following, where the treatments are based on the existence of one–isomer benzenoid series (Dias 1990a; 1990b; Brunvoll and Cyvin SJ 1990; Cyvin SJ, Brunvoll and Cyvin 1991d) and on some properties of extremal benzenoids (Cyvin SJ, Brunvoll and Cyvin 1991b; Brunvoll, Cyvin BN and Cyvin 1993b).

Consider first (Dias 1982a) C_6H_6 benzene, $C_{24}H_{12}$ coronene, $C_{54}H_{18}$ circumcoronene, $C_{96}H_{24}$ dicircumcoronene, etc. All these benzenoids are unique for each formula, i.e. they belong to a one–isomer series, and all of them are extremal. Let again g be the number of hexagons. Then (Brunvoll and Cyvin SJ 1990) $g = 3k^2+3k+1$ ($k=0,1,2, ...$), and eqn. (46) gives

$$N_{min}(3k^2+3k+1) = 6(k + 1)^2 ; \quad k = 0, 1, 2, ... \tag{3.48}$$

Table 3.3. Values of N_{min} and P_{min} for extremal benzenoids with g hexagons.

g	Extremal benzenoid(s)	N_{min}	P_{min}	g	Extremal benzenoid(s)	N_{min}	P_{min}
1		6	2	10		32	13
2		10	3				
3		13	5	11		35	14
4		16	6				
5		19	7	12		37	16
6		22	8				
7		24	10	13		40	16
8		27	11	14		42	18
9		30	12	15		45	19

Now it is relatively easy to find a combinatorial formula for the number of parallel edges, which is the same in all the three directions for one of the systems under consideration. It was found

$$P_{min}(3k^2+3k+1) = (k+1)(3k+2) ; \quad k = 0, 1, 2, \dots \tag{3.49}$$

Finally, on inserting from eqns. (48) and (49) into (47) it is attained at:

$$h_{min}(3k^2+3k+1) = (k+1)(9k+8) ; \quad k = 0, 1, 2, \dots \tag{3.50}$$

A corresponding analysis was performed for the one–isomer series (Dias 1984a; 1984b) $C_{10}H_8$ naphthalene, $C_{32}H_{14}$ ovalene, $C_{66}H_{20}$ circumovalene, $C_{112}H_{26}$ dicircumovalene, etc. In this case it was found (Brunvoll and Cyvin SJ 1990) $g = (k+1)(3k+2)$ for $k=0,1,2,....$, and consequently

$$N_{min}(3k^2+5k+2) = 6k^2 + 10k + 5 + \lceil(36k^2 + 60k + 21)^{1/2}\rceil \; ; \quad k = 0, 1, 2, ... \qquad (3.51)$$

Furthermore,

$$P_{min}(3k^2+5k+2) = 3k^2 + 7k + 3 \; ; \quad k = 0, 1, 2, ... \qquad (3.52)$$

and as the final result:

$$h_{min}(3k^2+5k+2) = 9k^2 + 17k + 8 + \lceil(36k^2 + 60k + 21)^{1/2}\rceil \; ; \quad k = 0, 1, 2, ... \qquad (3.53)$$

In the above references (Dias 1984a; 1984b; Brunvoll and Cyvin SJ 1990) also the following one–isomer series is treated: $C_{16}H_{10}$ pyrene, $C_{42}H_{16}$ circumpyrene, $C_{80}H_{22}$ dicircumpyrene, $C_{130}H_{28}$ tricircumpyrene, etc. In this case $g = (k+1)(3k+4)$ for $k=0,1,2,....$, which gives

$$N_{min}(3k^2+7k+4) = 6k^2 + 14k + 9 + \lceil(36k^2 + 84k + 45)^{1/2}\rceil \; ; \quad k = 0, 1, 2, ... \qquad (3.54)$$

Furthermore it was found

$$P_{min}(3k^2+7k+4) = 3(k + 1)(k + 2) \; ; \quad k = 0, 1, 2, ... \qquad (3.55)$$

and finally:

$$h_{min}(3k^2+7k+4) = 9k^2 + 23k + 15 + \lceil(36k^2 + 84k + 45)^{1/2}\rceil \; ; \quad k = 0, 1, 2, ... \qquad (3.56)$$

In Table 4 some values of $h_{min}(g)$ according to the above algorithm are collected. Those for $1 \leq g \leq 15$ emerge readily from the numbers of Table 3; those for $g > 15$ (non–consecutive values) were obtained from eqns. (50), (53) and (56).

3.5.4 Discussion and Depictions of Forms

It is a fact that the algorithm (Par. 3.5.3) does not generate all the g–tuple coronoids with $h = h_{min}(g)$. This is easily demonstrated by examples.

Examples

1. For $g = 3$ it is inferred from Table 3: $h_{min}(3) = 13 + 5 = 18$. In addition to the smallest triple coronoid (with $h = 18$) derived from phenalene according to the algorithm one

Table 3.4. Values of $h_{min}(g)$ for the (presumably) smallest g–polyhexes, and the lower bounds, $h^*(0)$, for these values; $h_{min}(g) \geq h^*(0)$.

g	h_{min}	$h^*(0)$	g	h_{min}	$h^*(0)$
0	1	1	19	78	75
1	8	8	24	95	92
2	13	13	30	116	113
3	18	17	37	140	136
4	22	21	44	163	159
5	26	25	52	190	185
6	30	29	61	220	215
7	34	33	70	249	244
8	38	36	80	282	276
9	42	40	91	318	311
10	45	44	102	353	347
11	49	47	114	392	385
12	53	51	127	434	426
13	56	54	140	475	467
14	60	58	154	520	512
15	64	61	169	568	559

may derive another such system from anthracene as shown below.

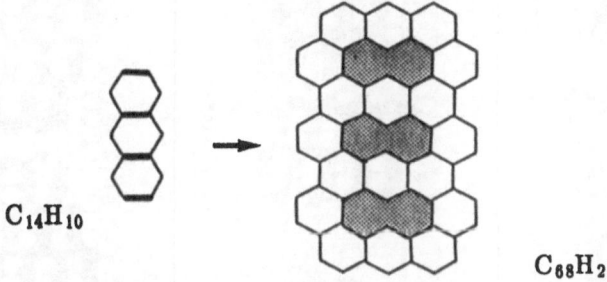

$C_{14}H_{10}$

$C_{68}H_{28}$

Notice that this derivation does not follow the algorithm because anthracene ($C_{14}H_{10}$) is not an extremal benzenoid. It has as its number of vertices $N = 14 > N_{min}(3)$, which equals 13 (cf. Table 3). On the other hand, the minimum number of parallel edges in anthracene (cf. the above depiction) is $P = 4 < P_{min}(3)$, which is 5. This is no paradox when it is remembered that $P_{min}(g)$ by definition always pertains to an extremal benzenoid. It follows that, for anthracene, $N + P = 18 = h_{min}(3)$.

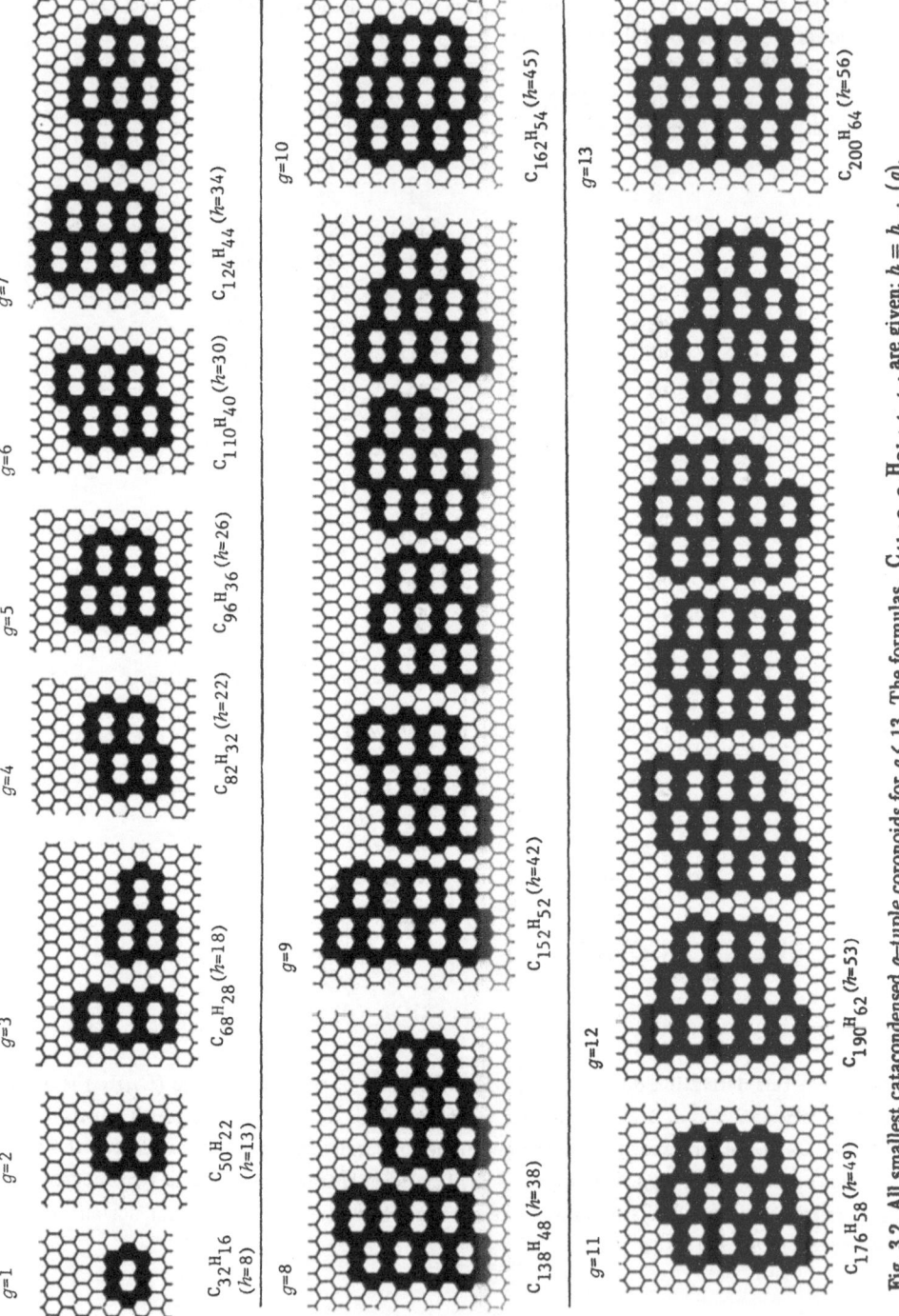

Fig. 3.2. All smallest catacondensed g-tuple coronoids for $g \leq 13$. The formulas, $C_{4h+2-2g} H_{2h+4-4g}$, are given; $h = h_{\min}(g)$.

$g=1$ $g=2$ $g=3$ $g=4$ $g=5$ $g=6$ $g=7$

$C_{32} H_{16} (h=8)$ $C_{50} H_{22} (h=13)$ $C_{68} H_{28} (h=18)$ $C_{82} H_{32} (h=22)$ $C_{96} H_{36} (h=26)$ $C_{110} H_{40} (h=30)$ $C_{124} H_{44} (h=34)$

$g=8$ $g=9$ $g=10$

$C_{138} H_{48} (h=38)$ $C_{152} H_{52} (h=42)$ $C_{162} H_{54} (h=45)$

$g=11$ $g=12$ $g=13$

$C_{176} H_{58} (h=49)$ $C_{190} H_{62} (h=53)$ $C_{200} H_{64} (h=56)$

2. In the second example of Par. 3.5.3 the unique smallest multiple coronoid for $g = 7$ was derived from coronene according to the algorithm under consideration. Another such system emerges from $C_{25}H_{13}$ benzo[hi]anthanthrene. This nonextremal benzenoid has 25 internal vertices (N) against $N_{min}(7) = 24$ (cf. Table 3). On the other hand, its minimum number of parallel edges (P) is 9, while $P_{min}(7) = 10$. Hence $N + P = 34 = h_{min}(7)$.

Figure 2 shows the forms of the smallest catacondensed g–tuple coronoids for $g \leq 13$, obtained from systematic generations by hand. All the systems fulfil the assumptions (Par. 3.5.2) of being naphthalenic with parallel holes. Those which emerge from the algorithm of Par. 3.5.3 (cf. especially Table 3) are all found in Fig. 2. In addition one "off–algorithm" system each for $g = 3$ and $g = 7$, which account for the above examples, are also found in Fig. 2, and finally one off–algorithm system for $g = 8$, one for $g = 9$, and four for $g = 12$.

3.5.5 Pericondensed Smallest Multiple Coronoids

It is no contradiction to the assumption of Par. 3.5.2 that there exist pericondensed triple coronoids among the smallest multiple coronoids. The situation was described in detail by Cyvin SJ and Brunvoll (1990). The two relevant systems, which have been detected, are depicted in Fig. 3 and elsewhere (Brunvoll, Cyvin BN and Cyvin 1990; Cyvin SJ, Brunvoll and Cyvin 1990a; 1991e). They have the same numbers of hexagons ($h = 18$) as the smallest catacondensed triple coronoids (see $C_{68}H_{28}$ in Fig. 2), but one internal vertex ($n_i = 1$) each. In fact these pericondensed systems have less vertices and less vertices of degree two (formula $C_{67}H_{27}$) than the catacondensed systems ($C_{68}H_{28}$) in question.

The existence of pericondensed systems among smallest g–polyhexes seems to be a unique property of triple coronoids ($g = 3$).

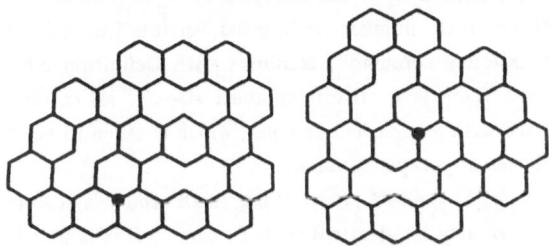

Fig. 3.3. Two $C_{67}H_{27}$ 3–polyhexes (triple coronoids) with $h = 18$ and $n_i = 1$. The internal vertices are indicated by black dots.

3.6 Perfect and Imperfect Extremal Coronoids

3.6.1 Introduction

Two g–polyhexes (with the same g) are said to be *isomers* (in a restricted sense) when they have the same set of invariants; they are characterized, e.g., by the same pair of (h, n_i) values or the same formula C_nH_s. Figures 2 and 3 provide examples of coronoid isomers.

In the present section the great deal of knowledge we have on benzenoid isomers is exploited at several occasions. For this subject the reader is referred to four reviews (Brunvoll and Cyvin SJ 1990; Cyvin SJ, Brunvoll and Cyvin 1991d; Brunvoll, Cyvin BN and Cyvin 1992b; Cyvin SJ, Cyvin and Brunvoll 1993e) and references cited therein. All quotations which are made to numbers of isomers can be found in Brunvoll and Cyvin SJ (1990). The most extensive depictions in Cyvin SJ, Brunvoll and Cyvin (1991d) shall also prove to be most useful, and especially those for extremal benzenoids. Many forms of such systems, some of which being referred to in the subsequent analyses, are found in the cited work, either depicted directly or specified as circumscribed systems. It is noted that the members of the so—called constant—isomer series (Dias 1990a; 1990b) are precisely the extremal benzenoids (Cyvin SJ, Cyvin and Brunvoll 1993e); cf. also Par. 3.5.3.

From the definition of extremal g—polyhexes (Par. 3.3.4) it is clear that there exists an extremal g—polyhex for every possible value of $h \geq h_{min}(g)$ when g is fixed. Let it be symbolized by $A_g(h)$. For a given pair of values (h, g) there may exist more than one extremal g—polyhex: $A_g(h)$, $A_g'(h)$, ..., which may be nonisomorphic.

Some introductory observations about extremal coronoids are reported below. They are followed by comments which tend to justify the inherent assumptions. However, the observations are not to be considered as rigorously proved.

Observations

1. Any extremal coronoid is naphthalenic.

2. If $A_g(h)$ is a perfect extremal coronoid, then all the larger extremal coronoids $A_g(h')$ for $h' > h$ are also perfect.

In connection with Observation 1, suppose that a coronoid C_g has a corona hole larger than naphthalene. Then we can imagine that a closer packing of the hexagons of C_g is possible by a partial filling of the corona hole so that the total number of internal vertices increases. In Observation 2 the crucial term ("perfect extremal coronoid") conforms with Definition 3.5 of Par. 3.3.4. It is reasonable to imagine that there is a critical smallest size for an extremal benzenoid, say **A**, so that **A** can be perforated with g naphthalene holes, which is taken to be the necessary condition for creating a perfect extremal g—tuple coronoid.

Consider two extremal coronoids $A_g(h)$ and $A_g'(h)$, along with their associated benzenoids, say **B** and **B'**, respectively. Here **B'** may be identical with **B**, namely if it is possible to perforate **B** = **B'** by g naphthalene holes in two different ways so as to create the two nonisomorphic systems $A_g(h)$ and $A_g'(h)$. In general **B** and **B'** must have the same invariants as indicated by: $B(H, N_i)$, $B'(H, N_i)$. This is a consequence of Observation 1. It follows that **B** and **B'** also have the same formula, say $C_N H_S$, and the same perimeter length, say N_e. The connections between the invariants of **B** and those of $A_g(h)$ with n_i internal vertices and the formula $C_n H_s$, are governed by eqns. (39) − (41).

A simple deduction from Observation 1 and the above discussion is given below. Consider two extremal coronoids $A_g(h)$ and $A_g'(h)$.

Corollary. If $A_g(h)$ is a perfect extremal coronoid, then any extremal coronoid $A_g{'}(h)$ is also perfect.

In this case the benzenoid associated with $A_g(h)$ is extremal. Let it be denoted by $A(H)$, where H is its number of hexagons. Since the associated benzenoid to $A_g{'}(h)$, say A', has the same invariants as $A(H)$ it must also be extremal and can be denoted by $A'(H)$ in order to indicate that it has the same number of hexagons as $A(H)$.

3.6.2 *Numbers of Hexagons and of Internal Vertices*

The numbers of internal vertices of a g–polyhex are restricted by the inequalities

$$0 \leq n_i \leq 2h + 1 - 6g - \lceil (12h - 3 + 24g)^{1/2} \rceil \tag{3.57}$$

According to the assumptions of Par. 3.5.2 the lowest value ($n_i = 0$) is realized for any g when $h = h_{min}(g)$. Then it must be realized for any h value because we can always extend a catacondensed g–polyhex by catacondensed appendages without limitation for the number of hexagons. It can also be assumed with confidence that all integer values $n_i \leq (n_i)_{max}(h, g)$ are realized. Here it should be recalled that the upper bound in (57) is not always realized as this $(n_i)_{max}$; cf. eqn. (44) with the accompanying discussion. However, the number of internal vertices in an extremal benzenoid or a g–tuple perfect extremal coronoid is equal to the right–hand side of (57), for which we write:

$$n_i^* = 2h + 1 - 6g - \lceil (12h - 3 + 24g)^{1/2} \rceil \tag{3.58}$$

An extremal g–polyhex is characterized by having the minimum number of hexagons for a given number of internal vertices: $h = h_{min}(n_i, g)$. Let it be mentioned by passing that this property is not suitable as a definition of extremal g–polyhexes; for certain n_i values which are not found in extremal g–polyhexes the condition $h = h_{min}$ is realized for nonextremal systems. Nevertheless, it is useful to study the inequalities for h in terms of n_i for different g values.

From an analysis of the kind as the one of Harary and Harborth (1976) it was found

$$H \geq 1 + \lceil (1/2)(N_i + 6^{1/2} N_i^{1/2}) \rceil \tag{3.59}$$

for the numbers of hexagons (H) and numbers of internal vertices (N_i) in benzenoids. Here the lower bound for H, say H_{min}, is always realized. Now it is inferred that, in a g–tuple coronoid, h cannot be smaller for a given n_i than it would be in a perforated benzenoid with $H = H_{min}$ and the appropriate N_i value, and when the coronoid is naphthalenic. On this basis we find the lower bound of the inequalities

$$1 + 3g + \lceil (n_i/2) + (1/2)(6n_i + 60g)^{1/2} \rceil \leq h < \infty \qquad (3.60)$$

which apply to g–polyhexes. This lower bound was found simply on inserting H and N_i from eqn. (39) into (59). The above equation (60) implies also the obvious fact that the number of hexagons is not limited upwards for any given pair of (n_i, g), since hexagons can always be added in catacondensed appendages, whereby the number of internal vertices does not increase.

Here again we have a relation which is not "perfect" inasmuch as the lower bound, as given by the left–hand side of eqn. (60), is not always realized. Therefore it is best to write:

$$h_{min} \geq 1 + 3g + \lceil (n_i/2) + (1/2)(6n_i + 60g)^{1/2} \rceil \qquad (3.61)$$

In benzenoids and g–tuple perfect extremal coronoids the number of hexagons is equal to the right–hand side of (61); for these cases we write in analogy with eqn. (58):

$$h^* = 1 + 3g + \lceil (n_i/2) + (1/2)(6n_i + 60g)^{1/2} \rceil \qquad (3.62)$$

3.6.3 Catacondensed Extremal Coronoids

It was stated (Par. 3.6.2) that $n_i = 0$ is realized in a g–polyhex for every g and h. It should also be clear that, if a catacondensed g–tuple extremal coronoid exists, then it can only occur for $h = h_{min}(g)$. It is always possible, namely, to add a hexagon to a catacondensed coronoid so that it becomes pericondensed. (The same property is also found for benzenoids with one exception: benzene. Accordingly, both benzene ($H = 1$) and naphthalene ($H = 2$) are catacondensed extremal benzenoids.) The above discussion, along with the last lines of Par. 3.5.5, lead to the following conjecture.

Conjecture: An extremal catacondensed g–polyhex exists for every value of $g \neq 3$.

Moreover, it was found that the catacondensed extremal g–tuple coronoids for $g = 1$ and 2 are perfect, while those for $g \geq 4$ are all supposed to be imperfect. Therefore, in a sense, the unique case of $g = 3$, for which there is no catacondensed extremal coronoid (cf. Par. 3.5.5), marks a borderline between the perfect and imperfect catacondensed extremal coronoids.

Now we are able to give a new approach to the smallest g–polyhexes (cf. Sect. 3.5 and especially Table 4). On inserting $n_i = 0$ in eqn. (62) the expression reduces to

$$h^*(0) = 1 + 3g + \lceil (15g)^{1/2} \rceil \qquad (3.63)$$

and represents a lower bound for $h_{min}(g)$. Values of $h^*(0)$ are included in Table 4 and are seen to fit nicely in with the $h_{min}(g)$ values obtained from the algorithm of Par. 3.5.3. In the following we shall analyse the smallest g–polyhexes for the lowest g values, starting from eqn. (63). Then the following expressions for the coefficients of C_nH_s in a catacondensed g–polyhex are of interest.

$$n(h, 0) = 4h + 2 - 2g , \quad s(h, 0) = 2h + 4 - 4g \qquad (3.64)$$

These relations emerge readily from $n(h, n_i)$ and $s(h, n_i)$ as found in Table 1; cf. also the legend of Fig.2.

For $g = 0$, 1 and 2, eqn. (63) gives $h^*(0) = 1$, 8 and 13, respectively. These values are equal to $h_{min}(g)$ according to the previous findings (cf. Table 4). The pertinent formulas, according to eqn. (64), come out as C_6H_6, $C_{32}H_{16}$ and $C_{50}H_{22}$, respectively. In the cases of $g = 1$ and 2 the corresponding associated benzenoids, characterized by formulas and the (H, N_i) pairs of invariants, are $C_{32}H_{14}(H=10, N_i=10)$ and $C_{50}H_{18}(H=17, N_i=10)$, as is readily found from eqns. (39) and (40). Here $C_{32}H_{14}$ represents the unique isomer of an extremal benzenoid, viz. ovalene. It is just possible to perforate ovalene by one naphthalene hole so as to generate the unique $C_{32}H_{16}$ isomer of single coronoids (see Fig 2) which is a perfect extremal system. With regard to $C_{50}H_{18}$ there are the following nine isomers (Dias 1984a; 1984b; 1986; Cyvin SJ, Brunvoll and Cyvin 1991d; Cyvin SJ, Cyvin and Brunvoll 1993e) to be inspected for the possibility to perforate them by two naphthalene holes.

$C_{50}H_{18}$

Here the heavy–line cycles indicate boundaries which no corona hole can trespass. It is seen that exactly one of these benzenoids, viz. circumperylene (the last system), can be perforated in the prescribed way, leading to the unique $C_{50}H_{22}$ isomer of double coronoids (see Fig. 2), which again is a perfect extremal system. So far we find that the present approach nicely reproduces the previous findings. The analysis is pursued to a few higher g values in the following.

For $g = 3$, $h^* = 17$. This corresponds to the formula $C_{64}H_{26}$ and the associated benzenoids $C_{64}H_{20}(H=23, N_i=30)$, of which there are four isomers (Dias 1984a; 1985a; 1985b; 1986; Cyvin SJ, Brunvoll and Cyvin 1991d):

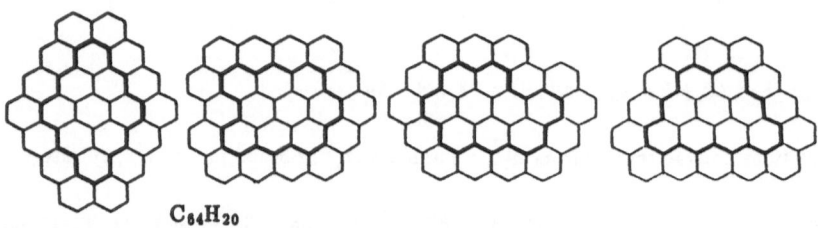

$$C_{64}H_{20}$$

Here it is found by inspection that three naphthalene holes cannot be introduced in any of these benzenoids. Hence it is inferred that no triple coronoid with the formula $C_{64}H_{26}$ can be constructed. We must therefore look for the possibilities with $h > h^*$. Then the first case to be inspected is $h = 18$, corresponding to $C_{68}H_{28}$ and the associated benzenoids $C_{68}H_{22}(H=24, N_i=30)$. These benzenoids are nonextremal and count 789 isomers (Stojmenović et al. 1986; Brunvoll and Cyvin SJ 1990), a too large number to be inspected by sight, even if the depictions were available. However, two of these isomers perforated by three naphthalene holes, have actually been constructed (see Fig. 2) as the smallest catacondensed triple coronoids, in consistency with the present analysis.

For $g = 4$, $h^* = 21$. One obtains as a starting point the formula $C_{78}H_{30}$, which corresponds to the associated benzenoids $C_{78}H_{22}(H=29, N_i=40)$. The actual four isomers (Cyvin SJ, Brunvoll and Cyvin 1991d) of these extremal benzenoids were inspected, but none of them could be perforated by four naphthalene holes. Consequently, we turn to $h = 22$, the formula $C_{82}H_{32}$ and associated benzenoids $C_{82}H_{24}(H=30, N_i=40)$. There are 1799 isomers of these nonextremal benzenoids. In accord with Fig. 2 it is claimed that only one of them can be perforated by four naphthalene holes in a unique constellation.

For $g = 5$, $h^* = 25$. Here the starting tentative formula is $C_{92}H_{34}$ and its associated benzenoid characterized by $C_{92}H_{24}(H=35, N_i=50)$. The actual nine isomers, which incidentally are circum–$C_{50}H_{18}$ (see above), were again inspected to find that none of them could be perforated by five naphthalene holes. In the next step, when $h = 26$ is assumed, the formula $C_{96}H_{34}$ emerges, and the benzenoids to be inspected are characterized by $C_{96}H_{26}(H=36, N_i=50)$. Here the number of isomers is as formidable as 6155, but again it is claimed, in accord with Fig. 2, that only one of them can give the pertinent coronoid system.

3.6.4 Extension to Pericondensed Extremal Coronoids

The approach of the preceding paragraph is extended to a systematic search for pericondensed extremal g–tuple coronoids. In the different steps of the strategy, which is outlined below, the possibilities to construct extremal (naphthalenic) coronoids $A_g(h,n_i)$ are investigated: the corresponding associated benzenoids, $A(H=h+2g,\ N_i=n_i+10g)$ or $B(H=h+2g,\ N_i=n_i+10g)$, are inspected for the possibility to perforate them by g naphthalene holes. Here the symbols A and B are used to identify extremal or nonextremal benzenoids, respectively. If a benzenoid A can be perforated by g naphthalene holes, then a perfect extremal g-tuple coronoid emerges. A detailed account on the strategy follows.

For the sake of brevity we shall refer to A or B as perforable (resp. nonperforable) when A or B can (resp. cannot) be perforated in the prescribed way, viz. by g naphthalene holes.

Algorithm

(i) Choose a g value, which is to be kept constant throughout the steps.

(ii) Start with $h_{min}(g) \equiv h_{min}$, and determine $n_i^*(h_{min},g)$ according to eqn. (58).

(iii) Inspect the extremal benzenoids $A(h_{min}+2g,\ n_i^*+10g)$ to see if they are perforable. If a perforable system is detected, one has already a perfect extremal coronoid $A_g(h_{min},n_i^*)$. Suppose that no such system exists.

(iv) Assume $n_i' < n_i^*$ successively, viz. $n_i^*-1,\ n_i^*-2,\ ....$, until a perforable $B(h_{min}+2g,\ n_i'+10g)$ occurs. Then $A_g(h_{min},n_i')$ is an imperfect extremal coronoid.

(v) Go to $h' > h_{min}$ successively, viz. $h_{min}+1,\ h_{min}+2,\ .....$. For each h' value, repeat the procedures (i) — (iii) and occasionally (iv), i.e. inspect whether $A(h'+2g,\ n_i^*+10g)$ and occasionally $B(h'+2g,\ n_i'+10g)$ are perforable.

The algorithm stops when a benzenoid $A(h+2g,\ n_i^*+10g)$ is perforable for the first time, i.e. for the smallest $h = h_{min}$ or $h = h'$. Then $A_g(h,n_i^*)$ represents the smallest perfect extremal g-tuple coronoid(s) for the chosen g value.

In an extremal g-tuple coronoid characterized by $A_g(h,n_i^*)$ the invariants $(h,\ n_i^*)$ obey eqn. (58). These invariants are also compatible with eqn. (62) in the sense that h^* becomes equal to h on inserting $n_i = n_i^*$. Now, in accord with Observation 2 of Par. 3.6.1, all the larger extremal g-tuple coronoids can be characterized by their invariants as $A_g(h,n_i^*)$, where h may be increased indeterminately, and n_i^* is found from eqn. (58) for each h value.

For $g = 1$ and 2 the above algorithm reproduces $A_1(h=8,\ n_i=0)$ and $A_2(h=13,\ n_i=0)$, viz. the two first systems of Fig. 2. Below the algorithm is applied to $g = 3$, 4, and partly to $g = 5$.

Triple Coronoids: (i) $g = 3$; (ii) $h_{min} = 18$, $n_i^* = 2$; (iii) $A = C_{66}H_{20}(H=24,\ N_i=32)$, the unique system:

$C_{66}H_{20}$ circumovalene

It is found by inspection that circumovalene is nonperforable (by three naphthalene holes). Notice that the heavy–line cycle in the above depiction cannot be trespassed by any corona hole. (iv) $n_i' = 1$, $B = C_{67}H_{21}(H=24,\ N_i=31)$ represents 43 isomers, which were generated and depicted probably for the first time. The depictions are not reproduced here, but they were checked in order to verify that exactly two of these isomers were perforable and led to the two $C_{67}H_{27}(h=18,\ n_i=1)$ extremal triple coronoids, which are shown in Fig. 3. (v) $h' = 19$, $n_i^* = 3$, $A = C_{69}H_{21}(H=25,\ N_i=33)$ represents 13 isomers (Cyvin SJ, Brunvoll and Cyvin 1991d), out of which the six perforable systems gave the six $C_{69}H_{27}(h=19,\ n_i=3)$ perfect extremal triple coronoids shown in Fig. 4.

It is a pleasing fact that the present results are consistent with previous findings: the two $C_{67}H_{27}$ isomers of triple coronoids which are treated in Par. 3.5.5; the six $C_{69}H_{27}$ isomers, which also were identified and depicted by Cyvin SJ, Brunvoll and Cyvin (1991e) from a generation of triple coronoids using certain principles entirely different from those of the present algorithm.

Quadruple Coronoids. (i) $g = 4$; (ii) $h_{min} = 22$, $n_i^* = 2$; (iii) $A = C_{80}H_{22}(H=30,\ N_i=42)$, the unique system:

C$_{80}$H$_{22}$ dicircumpyrene

This system is nonperforable. (iv a) $n_i' = 1$; $B = C_{81}H_{23}(H=30,\ N_i=41)$ represents 68 isomers. Again, these systems were generated and depicted, probably for the first time, and inspected with the result that all of them were nonperforable. (b) $n_i' = 0$, $B = C_{82}H_{24}(H=30,\ N_i=40)$ coincides with a part of the analysis in Par. 3.6.3 and leads to the $C_{82}H_{32}(h=22,\ n_i=0)$ catacondensed imperfect extremal quadruple coronoid (Fig. 2). (v) $h' = 23$, $n_i^* = 3$, $A = C_{83}H_{23}(H=31,\ N_i=43)$ represents 20 isomers (Cyvin SJ, Brunvoll and Cyvin 1991d), from which one perforable system was detected and gave a unique $C_{83}H_{31}(h=23,\ n_i=3)$ perfect extremal quadruple coronoid (see Fig. 4).

Quintuple Coronoids ($g = 5$). Whereas the above analysis for $g \leq 4$ supports the conjecture of Par. 3.6.3 we shall now resort to assuming that the conjecture also is valid for $g = 5$, so that the catacondensed $C_{96}H_{36}$ isomer (Fig. 2) is the only quintuple coronoid for $h = 26$. For $h = 27$ it was found that the $C_{98}H_{36}(n_i'=2)$ isomers are imperfect extremal systems, since both $A = C_{96}H_{26}(H=37,\ N_i=54)$ = dicircumcoronene, and the 72 benzenoid isomers $B = C_{97}H_{25}(H=37,\ N_i=53)$ are nonperforable. The smallest perfect extremal quintuple coronoid was detected as $C_{99}H_{35}(h=28,\ n_i=5)$ by perforating one of the 20 benzenoid isomers $A = C_{99}H_{25}(H=38,\ N_i=55)$.

In Fig. 4 the smallest perfect extremal g–tuple coronoids for $g = 1, 2, 3, 4$ and 5 are shown; they occur for $h = 8, 13, 19, 23$ and 28, respectively. The analysis was pursued, by perforating the appropriate extremal benzenoids, in order to generate the systems with a few higher h values for each g; the results are included in Fig. 4.

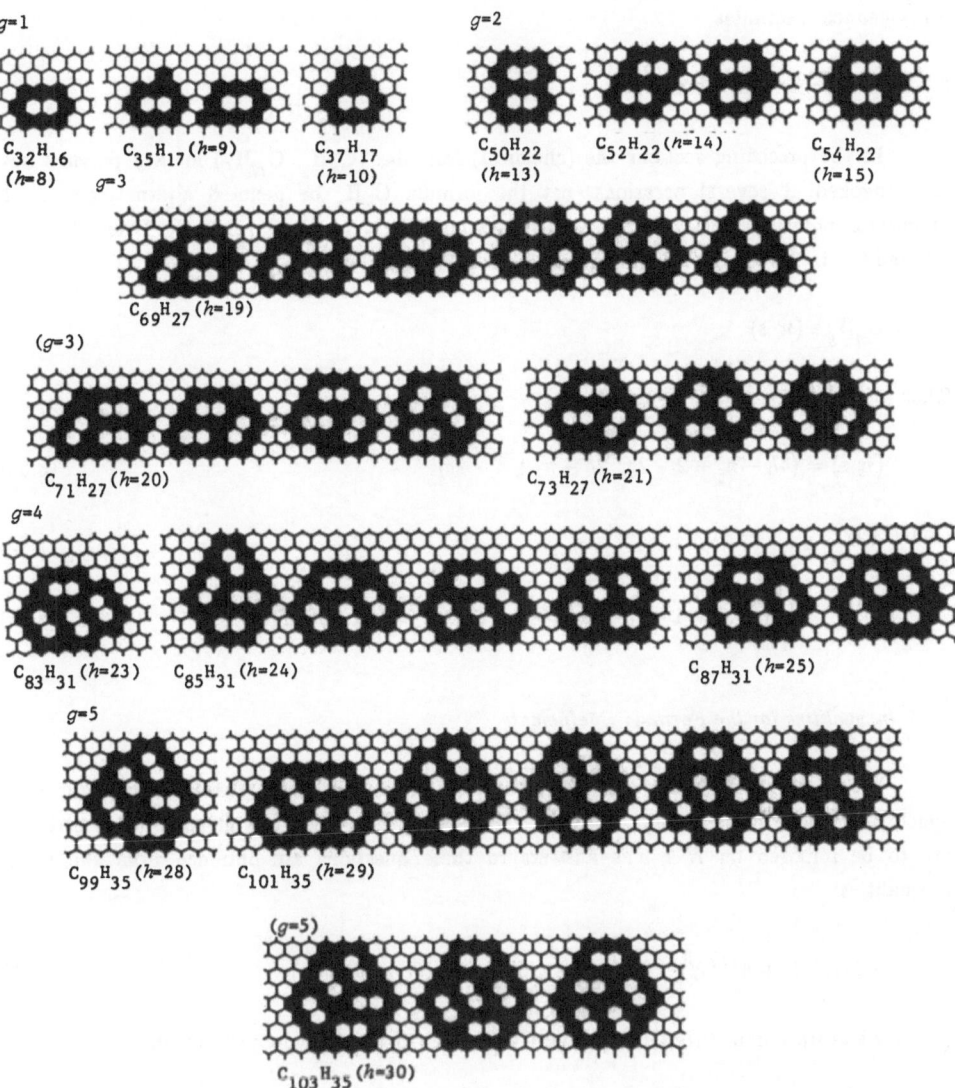

$g=1$ $C_{32}H_{16}$ $(h=8)$ $C_{35}H_{17}$ $(h=9)$ $C_{37}H_{17}$ $(h=10)$

$g=2$ $C_{50}H_{22}$ $(h=13)$ $C_{52}H_{22}$ $(h=14)$ $C_{54}H_{22}$ $(h=15)$

$g=3$ $C_{69}H_{27}$ $(h=19)$

$(g=3)$ $C_{71}H_{27}$ $(h=20)$ $C_{73}H_{27}$ $(h=21)$

$g=4$ $C_{83}H_{31}$ $(h=23)$ $C_{85}H_{31}$ $(h=24)$ $C_{87}H_{31}$ $(h=25)$

$g=5$ $C_{99}H_{35}$ $(h=28)$ $C_{101}H_{35}$ $(h=29)$

$(g=5)$ $C_{103}H_{35}$ $(h=30)$

Fig. 3.4. The perfect extremal coronoids for some of the smallest h values when $1 \leq g \leq 5$.

3.7 Chemical Formulas

3.7.1 *Introduction and Notation*

In the preceding sections the (chemical) formulas (C_nH_s, C_NH_S) for a g–polyhex have been invoked at several occasions. Let the formula C_nH_s be denoted alternatively in the "semicolon notation" as (Cyvin SJ and Brunvoll 1991; Cyvin SJ 1991c; 1992c; Brunvoll, Cyvin BN and Cyvin 1992b; Cyvin SJ, Cyvin and Brunvoll 1993e)

$$C_nH_s \equiv (n; s)$$

Then (cf. Table 1):

$$(n; s) = (4h - n_i + 2 - 2g; \ 2h - n_i + 4 - 4g) \tag{3.65}$$

Conversely,

$$(h, n_i) = (\tfrac{1}{2}(n - s) + 1 - g, \ n - 2s + 6 - 6g) \tag{3.66}$$

3.7.2 *Inequalities for the Formula Coefficients*

Let (N; S) represent the possible formulas for benzenoids (0–polyhexes). When N is given, which restrictions must be imposed on S ? And conversely: when S is given, which restrictions are to be imposed on N ? The answers to these questions are obtained from the known inequalities:

$$2\lceil (1/2)(N + 6^{1/2}N^{1/2}) \rceil - N \leq S \leq n + 2 - 2\lceil (1/4)(N - 2) \rceil \tag{3.67}$$

(Harary and Harborth 1976; Gutman and Cyvin 1989; Brunvoll and Cyvin SJ 1990) and

$$S - 6 + 2\lceil S/2 \rceil \leq N \leq S + 2\lfloor (1/12)(S^2 - 6S) \rfloor \tag{3.68}$$

(Brunvoll, Cyvin BN and Cyvin 1992b). In addition, the coefficients of (N; S) should have the same parity, i.e. either both N and S should be even, or both of them should be odd. Both equations (67) and (68) are perfect in the sense that the upper and lower bounds are realized for all values of N and S, respectively, which are allowed in benzenoids.

The above inequalities (67) and (68) were generalized to g–polyhexes with the following results:

$$2\lceil (1/2)(n + 6^{1/2}n^{1/2})\rceil - n + 2g \leq s \leq n + 2 - 2g - 2\lceil (1/4)(n - 2 + 2g)\rceil \tag{3.69}$$

$$s - 6 + 6g + 2\lceil s/2\rceil \leq n \leq s + 2\lfloor (1/12)(s - 2g)^2 - (s/2)\rfloor \tag{3.70}$$

In addition, n and s should again have the same parity. The relations (69) and (70) are not perfect inasmuch as the upper and lower bounds are not always realized in g–polyhexes. However, this is the case in general for $g = 0$, 1 and 2. These properties are similar to those of eqns. (57) and (60).

Examples: Consider the C_nH_s formulas for *2*-polyhexes. (a) $n = 70$; eqn. (69) gives $26 \leq s \leq 32$, hence the C_{70} formulas for double coronoids are $C_{70}H_{26}$, $C_{70}H_{28}$, $C_{70}H_{30}$ and $C_{70}H_{32}$. (b) $s = 23$; eqn. (70) gives $53 \leq n \leq 59$, hence the H_{23} formulas for double coronoids are $C_{53}H_{23}$, $C_{55}H_{23}$, $C_{57}H_{23}$ and $C_{59}H_{23}$.

3.7.3 *Table of Formulas*

The formulas of g–polyhexes for any g are found among the formulas C_NH_S of *0*–polyhexes (benzenoids). Table 5 shows a part of this formula table. It is presented in the same style as in Cyvin SJ and Cyvin (1993), where a more complete table of the C_NH_S benzenoid formulas in the coordinate system (H, N_i) is found. The opening statement of the present paragraph is a corollary of a more precise statement as expressed below.

Observation: The set of formulas C_nH_s for g–tuple coronoids ($g \geq 1$) forms a subset of the formulas for $(g-1)$–polyhexes.

Therefore, if C_nH_s is a formula for a g–tuple coronoid C_g ($g > 1$), then C_nH_s is also a formula for C_{g-1}; furthermore, any C_nH_s formula for $C_g(g \geq 1)$ is also a formula for a benzenoid. In Table 5 the regions of the formulas for *1*–polyhexes and *2*–polyhexes are indicated by heavy lines.

Table 5 is an alternative to the "periodic table for polycyclic aromatic hydrocarbons", which was devised by Dias (1982a; 1986; 1987; 1988; 1990c; 1990f); see also Brunvoll, Cyvin BN and Cyvin (1992b), Cyvin SJ, Cyvin and Brunvoll (1993e). It has been referred to as the *Dias periodic table* (Hall 1988; Cyvin SJ and Cyvin 1993). Here the C_nH_s formulas are arranged as an array in the coordinates (d_s, n_i). Dias (1982a; 1984b) was aware of the properties of subsets as expressed in the above observation. A detailed description of his identification of the smallest multiple coronoid formulas is given elsewhere (Cyvin SJ and Brunvoll 1990). Another discussion in Cyvin SJ and Brunvoll (1989) states that Dias seemed to have recognized that the coronoids, in a sense, are misplaced in the Dias periodic table. As Hall (1988) has put it: "They [molecules with holes] could be classed according to their chemical formula but this would bring together molecules with different chemical behaviour and some of the advantage of the table would be lost". In an attempt to overcome this difficulty Cyvin SJ and Brunvoll (1989) proposed separate

Table 3.5. Formulas for g–polyhexes.

$h+g$	$\alpha = 6$	7	8	9	10	11	12	13	14	15
	$\beta = n_i + 6g$									
7	$C_{24}H_{12}$									
8	$C_{28}H_{14}$	$C_{27}H_{13}$								
9	$C_{32}H_{16}$	$C_{31}H_{15}$	$C_{30}H_{14}$							
10	$C_{36}H_{18}$	$C_{35}H_{17}$	$C_{34}H_{16}$	$C_{33}H_{15}$						
11	$C_{40}H_{20}$	$C_{39}H_{19}$	$C_{38}H_{18}$	$C_{37}H_{17}$	$C_{36}H_{16}$	$C_{35}H_{15}$				
12	$C_{44}H_{22}$	$C_{43}H_{21}$	$C_{42}H_{20}$	$C_{41}H_{19}$	$C_{40}H_{18}$	$C_{39}H_{17}$	$C_{38}H_{16}$			
13	$C_{48}H_{24}$	$C_{47}H_{23}$	$C_{46}H_{22}$	$C_{45}H_{21}$	$C_{44}H_{20}$	$C_{43}H_{19}$	$C_{42}H_{18}$	$C_{41}H_{17}$		
14	$C_{52}H_{26}$	$C_{51}H_{25}$	$C_{50}H_{24}$	$C_{49}H_{23}$	$C_{48}H_{22}$	$C_{47}H_{21}$	$C_{46}H_{20}$	$C_{45}H_{19}$	$C_{44}H_{18}$	$C_{43}H_{17}$
15	$C_{56}H_{28}$	$C_{55}H_{27}$	$C_{54}H_{26}$	$C_{53}H_{25}$	$C_{52}H_{24}$	$C_{51}H_{23}$	$C_{50}H_{22}$	$C_{49}H_{21}$	$C_{48}H_{20}$	$C_{47}H_{19}$
16	$C_{60}H_{30}$	$C_{59}H_{29}$	$C_{58}H_{28}$	$C_{57}H_{27}$	$C_{56}H_{26}$	$C_{55}H_{25}$	$C_{54}H_{24}$	$C_{53}H_{23}$	$C_{52}H_{22}$	$C_{51}H_{21}$
17	$C_{64}H_{32}$	$C_{63}H_{31}$	$C_{62}H_{30}$	$C_{61}H_{29}$	$C_{60}H_{28}$	$C_{59}H_{27}$	$C_{58}H_{26}$	$C_{57}H_{25}$	$C_{56}H_{24}$	$C_{55}H_{23}$
18	$C_{68}H_{34}$	$C_{67}H_{33}$	$C_{66}H_{32}$	$C_{65}H_{31}$	$C_{64}H_{30}$	$C_{63}H_{29}$	$C_{62}H_{28}$	$C_{61}H_{27}$	$C_{60}H_{26}$	$C_{59}H_{25}$
19	$C_{72}H_{36}$	$C_{71}H_{35}$	$C_{70}H_{34}$	$C_{69}H_{33}$	$C_{68}H_{32}$	$C_{67}H_{31}$	$C_{66}H_{30}$	$C_{65}H_{29}$	$C_{64}H_{28}$	$C_{63}H_{27}$
20	$C_{76}H_{38}$	$C_{75}H_{37}$	$C_{74}H_{36}$	$C_{73}H_{35}$	$C_{72}H_{34}$	$C_{71}H_{33}$	$C_{70}H_{32}$	$C_{69}H_{31}$	$C_{68}H_{30}$	$C_{67}H_{29}$
21	$C_{80}H_{40}$	$C_{79}H_{39}$	$C_{78}H_{38}$	$C_{77}H_{37}$	$C_{76}H_{36}$	$C_{75}H_{35}$	$C_{74}H_{34}$	$C_{73}H_{33}$	$C_{72}H_{32}$	$C_{71}H_{31}$
22	$C_{84}H_{42}$	$C_{83}H_{41}$	$C_{82}H_{40}$	$C_{81}H_{39}$	$C_{80}H_{38}$	$C_{79}H_{37}$	$C_{78}H_{36}$	$C_{77}H_{35}$	$C_{76}H_{34}$	$C_{75}H_{33}$

periodic tables for g-polyhexes with different given values of g, and they displayed the table for single coronoids ($g = 1$) as an example. This implies a coordinate shift which, as we shall see, can be handled in terms of g-dependent invariants. We shall presently refer to them as *global invariants* and introduce them in relation to the formula table in the style of Table 5 rather than the Dias periodic table.

Let the global invariants α and β be defined by

$$\alpha = h + g \ , \ \beta = n_i + 6g \tag{3.71}$$

Then a formula $C_n H_s$ is expressed by

$$(n; s) = (4\alpha - \beta + 2; \ 2\alpha - \beta + 4) \tag{3.72}$$

where the parameter g does not appear explicitly; it is "hidden" in the global invariants. Now the idea is to use these invariants, viz. α and β, as coordinates for the $C_n H_s$ formulas. By this approach it is achieved that the coordinates (α, β) apply to any g-polyhex; cf. Table 5 for an illustration.

It follows straightforwardly from eqn. (72) that the table of formulas for g-polyhexes can be built up recursively by the scheme:

$$C_n H_s(\alpha, \beta) \longrightarrow C_{n-1} H_{s-1}(\alpha, \beta+1)$$
$$\downarrow$$
$$C_{n+4} H_{s+2}(\alpha+1, \beta)$$

The initial formula, situated at the upper-left corner of the table, depends on the g value.

Furthermore, one should observe the termination of each row in the right-hand direction. This termination results in the *staircase boundary*. It is determined by the formulas situated at the extreme-right of each row, namely the formulas which pertain to the extremal g-polyhexes, as should be clear from the definition of these systems (Par. 3.3.4). When perfect extremal g-tuple coronoids are involved, then the corresponding staircase boundary reflects a part of the staircase boundary for benzenoids. It is specifically the part which corresponds to the benzenoids associated with the g-tuple coronoids in question. Thus, for instance, the staircase boundary for single coronoids has the same shape as the one for benzenoids when starting from $C_{32}H_{14}$ ovalene (cf. Table 5, where the start of this staircase boundary for benzenoids is indicated by thin lines). A staircase boundary of this kind, determined by perfect extremal coronoids or extremal benzenoids, shall be referred to as a *perfect staircase* (boundary). If imperfect extremal coronoids are involved we shall call it an *imperfect staircase*.

A perfect extremal coronoid or an extremal benzenoid is characterized by (h^*, n_i^*) in the notation of eqns. (58) and (62). Introduce

$$\alpha^* = h^* + g \ , \ \beta^* = n_i^* + 6g \tag{3.73}$$

according to eqn. (71). Then, by means of eqns. (58) and (62) it is found

$$\beta^* = 2\alpha + 1 - 2g - \lceil (12\alpha - 3 + 12g)^{1/2} \rceil \tag{3.74}$$

and

$$\alpha^* = 1 + g + \lceil (\beta/2) + (1/2)(6\beta + 24g)^{1/2} \rceil \tag{3.75}$$

respectively. From these relations it is apparent that β^* decreases when g increases for a given α, while α^* increases when g increases for a given β. Herefrom the observation at the beginning of this paragraph is explained for perfect staircases, but then it must obviously be sound for imperfect staircases as well. It should be noted, however, that this deduction is based on certain unproved presumptions, which were made in preceding sections of this chapter.

In the following we shall specify the $C_n H_s$ formulas which actually exist for the smallest g–tuple coronoids when $g = 3$, 4 and 5; this tabulation is a summary of some of the results deduced in Par. 3.6.3 and Par. 3.6.4.

For $g = 3$:

α	h	$\beta = 18$ $n_i = 0$	19 1	20 2	21 3	22 4	23 5
20	17	$(C_{64}H_{26})$					
21	18	$C_{68}H_{28}$	$C_{67}H_{27}$	$(C_{66}H_{26})$			
22	19	$C_{72}H_{30}$	$C_{71}H_{29}$	$C_{70}H_{28}$	$C_{69}H_{27}$		
23	20	$C_{76}H_{32}$	$C_{75}H_{31}$	$C_{74}H_{30}$	$C_{73}H_{29}$	$C_{72}H_{28}$	$C_{71}H_{27}$

For $g = 4$:

α	h	$\beta = 24$ $n_i = 0$	25 1	26 2	27 3	28 4	29 5
25	21	$(C_{78}H_{30})$					
26	22	$C_{82}H_{32}$	$(C_{81}H_{31})$	$(C_{80}H_{30})$			
27	23	$C_{86}H_{34}$	$C_{85}H_{33}$	$C_{84}H_{32}$	$C_{83}H_{31}$		
28	24	$C_{90}H_{36}$	$C_{89}H_{35}$	$C_{88}H_{34}$	$C_{87}H_{33}$	$C_{86}H_{32}$	$C_{85}H_{31}$

For $g = 5$:

α	h	$\beta = 30$ $n_i = 0$	31 1	32 2	33 3	34 4	35 5
30	25	$(C_{92}H_{34})$					
31	26	$C_{96}H_{36}$	$(C_{95}H_{35})$	$(C_{94}H_{34})$			
32	27	$C_{100}H_{38}$	$C_{99}H_{37}$	$C_{98}H_{36}$	$(C_{97}H_{35})$	$(C_{96}H_{34})$	
33	28	$C_{104}H_{40}$	$C_{103}H_{39}$	$C_{102}H_{38}$	$C_{101}H_{37}$	$C_{100}H_{36}$	$C_{99}H_{35}$

In the above diagrams the staircase boundaries are indicated by heavy lines. For the portions where they are imperfect they are augmented by stippled lines to the shapes which conform with perfect boundaries. Hence it should be understood that the parenthesized formulas actually do not exist for the coronoids in question. The heavy formulas pertain to perfect extremal coronoids.

Figure 5 shows a mapping of C_nH_s formulas for g–polyhexes in terms of a dot diagram, where each formula is represented by a dot. The same representation has been employed (for benzenoids and helicenes) by Cyvin SJ and Cyvin (1993). From Fig. 5 a quite large portion of the formula table can be retrieved when it only is remembered that the table starts with C_6H_6 at the top. The staircase boundaries from Table 5 and the above diagrams are all reproduced in Fig. 5, which in addition summarizes an original analysis for 6–tuple coronoids. In that analysis the smallest perfect extremal systems were identified as having the formula $C_{115}H_{39}$ and $h = 33$; the largest imperfect extremal systems were established as $C_{113}H_{39}$ ($h = 32$); examples of $C_{112}H_{40}$ ($h = 31$) systems were generated and supposed to be imperfect extremal; finally the catacondensed $C_{110}H_{40}$ 6–tuple coronoid (cf. Fig. 2) is also supposed to be an imperfect extremal system.

3.8 Numbers of Isomers

3.8.1 *Definition and Notation*

It is of interest to enumerate the g–polyhex isomers (in the sense of Par. 3.6.1) for different formulas C_nH_s. The following definition is inspired from a notation for benzenoids, which has been invented (Cyvin SJ and Brunvoll 1991; Cyvin SJ 1991c; 1992c; Cyvin SJ, Cyvin and Brunvoll 1993e).

Definition 3.6: The *cardinality* (of genus g) of a formula C_nH_s, written

$$|C_nH_s|_g \equiv |n;\ s|_g$$

is the number of nonisomorphic g–polyhexes which are compatible with that formula. In other words, $|C_nH_s|_g$ is the number of g–polyhex C_nH_s isomers.

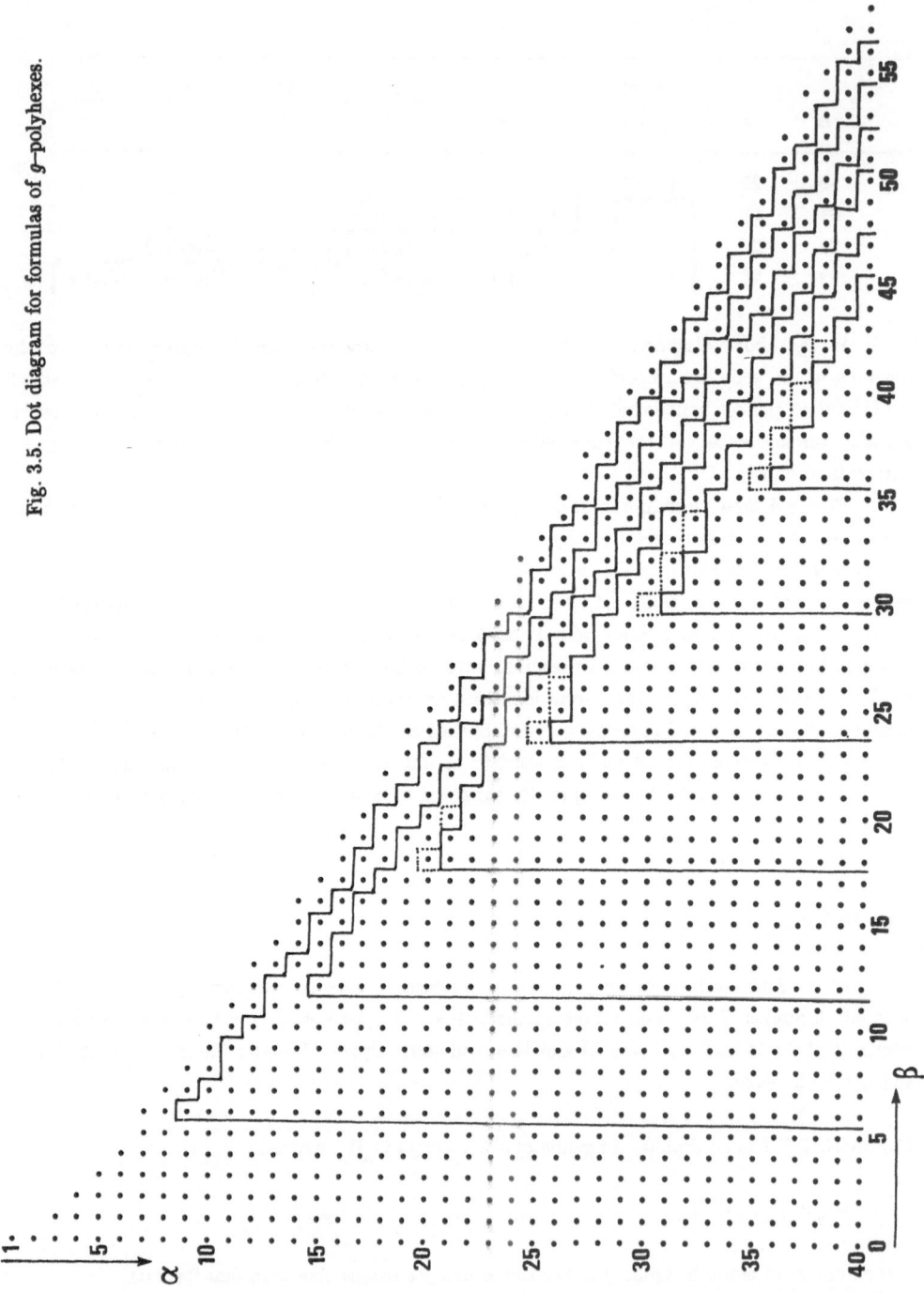

Fig. 3.5. Dot diagram for formulas of *g*-polyhexes.

3.8.2 *Numerical Values*

The definition of cardinalities of genus g (Par. 3.8.1) is illustrated below.

Examples

$|C_{50}H_{22}|_0 = 18396$	(Stojmenović et al. 1986; Brunvoll and Cyvin 1990)
$|C_{50}H_{22}|_1 = 1969$	(Cyvin SJ, Brunvoll and Cyvin 1991a)
$|C_{50}H_{22}|_2 = 1$	(cf. Fig. 2)
$|C_{50}H_{22}|_3 = 0$	

Very much work has been done on the computations of cardinalities of genus zero (viz. the numbers of C_NH_S benzenoid isomers); the reader may consult three extensive reviews (Brunvoll and Cyvin SJ 1990; Brunvoll, Cyvin BN and Cyvin 1992b; Cyvin SJ, Cyvin and Brunvoll 1993e) and references cited therein. These numbers fall outside the main scope of the present book.

With regard to the cardinalities of genus one (viz. the numbers of single coronoid C_nH_s isomers) the extensive enumerations in Volume I are not especially oriented towards these numbers. Nevertheless, the last chapter therein (I–9) has a clear relevance to the topic in question, although the different C_nH_s formulas are not specified therein. On the other hand, the numbers of C_nH_s single coronoid isomers are treated as a main topic in one of the subsequent chapters of the present book.

Table 6 shows the numbers of isomers for some classes of multiple coronoids, most of them taken from the enumerations of Cyvin SJ and Brunvoll (1990). The supplementary values for $g > 2$ in Table 6 are present results, which can be taken out from Figs. 2 and 4; in addition, it is inferred from Fig. 2: $|C_{110}H_{40}|_6 = |C_{162}H_{54}|_{11} = |C_{200}H_{64}|_{13} = 1$, $|C_{124}H_{44}|_7 = |C_{138}H_{48}|_8 = 2$, $|C_{152}H_{52}|_9 = |C_{190}H_{62}|_{12} = 5$. In Table 6 the separate numbers for Kekuléan and non–Kekuléan systems are included; in Fig. 2 the systems are catacondensed, and therefore all of them are Kekuléan.

Table 3.6. Numbers of g–polyhex (g–tuple coronoid) isomers.

g	h	n_i	Formula	Kekuléan	non–Kekuléan	Total
2	13	0	$C_{50}H_{22}$	1^a	0	1^a
	14	0	$C_{54}H_{24}$	5^a	0	5^a
		1	$C_{53}H_{23}$	0	4^a	4^a
		2	$C_{52}H_{22}$	2^a	0	2^a
	15	0	$C_{58}H_{26}$	33^a	0	33^a
		1	$C_{57}H_{25}$	0	52^a	52^a
		2	$C_{56}H_{24}$	38^a	10^a	48^a
		3	$C_{55}H_{23}$	0	15^a	15^a
		4	$C_{54}H_{22}$	1^a	0	1^a
	16	0	$C_{62}H_{28}$	211^a	0	211^a
		1	$C_{61}H_{27}$	0	434^a	434^a
		2	$C_{60}H_{26}$	379^a	127^a	506^a
		3	$C_{59}H_{25}$	0	334^a	334^a
		4	$C_{58}H_{24}$	88^a	33^a	121^a
		5	$C_{57}H_{23}$	0	12^a	12^a
	17	7	$C_{59}H_{23}$	0	4	4
3	18	0	$C_{68}H_{28}$	2^a	0	2^a
		1	$C_{67}H_{27}$	0	2^a	2^a
	19	0	$C_{72}H_{30}$	17^a	0	17^a
		1	$C_{71}H_{29}$	0	26^a	26^a
		2	$C_{70}H_{28}$	16^a	6^a	22^a
		3	$C_{69}H_{27}$	0	6^a	6^a
	20	0	$C_{76}H_{32}$	145^a	0	145^a
		1	$C_{75}H_{31}$	0	328^a	328^a
		2	$C_{74}H_{30}$	284^a	107^a	391^a
		3	$C_{73}H_{29}$	0	243^a	243^a
		4	$C_{72}H_{28}$	67^a	36^a	103^a
		5	$C_{71}H_{27}$	0	4^a	4^a

Table 3.6 (continued).

g	h	n_i	Formula	Kekuléan	non–Kekuléan	Total
3	21	7	$C_{73}H_{27}$	0	3	3
4	22	0	$C_{82}H_{32}$	1	0	1
	23	3	$C_{83}H_{31}$	0	1	1
	24	5	$C_{85}H_{31}$	0	4	4
	25	7	$C_{87}H_{31}$	0	2	2
5	26	0	$C_{96}H_{36}$	1	0	1
	28	5	$C_{99}H_{35}$	0	1	1
	29	7	$C_{101}H_{35}$	0	5	5
	30	9	$C_{103}H_{35}$	0	3	3

[a] Cyvin SJ, Brunvoll J (1990). Chem Phys Letters 170: 364

Chapter 4

INVARIANTS OF SINGLE CORONOIDS

4.1 Introduction

In this chapter the findings for invariants of g–polyhexes (cf. Sect. 3.2) are summarized for the special case of single coronoids ($g = 1$) and extended in a more detailed treatment. It comes as a supplement to the treatment of invariants of single coronoids in Vol.I–3.2.2. In particular, the upper and lower bounds of several invariants are treated extensively. This leads naturally to a treatment of extremal single coronoids (cf. Par. 3.3.4), which invariably are perfect.

4.2 Summary of Invariants and Relations Between Them

4.2.1 *Summary of Relations*

For $g = 1$ eqn. (3.1) assumes a particularly simple form without any constant term, viz.

$$h = m - n \tag{4.1}$$

In fact, nearly all the constant terms disappear from Table 3.1 when $g = 1$ is inserted; the only exceptions occur for $d_s(h, n_i)$ and $d_s(n, s)$. For the sake of convenience the special case of Table 3.1 for single coronoids is reproduced here (Table 1), but the invariant d_s is replaced by a related invariant denoted ν (see below), whereby the constant terms are avoided. Another especially simple relation, which emerges for single coronoids, is (cf. Table 1)

$$s = t \tag{4.2}$$

It was stated explicitly by Dias (1082a).

For the outer perimeter, eqns. (3.9) and (3.10) are immediately applicable to single coronoids; for the inner perimeter one has

$$n_b{}'' = s'' + t'' = 2s'' + 6 \tag{4.3}$$

and

$$t'' - s'' = 6 \tag{4.4}$$

Table 4.1. Invariants of single coronoids.

Invariant	Function of (h, n_i)	Function of (n, s)
h	h	$(1/2)(n - s)$
n	$4h - n_i$	n
m	$5h - n_i$	$(1/2)(3n - s)$
n_i	n_i	$n - 2s$
n_b	$4h - 2n_i$	$2s$
$s \equiv n_2$	$2h - n_i$	s
t	$2h - n_i$	s
n_3	$2h$	$n - s$
m_i	$h + n_i$	$(1/2)(3n - 5s)$
ν	$n_i - h$	$(1/2)(n - 3s)$

cf. eqns. (3.18) and (3.19), respectively.

Below we summarize additional formulas for single coronoids, which are consistent with the expressions of Sect. 3.3. The corona hole (cf. Par. 3.3.1) for a single coronoid is represented by a benzenoid B^O with the invariants h^o, n_i^o, n^o, s^o, etc. Then

$$(n^o; s^o) = (4h^o - n_i^o + 2; \; 2h^o - n_i^o + 4) \tag{4.5}$$

Conversely,

$$(h^o, n_i^o) = (\tfrac{1}{2}(n^o - s^o) + 1, \; n^o - 2s^o + 6) \tag{4.6}$$

Furthermore:

$$t^o = 2h^o - n_i^o - 2 = s^o - 6 \tag{4.7}$$

$$n_e^o = 4h^o - 2n_i^o + 2 = 2s^o - 6 \tag{4.8}$$

Let $B(H, N_i) = B(N; S)$ be the associated benzenoid with the single coronoid in question, and let its perimeter length be denoted by N_e. Then:

$$(H, N_i) = (h + h^o, \; n_i + n^o) = (h + h^o, \; 4h^o + n_i - n_i^o + 2) \tag{4.9}$$

$$(N; S) = (n + n^o - 2s^o + 6; \; s - s^o + 6) \tag{4.10}$$

$$N_e = 4(h - h^o) - 2(n_i - n_i^o) - 2 = 2(s - s^o) + 6 \tag{4.11}$$

A naphthalenic single coronoid (cf. Par. 3.3.3) has exactly one naphthalene hole. Figure 1 summarizes the characteristic features of this hole and the corresponding benzenoid (naphthalene). The special cases of Eqs. (9) – (11) for naphthalenic single coronoids read:

$$(H, N_i) = (h + 2, \; n_i + 10) \tag{4.12}$$

$$(N, S) = (n; \; s - 2) \tag{4.13}$$

$$N_e = 4h - 2n_i - 10 = 2s - 10 \tag{4.14}$$

Hence

$$N_e = n_b - 10 \tag{4.15}$$

One has also

$$M = m + 1 \tag{4.16}$$

where M is the number of edges in B.

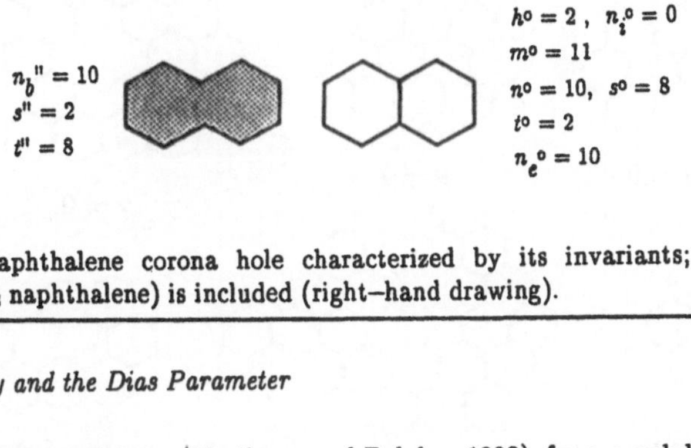

$$n_b'' = 10$$
$$s'' = 2$$
$$t'' = 8$$

$$h^o = 2, \; n_i^o = 0$$
$$m^o = 11$$
$$n^o = 10, \; s^o = 8$$
$$t^o = 2$$
$$n_e^o = 10$$

Fig. 4.1. The naphthalene corona hole characterized by its invariants; the corresponding benzenoid ($C_{10}H_8$ naphthalene) is included (right–hand drawing).

4.2.2 Connectivity and the Dias Parameter

Define the *connectivity*, μ (Morikawa and Balaban 1992), for a g–polyhex by

$$\mu = m_i - n_3 + 1 \tag{4.17}$$

This is Euler's relation applied to the system of internal edges; cf. eqns. (3.1) and (3.2). On inserting the functions $m_i(h, n_i)$ and $n_3(h)$ from Table 3.1 into eqn. (17) one arrives at

$$\mu = n_i - h + 2 - g = -d_s \tag{4.18}$$

which shows the connection of μ with the Dias parameter (d_s).

Let the *reduced connectivity*, ν, be defined by

$$\nu = \mu - 1 \qquad (4.19)$$

Then the following functions for this invariant (ν) are obtained

$$\nu = n_i - h + 1 - g = (1/2)(n - 3s) + 6 - 6g \qquad (4.20)$$

The special cases for $g = 1$ are entered in Table 1.

Figure 2 exemplifies the reduced connectivity for some $0-$ and $1-$polyhexes, actually the same systems which were used previously for the exemplification of the Dias parameter in Cyvin SJ and Brunvoll (1989).

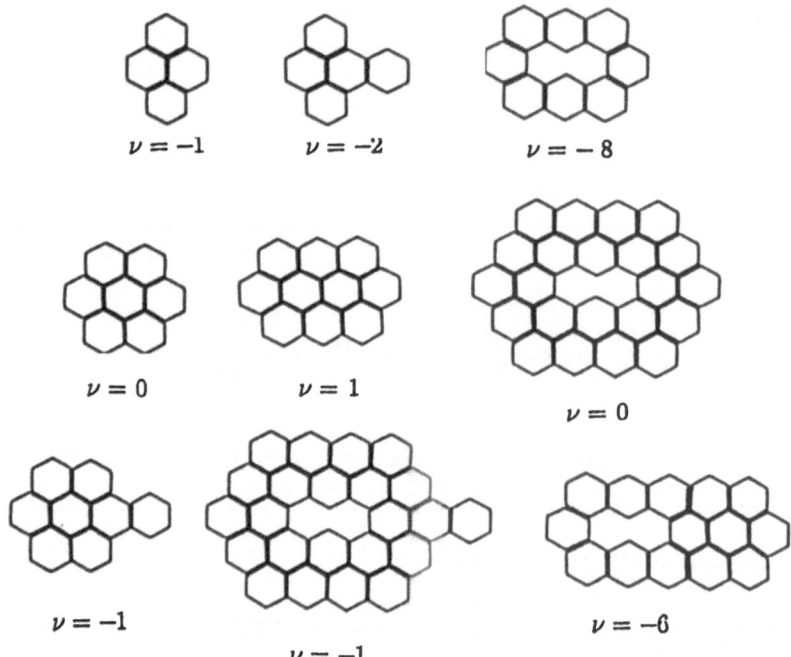

Fig. 4.2. Five benzenoids and four single coronoids with their reduced connectivities (ν) indicated. The internal edges are given as heavy lines. Formulas (from left to right): top row $C_{16}H_{10}$, $C_{20}H_{12}$, $C_{32}H_{16}$; middle row $C_{24}H_{12}$, $C_{32}H_{14}$, $C_{66}H_{22}$; bottom row $C_{28}H_{14}$, $C_{70}H_{24}$, $C_{48}H_{20}$.

4.3 Maximum Number of Internal Vertices, and Minimum Number of Hexagons

4.3.1 *Maximum Number of Internal Vertices, and Extremal Single Coronoids*

An extremal single coronoid (cf. Par. 3.3.4, especially Definition 3.4), A, is defined by having the maximum number of internal vertices (n_i) for a given number of hexagons (h): $n_i = (n_i)_{max}(h)$. This upper bound for n_i is known to be

$$(n_i)_{max} = 2h - 5 - \lceil (12h + 21)^{1/2} \rceil \tag{4.21}$$

as obtained from eqn. (3.44) on inserting $g = 1$. The expression (21) is also obtained from

$$(N_i)_{max} = 2H + 1 - \lceil (12H - 3)^{1/2} \rceil \tag{4.22}$$

for benzenoids; cf. eqn. (3.43) and the accompanying text. One only has to insert $H = h + 2$ and $(N_i)_{max} = (n_i)_{max} + 10$ in consistency with eqn. (12).

A coronoid A has a naphthalene hole; in other words, it is naphthalenic (cf. Definition 3.3 in Par. 3.3.3, and Observation 1 of Par. 3.6.1). Furthermore, any A is a perfect extremal coronoid; it is a perforated extremal benzenoid (cf. Definition 3.5). In this connection, notice the sign of equality in eqn. (21), in contrast to the form of eqn. (3.44). The above properties were used in the derivation of (21).

It is possible to give a firm justification to the effect that A is naphthalenic, in addition to the arguments in Vol.I–3.2.3, which are relevant in this connection. Let a coronoid (not necessarily single, in fact) be denoted by C. It is claimed that a corona hole of C, which holds more than two hexagons, can always be partially filled by hexagons from the outer perimeter of C so that the net number of internal vertices (n_i) increases. This property, if it could be demonstrated, would clearly prove that C is not extremal. Now, a hexagon can always be added into a corona hole so that n_i increases at least by two. An increment of more than two can always be achieved when the additions are continued. On the other hand, a hexagon from the outer perimeter can always be chosen so that n_i decreases at most by two when it is deleted. Below we show a *perforated snowflake* (Cyvin SJ, Bergan and Cyvin 1987; Cyvin SJ, Cyvin, Brunvoll and Bergan 1987; Cyvin SJ, Brunvoll and Cyvin 1989a; Cyvin SJ, Brunvoll, Cyvin, Točić and Kovačević 1989), which exemplifies the most unfavourable situation with regard to the described process.

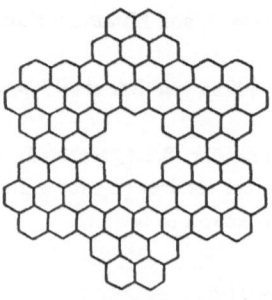

Here the deletion of a hexagon from the perimeter causes n_i to decrease at least by two, while an addition anywhere into the corona hole causes n_i to increase by two; see the left–hand drawing below, where the grey hexagon is the one added into the corona hole. But already in the second move (see the right–hand drawing below) it is possible to delete a hexagon (from the bottom of the left–hand drawing) with a decrease in n_i by only one, and to cover an inner cove with it so that the increase in n_i is three. The net increase in n_i is not less than two, demonstrating that the perforated snowflake certainly is not an extremal coronoid.

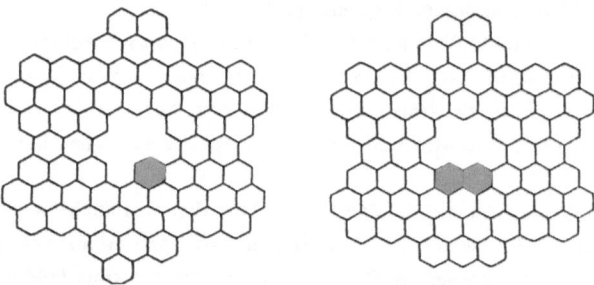

In Fig. 3 a diagram of the function $(n_i)_{max}(h)$ in consistency with eqn. (21) is shown. It is represented by the circles and the full–drawn curve through them.

4.3.2 *Minimum Number of Hexagons*

The expression of h_{min} as a function of n_i for single coronoids is obtained from eqn. (3.61) on inserting $g = 1$;

$$h_{min} = 4 + \lceil (n_i/2) + (1/2)(6n_i + 60)^{1/2} \rceil \tag{4.23}$$

Here again the sign of equality is appropriate. This expression is also obtainable directly from

$$H_{min} = 1 + \lceil (1/2)(N_i + 6^{1/2} N_i^{1/2}) \rceil \tag{4.24}$$

for benzenoids; cf. eqn. (3.59), in the same way as eqn. (21) could be derived from (22).

In any extremal single coronoid, A, one has clearly $h = h_{min}(n_i)$. However, this does not give the full information on the desired function, since A systems do not exist for every n_i; see below. These features are elucidated by Fig. 3, which includes the function $h_{min}(n_i)$ in consistency with eqn. (23). This function is represented by the curve drawn through the pertinent points (indicated by asterisks), stippled when it deviates from $(n_i)_{max}(h)$.

4.3.3 *Spiral Walk*

The process used by Harary and Harborth (1976) in order to generate one extremal benzenoid for every number of hexagons has been referred to as the *spiral walk* (Brunvoll, Cyvin BN and Cyvin 1993b). An application of the spiral walk to single coronoids is illustrated in Fig. 4. Starting from the smallest coronoid (which has $h = 8$) new hexagons are added, one at a time, in a spiral fashion as is indicated by the numerals in the figure. This spiral walk generates exactly one extremal single coronoid $A(h)$ for each $h = 8, 9, 10, 11, \ldots$. In all cases but two, however, there exist more $A(h)$ isomers for a given h. The two exceptions occur for $h = 8$ and $h = 10$; Figure 3.4 shows the pertinent two unique A isomers, viz. $C_{32}H_{16}$ and $C_{37}H_{17}$, respectively.

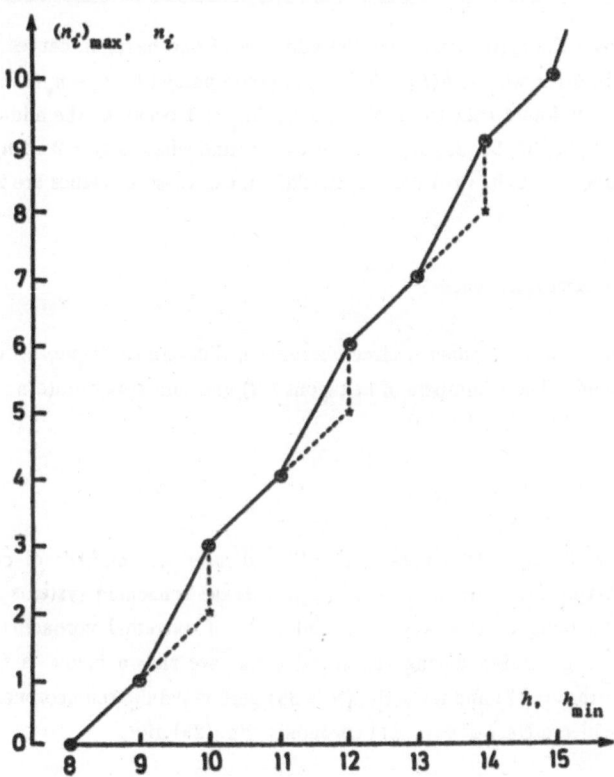

Fig. 4.3. Single coronoids: diagram of $(n_i)_{max}(h)$ on the full–drawn curve and of $h_{min}(n_i)$ on the stippled curve and parts of the full–drawn curve.

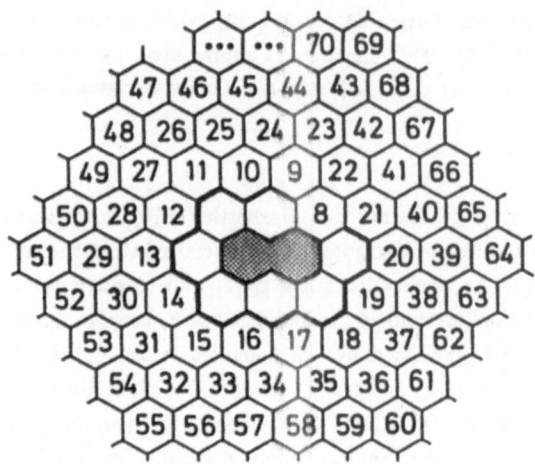

Fig. 4.4. The spiral walk for generating extremal single coronoids.

It is clear from the spiral walk that the addition of one hexagon causes n_i to increase by either one or two. In other words, $A(h) \to A(h+1)$ is accompanied by $n_i \to n_i + \Delta n_i$, where $\Delta n_i = 1$ or 2. From Fig. 4 it is found that the first cases of $\Delta n_i = 1$ occur at the additions of hexagons Nos. 9, 11, 13, 15, 18, 20, 23, 26, 29, 32, ... For every time when $\Delta n_i = 2$ an n_i value is skipped and consequently does not exist in A coronoids. The first of these n_i values are 2, 5, 8, 11, 13, 16, 19, 21, 24, 26, ...

4.3.4 Perforated Polycircumcoronenes

The systems benzene, coronene, circumcoronene, dicircumcoronene, ... constitute a class of extremal benzenoids. Their numbers of hexagons (H) and numbers of internal vertices (N_i) are given by (Brunvoll and Cyvin SJ 1990)

$$(H, N_i) = (3K^2 + 3K + 1, \ 6K^2) \tag{4.25}$$

where $K = 0, 1, 2, 3, ...$. This parameter (K) indicates the number of circumscribings of benzene. When starting from circumcoronene each of these benzenoid systems can be perforated by one naphthalene hole. In this way a special class of extremal coronoids is created. Two members, which are generated during the spiral walk, are shown below: a $C_{54}H_{20}$ perforated circumcoronene (with $h = 17$) and a $C_{96}H_{26}$ ($h = 35$) perforated dicircumcoronene.

For the considered class of extremal coronoids eqn. (25) gives

$$(h, n_i) = (3k^2 + 9k + 5, \ 6k^2 + 12k - 4) \tag{4.26}$$

simply by inserting for (H, N_i) from eqn. (12) and simultaneously making the substitution $K = k + 1$. Then the two systems below ($C_{54}H_{20}$ and $C_{96}H_{26}$) correspond to $k = 1$ and $k = 2$, respectively.

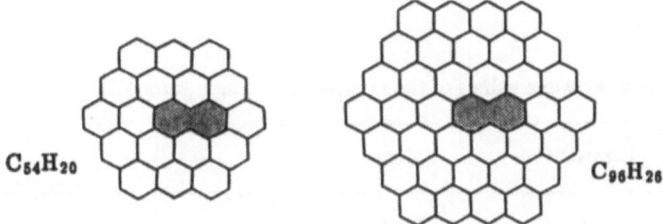

$$C_{54}H_{20} \qquad\qquad\qquad\qquad C_{96}H_{26}$$

From (26) one obtains

$$k = (1/2)[(1/3)(12h + 21)^{1/2} - 3] = (1/6)(6n_i + 60)^{1/2} - 1 \tag{4.27}$$

These relations were used to eliminate k from the equations which are implied in (26). Then it was obtained, in the first place,

$$n_i = 2h - 5 - (12h + 21)^{1/2} \tag{4.28}$$

which is seen to be consistent with eqn. (21). Next, it was obtained

$$h = 4 + (n_i/2) + (1/2)(6n_i + 60)^{1/2} \tag{4.29}$$

in consistency with eqn. (23). The two expressions (28) and (29) are the special cases of (21) and (23), respectively, where the square roots are integers already before the ceiling functions are taken.

4.4 Possible Values of the Invariants

The invariants of single coronoids have integer values restricted to certain domains. The smallest (single) coronoid has $h = 8$, and also all values $h > 8$ are obviously possible. For n_i, zero and all positive integers are possible, and specifically $n_i = 0, 1, 2, \ldots, (n_i)_{max}$ for every h. Now, with the aid of eqn. (21) and the expressions of Table 1 the domains of all the other invariants under consideration can be worked out. Here we give a complete account of the possible values for the invariants which are listed in Table 1. This material supplements the relevant information of Vol. I–3.2.3.

The number of vertices may be $n = 32$ or all integers $n \geq 35$. Similarly for the number of edges, $m = 40, 44, 45$ or $m \geq 47$. For the invariants $s = t$ one finds $s \geq 16$, $t \geq 16$, while the combined perimeter length (for the outer and inner perimeter) may assume all even integers $n_b \geq$

32. Also n_3 can only have even integer values, and $n_3 \geq 16$. The possible numbers of internal edges are given by $m_i \geq 8$. Finally, the invariant ν (as well as d_s and μ) may assume any positive and negative integer values or zero.

A survey of the possible values for the invariants considered here (see Table 1) is given in Table 2.

Table 4.2. Possible values of the invariants in single coronoids.

Invariant	Values
h	8,9,10,11,....
n	32, 35,36,37,38,....
m	40, 44,45, 47,48,49,50,....
n_i	0,1,2,3,....
n_b	32, 34, 36, 38,....
s	16,17,18,19,....
t	16,17,18,19,....
n_3	16, 18, 20, 22,....
m_i	8,9,10,11,....
ν	0,$\pm$1,$\pm$2,$\pm$3,....

4.5 Upper and Lower Bounds for Some Invariants

4.5.1 General

Assume that an invariant x for single coronoids is selected. Then it is of interest to determine the upper and lower bounds of x, if they exist, in terms of an other invariant, say y. Such an analysis may result in inequalities of the kind $f(y) \leq x \leq g(y)$, where $f(y) = x_{min}$ and $g(y) = x_{max}$ are functions to be determined.

The functions $(n_i)_{max}(h)$ and $h_{min}(n_i)$ are special cases, which were encountered in eqn. (21) and eqn. (23), respectively. They are relevant for the treatment in the next paragraph.

4.5.2 Functions of the Number of Hexagons and of the Number of Internal Vertices

The inequalities for n_i and h in terms of h and n_i, respectively, as given in eqns. (3.57) and (3.60), were specialized for single coronoids by inserting $g = 1$. The results are found among the relations of Table 3. All the other inequalities in this table were easily deduced from the two mentioned relations and those of Table 1. Some of the inequalities in terms of h, viz. those for n, m, n_i and n_b, are given in Vol. I–3.2.3.

All the upper and lower bounds under consideration (when they exist) are realized in single coronoids. With regard to the functions of h, also all the intermediate integer values are realized for every h with the exception for n_b. In this special case it is guaranteed that the upper and lower bounds come out as even integers, and only the even integers between them should be taken. In other words, the n_b values should be taken in intervals by 2, as is indicated in Table 3.

Table 4.3. Inequalities for invariants of single coronoids in terms of h and n_i.*

Inequalities	Interval
$2h + 5 + \lceil (12h+21)^{1/2} \rceil \leq n \leq 4h$	1
$3h + 5 + \lceil (12h+21)^{1/2} \rceil \leq m \leq 5h$	1
$0 \leq n_i \leq 2h - 5 - \lceil (12h+21)^{1/2} \rceil$	1
$10 + 2\lceil (12h+21)^{1/2} \rceil \leq n_b \leq 4h$	2
$5 + \lceil (12h+21)^{1/2} \rceil \leq s \leq 2h$	1
$h \leq m_i \leq 3h - 5 - \lceil (12h+21)^{1/2} \rceil$	1
$-h \leq \nu \leq h - 5 - \lceil (12h+21)^{1/2} \rceil$	1
$4 + \lceil (n_i/2) + (1/2)(6n_i+60)^{1/2} \rceil \leq h$	1
$16 - n_i + 4\lceil (n_i/2) + (1/2)(6n_i+60)^{1/2} \rceil \leq n$	4
$20 - n_i + 5\lceil (n_i/2) + (1/2)(6n_i+60)^{1/2} \rceil \leq m$	5
$16 - 2n_i + 4\lceil (n_i/2) + (1/2)(6n_i+60)^{1/2} \rceil \leq n_b$	4
$8 - n_i + 2\lceil (n_i/2) + (1/2)(6n_i+60)^{1/2} \rceil \leq s$	2
$8 + 2\lceil (n_i/2) + (1/2)(6n_i+60)^{1/2} \rceil \leq n_3$	2
$n_i + 4 + \lceil (n_i/2) + (1/2)(6n_i+60)^{1/2} \rceil \leq m_i$	1
$\nu \leq n_i - 4 - \lceil (n_i/2) + (1/2)(6n_i+60)^{1/2} \rceil$	1

* In terms of n_i: h, n, m, n_b, s and n_3 have no upper bounds; ν has no lower bound.

Example: For $h = 10$, $34 \leq n_b \leq 40$. Hence n_b may assume the values 34, 36, 38 and 40 in single coronoids with ten hexagons.

In general, the interval (cf. the last column in Table 3) is the absolute magnitude of the coefficient preceding the respective ceiling function. Also in the cases of the functions of n_i the values of the different invariants are to be taken in certain intervals, as indicated in Table 3.

Examples: For $n_i = 2$, $38 \leq n$ and $48 \leq m$, with the intervals of 4 and 5, respectively. Hence $n = 38, 42, 46, 50, \ldots$, and $m = 48, 53, 58, 63, \ldots$ for single coronoids with two internal vertices.

4.5.3 *Functions of Invariants Other Than the Number of Hexagons and the Number of Internal Vertices*

Disposition. The invariant $x = h$ is selected. Then the lower bounds, $h_{min}(y)$, and subsequently the upper bounds, $h_{max}(y)$, are determined, where $y = n$, m or s. The corre— sponding inequalities are tabulated. The table is augmented by the inequalities $f(n_b) \leq h \leq g(n_b)$ in terms of the number of boundary vertices (or the combined perimeter length), n_b. These last inequalities are easily accessible through the simple connection between s and n_b (cf. Table 1). Examples of the application of some of the inequalities are provided.

Lower Bounds. From some of the relations of Table 1 it is obtained:

$$h = (n + n_i)/4 \tag{4.30}$$
$$h = (m + n_i)/5 \tag{4.31}$$
$$h = (s + n_i)/2 \tag{4.32}$$

From these equations it is clear that $h = h_{min}(y)$, where $y = n$, m, s, is fulfilled for any catacondensed single coronoid ($n_i = 0$). For these systems, $n \equiv 32 \pmod 4$, $m \equiv 40 \pmod 5$ and $s \equiv 16 \pmod 2$. Clearly, in order to account for all the possible values of y one must also allow n_i to have some of the smallest nonvanishing values. The cases for $y = n$, m and s are treated separately below.

In single coronoids with $h = h_{min}(n)$ one finds $n_i = 0$, 1, 2 or 3. Specifically one finds the smallest n_i value which makes h to be an integer in eqn. (4.30), i.e. $(n + n_i) \equiv 0 \pmod 4$. In precise terms,

$$h_{min} = \begin{cases} n/4; & n = 32 + 4j \\ (n+1)/4; & n = 35 + 4j \\ (n+2)/4; & n = 38 + 4j \\ (n+3)/4; & n = 37 + 4j \end{cases} \tag{4.33}$$

where $j = 0, 1, 2, 3, \ldots$. The following compressed form is clearly valid for all n values.

$$h_{min} = \lceil n/4 \rceil \tag{4.34}$$

In an analogous way, $h = h_{min}(m)$ is found for $n_i = 0$, 1, 2, 3 or 4, and specifically:

$$h_{min} = \begin{cases} m/5; & m = 40 + 5j \\ (m+1)/5; & m = 44 + 5j \\ (m+2)/5; & m = 48 + 5j \\ (m+3)/5; & m = 47 + 5j \\ (m+4)/5; & m = 51 + 5j \end{cases} \tag{4.35}$$

where again $j = 0, 1, 2, 3, ...$. In compressed form:

$$h_{min} = \lceil m/5 \rceil \tag{4.36}$$

Finally, one has $h = h_{min}(s)$ for $n_i = 0$ or 1, depending on whether s is even or odd, respectively. We may write:

$$h_{min} = \begin{cases} s/2 \, ; & s = 16, 18, 20, 22, ... \\ (s+1)/2; & s = 17, 19, 21, 23, ... \end{cases} \tag{4.37}$$

and in compressed form:

$$h_{min} = \lceil s/2 \rceil \tag{4.38}$$

Upper Bounds. For a benzenoid $B(N;S)$ with H hexagons and M edges the following upper bounds are known already from the analysis of Harary and Harborth (1976).

$$H_{max} = N + 1 - \lceil (1/2)(N + 6^{1/2}N^{1/2}) \rceil \tag{4.39}$$

$$H_{max} = M - \lceil (2/3)(M-1) + (1/3)(4M+1)^{1/2} \rceil \tag{4.40}$$

The expression for $H_{max}(S)$ is a more recent deduction (Brunvoll, Cyvin BN, Cyvin, Gutman et al. 1989; Cyvin SJ 1992c; Brunvoll, Cyvin BN and Cyvin 1992b):

$$H_{max} = 1 + \lfloor (1/12)(S^2 - 6S) \rfloor \tag{4.41}$$

An extremal coronoid has simultaneously $h = h_{max}(n)$, $h = h_{max}(m)$ and $h = h_{max}(s)$. These maximum values are therefore readily obtained from eqns. (39)–(41) by means of the appropriate transformations according to (12), (13) and (16). However, in order to cover all the possible values of $y = n, m, s$ one must take into account some nonextremal systems too. Nevertheless, the procedure described above leads to generally valid expressions of $h_{max}(y)$ for single coronoids. It is not implied in this statement that all the nonextremal single coronoids C with $h = h_{max}(y)$ are naphthalenic; it is sufficient that there always is found (at least) one naphthalenic isomer of C.

The results are given in the following. For $h_{max}(n)$ one finds

$$h_{max} = n - 1 - \lceil (1/2)(n + 6^{1/2}n^{1/2}) \rceil \tag{4.42}$$

Next, for $h_{max}(m)$,

$$h_{max} = m - 1 - \lceil (2m/3) + (1/3)(4m+5)^{1/2} \rceil \qquad (4.43)$$

and finally for $h_{max}(s)$:

$$h_{max} = \lfloor (1/12)(s^2 - 10s + 4) \rfloor \qquad (4.44)$$

Inequalities. The inequalities on the basis of eqns. (34), (36), (38) and (42)–(44) are entered in Table 4. Here the upper and lower bounds are always realized in single coronoids, as well as all intermediate integer values; hence the intervals 1.

Table 4.4. Inequalities for h of single coronoids in terms of selected invariants.

Inequalities	Interval
$\lceil n/4 \rceil \leq h \leq n - 1 - \lceil (1/2)(n + 6^{1/2}n^{1/2}) \rceil$	1
$\lceil m/5 \rceil \leq h \leq m - 1 - \lceil (2m/3) + (1/3)(4m+5)^{1/2} \rceil$	1
$\lceil n_b/4 \rceil \leq h \leq \lfloor (1/48)(n_b^2 - 20n_b + 16) \rfloor$	1
$\lceil s/2 \rceil \leq h \leq \lfloor (1/12)(s^2 - 10s + 4) \rfloor$	1

Inclusion of the Number of Boundary Vertices. Table 4 is supplemented by the inequalities in terms of n_b, which are readily obtained through $s = n_b/2$. The upper bound of these inequalities, viz. $h_{max}(n_b)$, can alternatively be obtained directly from (Cyvin SJ, Brunvoll, Cyvin and Tošić 1990)

$$H_{max} = \lfloor (1/48)(N_e^2 + 12) \rfloor \qquad (4.45)$$

which holds for benzenoids. One only has to insert $H_{max} = h_{max} + 2$ and $N_e = n_b - 10$ in accord with eqns. (12) and (15). The result is consistent with the pertinent expression in Table 4.

Examples. (a) For $n = 48$, $12 \leq h \leq 14$ ($h = 12, 13, 14$); (b) for $m = 60$, $12 \leq h \leq 13$ ($h = 12, 13$); (c) for $s = 24$ or $n_b = 48$, $12 \leq h \leq 28$ ($h = 12, 13, 14, 15, ..., 28$).

4.6 Minimum Number of Vertices of Degree Two, and Maximum Number of Hexagons

4.6.1 *Minimum Number of Vertices of Degree Two*

For a given h in a single coronoid, the minimum number of vertices of degree two (s) is

$$s_{min} = 5 + \lceil (12h + 21)^{1/2} \rceil \qquad (4.46)$$

cf. Table 3. The property $s = s_{min}(h)$ may serve as a definition for the extremal single coronoids. The expression of $s_{min}(h)$ is obtainable alternatively from the corresponding expression for benzenoids, viz. (Gutman and Cyvin 1989; Brunvoll, Cyvin BN and Cyvin 1992b)

$$S_{min} = 3 + \lceil (12H - 3)^{1/2} \rceil \qquad (4.47)$$

Here one only has to make the substitutions $S_{min} = s_{min} - 2$ and $H = h + 2$ in order to attain at eqn. (46); the transformations conform with eqns. (13) and (12), respectively.

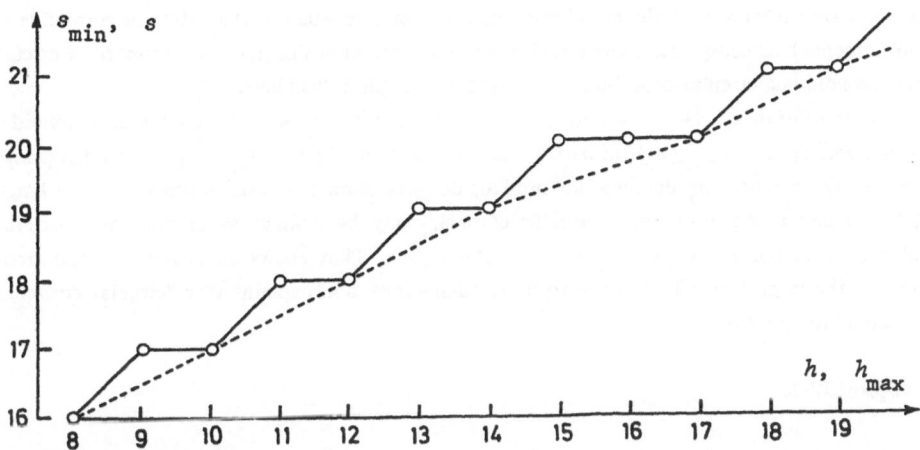

Fig. 4.5. Single coronoids: diagram of $s_{min}(h)$ on the full–drawn curve and of $h_{max}(s)$ on the stippled curve.

Figure 5 shows a diagram of the function $s_{min}(h)$, represented by the circles with the full–drawn curve through them.

4.6.2 *Maximum Number of Hexagons, and Circular Single Coronoids*

For a given s in a single coronoid, the maximum number of hexagons (h) is given by eqn. (44). Figure 5 includes a diagram of the function $h_{max}(s)$, represented by the circles on the stippled curve.

A single coronoid which has $h = h_{max}$ for a given s is clearly an extremal coronoid, A. However, the property $h = h_{max}(s)$ cannot serve as a definition for extremal single coronoids. As a matter of fact, eqn. (44) covers only selected values of h out of those which are possible for single coronoids and for which A systems exist. The A systems are represented by the full–drawn curve of Fig. 5. It is observed that $h = h_{max}(s)$ is not always fulfilled for these systems; this occurs for $h = 9, 11, 13, 15, 16, 18, ...$ (cf. Fig. 5).

A *circular* (single) *coronoid* (Cyvin SJ 1991b) may be defined in analogy with a circular benzenoid (Cyvin 1992c; Brunvoll, Cyvin BN and Cyvin 1992b; Cyvin SJ, Cyvin and Brunvoll 1993e).

Definition 4.1: A circular coronoid is defined by having the maximum number of hexagons (h) for a given number of vertices of degree two (s): $h = h_{max}(s)$.

The circular single coronoids are represented by the dots on the stippled curve in Fig. 5. It is an important feature that the circular coronoids form a subclass of the extremal coronoids (of the same genus). Consequently, the circular coronoids are naphthalenic. Furthermore, a circular single coronoid is a circular benzenoid perforated by a naphthalene hole.

In Definition 4.1 the invariant s can be replaced by n_b so that a circular coronoid is characterized by $h = h_{max}(n_b)$. Moreover, one has also clearly $h = h_{max}(n_b')$ as an alternative characterization, where n_b' denotes the numbers of vertices on the outer perimeter or the length of this perimeter. Accordingly, a circular coronoid may be defined by having the maximum number of hexagons for a given outer perimeter length. That shows an analogy to the circle, which has the largest area in relation to its circumference, and explains why "circular coronoid" is a rational designation.

4.6.3 *Spiral Walk*

The spiral walk (cf. Par. 4.3.3, and in particular Fig. 4) generates exactly one circular single coronoid for every s value. Specifically, the first circular single coronoids are generated by the additions of hexagons Nos. 8, 10, 12, 14, 17, 19, 22, 25, 28, 31, ... The increment $\Delta n_i = 1$ (cf. Par. 4.3.3) occurs always by the addition of one hexagon immediately after the generation of a circular coronoid.

A closer inspection of the spiral walk is elucidating. It is observed that s, as well as n_b and n_b', are unchanged whenever the addition of a hexagon is accompanied by $\Delta n_i = 2$. Only when $\Delta n_i = 1$, s increases by one unit, while n_b and n_b' increase by two units each.

After the above discussion it is clear that the generation of circular single coronoids for h

> 8 can be described in terms of additions of segments, viz. single linear chains of hexagons. The first segments, specified by the numbers of hexagons with reference to Fig. 4, given in curled brackets, are: {9,10}, {11,12}, {13,14}, {15,16,17}, {18,19}, {20,21,22}, {23,24,25}, {26,27,28}, {29,30,31}, {32,33,34,35}, ...

4.6.4 Detailed Analysis

Here we give the first detailed analysis of the circular single coronoids, O (but see also the work by Cyvin SJ 1991b).

Depictions. Figure 6 shows the smallest O systems arranged systematically by a pair of coordinates $\{\epsilon, k\}$.

Here $\epsilon = 0, 1, 2, 3, 4, 5$, and each of these values is associated with a characteristic shape of the circular coronoid. These shapes were already known at least to Balaban (1971), who investigated their important role in the studies of annulenes; cf. also a later work in this area (Cyvin SJ, Brunvoll and Gutman 1990). The same shapes are also encountered under the studies of certain primitive single coronoids called hollow hexagons (Cyvin SJ, Brunvoll and Cyvin 1989d; Cyvin SJ, Brunvoll, Cyvin, Bergan and Brendsdal 1991), where the extremal property $h^o = h^o_{max}(h)$ for the corona hole is of interest. A detailed treatment of this topic is offered in Vol. I–4.6.

The second coordinate, k, for circular single coronoids (cf. Fig. 6) indicates the number of circumscribings. For $\epsilon = 0$ and 1, $k = 1, 2, 3, 4, ...$; for $\epsilon = 2, 3, 4$ and 5, $k = 0, 1, 2, 3, ...$.

The depictions of circular single coronoids in Fig. 6 are augmented by two degenerate coronoids, which fit into the scheme when attributed to $k = 0$ for $\epsilon = 0$ and 1. In fact, the scheme can be extended further by the following degenerate coronoids (starting from [10]annulene).

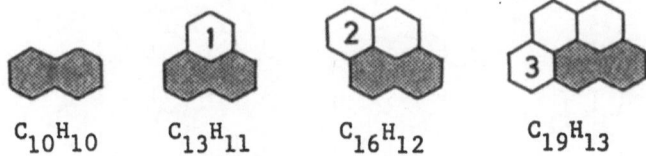

$$C_{10}H_{10} \qquad C_{13}H_{11} \qquad C_{16}H_{12} \qquad C_{19}H_{13}$$

These systems correspond to $k = -1$ and $\epsilon = 2, 3, 4, 5$.

Explicit Formula. In a given column of Fig. 6 it is found that the last added segments have the lengths (in terms of their numbers of hexagons) $k + 1$ and four times $k + 2$, when going from the next–to–top system to the bottom system. Then a segment of the length $k + 3$ should be added to the bottom system in order to move to the top of the next column. The systems of the top row ($\epsilon = 0$) have clearly $s = 6k + 14$. Their numbers of hexagons are found to be:

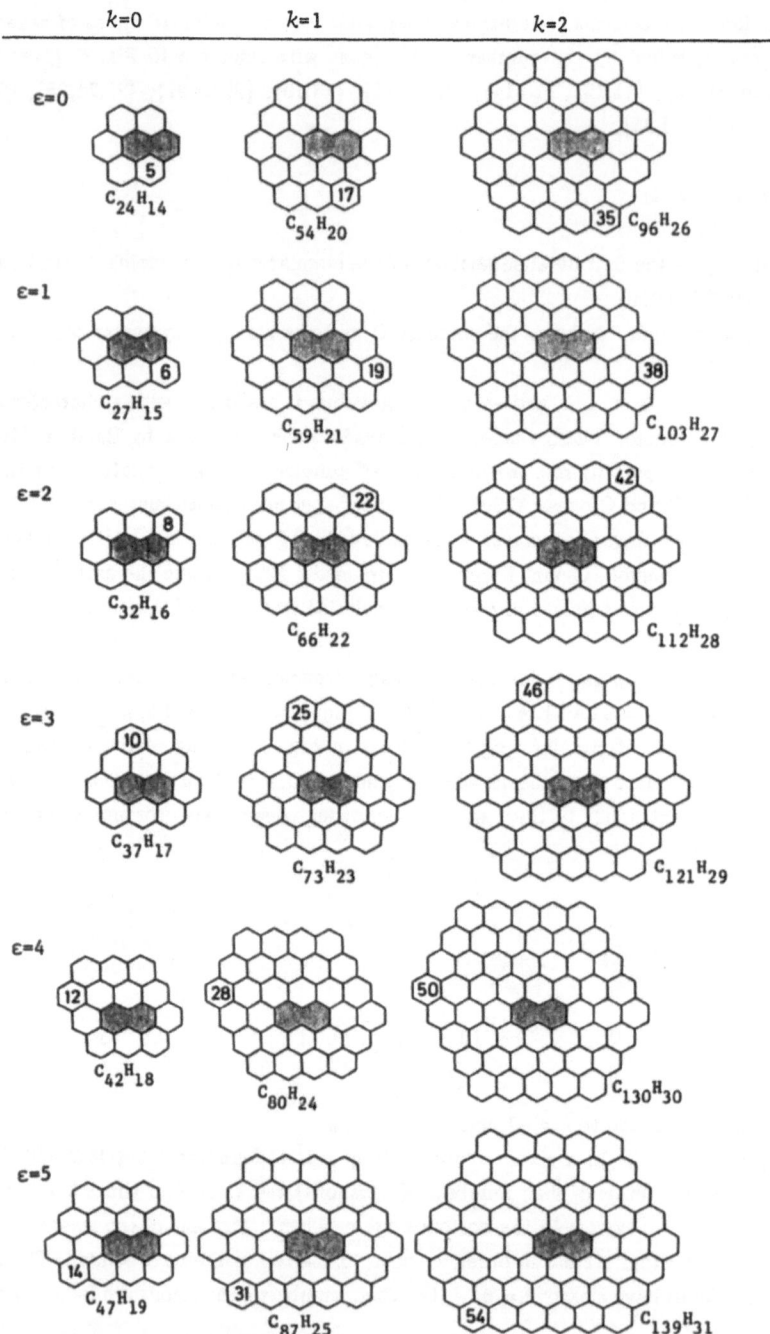

Fig. 4.6. The smallest circular single coronoids, augmented by two degenerate coronoids. The inscribed numerals indicate the number of last added hexagon during the spiral walk.

$$h_{max}(6k+14) = 5 + \sum_{i=0}^{k-1} [i + 1 + 4(i+2) + i + 3]$$

$$= 5 + 6 \sum_{i=0}^{k-1} (i + 2) = 3k^2 + 9k + 5 \tag{4.48}$$

This was a re–derivation of the number of hexagons in polycircumcoronenes perforated by one naphthalene hole each; see eqn. (26). Now the expressions for $h_{max}(s)$ in general are easily obtained with the following result.

$$h_{max}(6k+14) = 3k^2 + 9k + 5 \tag{4.49a}$$
$$h_{max}(6k+15) = 3k^2 + 10k + 6 \tag{4.49b}$$
$$h_{max}(6k+16) = 3k^2 + 11k + 8 \tag{4.49c}$$
$$h_{max}(6k+17) = 3k^2 + 12k + 10 \tag{4.49d}$$
$$h_{max}(6k+18) = 3k^2 + 13k + 12 \tag{4.49e}$$
$$h_{max}(6k+19) = 3k^2 + 14k + 14 \tag{4.49f}$$

These six relations pertain to $\epsilon = 0, 1, 2, 3, 4, 5$, respectively. They can be written in the compact form:

$$h_{max}(6k+14+\epsilon) = 3k^2 + 9k + 5 + (k+2)\epsilon - \lceil \epsilon/6 \rceil \tag{4.50}$$

Below we shall indicate an alternative derivation of eqn. (50); it is based on our knowledge of circular benzenoids, say $O(H,N_i) = O(N;S)$. By definition, they have the property $H = H_{max}(S)$. The formula for $H_{max}(S)$ as

$$H_{max}(6K+6+\epsilon) = 3K^2 + 3K + 1 + (K+1)\epsilon - \lceil \epsilon/6 \rceil \tag{4.51}$$

is a straightforward deduction from previous analyses of the O systems (Cyvin SJ 1992c; Cyvin SJ, Cyvin and Brunvoll 1993e), although it has not been given previously in exactly the form (51). Here again the transformations of eqns. (12) and (13) are to be used. On inserting $S = s - 2$ one obtains

$$s = 6K + 8 + e \tag{4.52}$$

where $K = k + 1$ as in the treatment of perforated polycircumcoronenes (Par. 4.3.4), which actually represent a special case ($\epsilon = 0$) of the O systems. The parameter ϵ in eqn. (51) may be identified with the same symbol in (50), and indeed the substitution of K by k in eqn. (52) leads to the same expression of s as in the argument on the left–hand side of (50). Now the substitution $H_{max} = h_{max} + 2$ is to be executed; it renders eqn. (51) into the form

$$h_{max} = 3K^2 + 3K - 1 + (K+1)\epsilon - \lceil \epsilon/6 \rceil \tag{4.53}$$

where it only remains to insert $K = k + 1$ in order to attain at eqn. (50).

Equivalence Between Two Expressions. It is instructive to establish the equivalence of eqn. (50) with (44). On inserting $s(\epsilon,k)$ from eqn. (50) into (44) it is obtained

$$
\begin{aligned}
h_{max} &= \lfloor (1/12)(6k + 14 + \epsilon)^2 - (5/6)(6k + 14 + \epsilon) + (1/3) \rfloor \\
&= 3k^2 + 9k + 5 + (k+1)\epsilon + \lfloor (1/12)(\epsilon^2 + 6\epsilon) \rfloor
\end{aligned} \tag{4.54}
$$

In order to establish the identity of this result with the right–hand side of eqn. (50) one has to prove that

$$\lfloor (1/12)(\epsilon^2 + 6\epsilon) \rfloor = \epsilon - \lceil \epsilon/6 \rceil \tag{4.55}$$

Here $\epsilon = 0$ makes both sides to vanish. But the relation (55) is not valid for arbitrary ϵ values; it fails for all $\epsilon \geq 6$. However, it is found for the values of interest, viz. $\epsilon = 1, 2, 3, 4, 5$, that both sides of (55) are equal to 0, 1, 2, 3, 4, respectively. That completes the proof of the equivalence between eqns. (44) and (50).

Concluding Remark. According to Cyvin SJ, Cyvin and Brunvoll (1993e) eqns. (44) and (50) may be said to belong to the *Harary–Harborth picture* and *Balaban picture*, respectively; cf. also Cyvin SJ and Brunvoll (1991). These designations are rational since eqn. (44) is based on the analysis of Harary and Harborth (1976), while the derivation of (50) is closely related to Balaban (1971).

Chapter 5

CHEMICAL FORMULAS OF SINGLE CORONOIDS

5.1 Introduction

Parts of this chapter, like the preceding chapter, is a specialization of relations from Chapt. 3 for $g = 1$, but also other aspects are treated. In particular, a concise treatise of the processes referred to as circumscribing and excising is offered. These processes are especially important when it comes to extremal polyhexes; the extremal single coronoids is another main subject of this chapter. All the treatments in this chapter are (more or less) oriented towards expressions in terms of the coefficients of the $C_n H_s$ formulas, viz. n and s, as is reflected in the title.

5.2 Terminology

It is recalled (cf. Par. 3.7.1) that a (chemical) formula $C_n H_s$ for a g–polyhex can be written $(n; s)$. Then one has for single coronoids:

$$(n; s) = (4h - n_i;\ 2h - n_i) \tag{5.1}$$

$$(h, n_i) = (\tfrac{1}{2}(n-s),\ n-2s) \tag{5.2}$$

cf. eqns. (3.65), (3.66) and Table 4.1.

Let the set of g–polyhexes which are compatible with the formula $C_n H_s$ be identified by the symbol

$$\{C_n H_s\} \equiv \{n;\ s\}$$

and denote a g–tuple coronoid system by C_g. Then the expression $C_g \in \{C_n H_s\}$ or $C_g \in \{n;\ s\}$ may be used to indicate that C_g has the formula $C_n H_s$ and therefore may be written $C_g(n;s)$. There may exist a coronoid $C_g{}'(n;s)$ which is not isomorphic with C_g and has the same formula $C_n H_s$: $C_g{}' \in \{C_n H_s\}$. Then C_g and $C_g{}'$ are two (nonisomorphic) g–tuple coronoid isomers of $C_n H_s$. Here "isomer" is used in a restricted sense according to the definition of this term in Par. 3.6.1.

The above formalism is also applicable to benzenoids, B: $B \in \{C_N H_S\}$ indicates that B has the formula $C_N H_S$.

Examples

1. If $C_g \in \{C_{32}H_{16}\}$, then C_g must be the smallest coronoid, which is single (see the first silhouette in Fig. 3.4).

2. If P_g is a *g*–polyhex, and $P_g \in \{C_{50}H_{22}\}$, then $g = 0$, 1 or 2; i.e. P_g is either a benzenoid, single coronoid or double coronoid (cf., e.g., Table 3.5).

3. Assume that C is a single coronoid, and $C \in \{C_{36}H_{18}\}$. Then C is one of the three $C_{36}H_{18}$ single coronoid isomers:

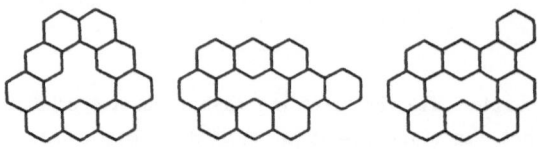

5.3 Inequalities in Terms of the Formula Coefficients

5.3.1 *When is a Given Formula Compatible With a Single Coronoid?*

Let $(n; s)$ represent the possible formulas for single coronoids. The possible values of n and s, independently of each other, are specified in Table 4.2.

When n is Given, Which are the Possible Values of s? The answer to this question is implied in the formula apparatus of Chapt. 4: pick out the inequalities $h_{min}(n) \le h \le h_{max}(n)$ from Table 4.4; insert $h(n,s)$ from Table 4.1; solve the inequalities with respect to s. During the last step the inequality has to be multiplied by two. The result is

$$2 - n + 2\lceil (1/2)(n + 6^{1/2}n^{1/2}) \rceil \le s \le n - 2\lceil n/4 \rceil \tag{5.3}$$

associated with the interval of 2. Eqn. (3) is the special case of eqn. (3.69) for $g = 1$. In the present case both the upper and lower bounds are realized for all possible values of n, and one has clearly $s_{max}(n) \equiv s_{min}(n)$ (mod 2). As indicated by the interval, the intermediate s values should be taken in steps by two. In consequence, n and s have always the same parity; i.e., either both of these invariants are even, or both of them are odd.

Example: Which C_{60} formulas are possible for single coronoids? For $n = 60$, eqn. (3) gives $22 \le s \le 30$. The answer is — $C_{60}H_{22}$, $C_{60}H_{24}$, $C_{60}H_{26}$, $C_{60}H_{28}$, $C_{60}H_{30}$.

A direct approach to the derivation of $s_{max}(n)$, analogous to the treatment of lower bounds in Par. 4.5.3, starts with

$$s = (n - n_i)/2 \tag{5.4}$$

as obtained from the relations of Table 4.1. It is clear that $s = s_{max}(n)$ is fulfilled for the catacondensed systems ($n_i = 0$). These systems cover the cases with $n \equiv 32 \pmod 4$. In order to take all the possible n values into account one must also allow for $n_i = 1$, 2 and 3, in which cases $n \equiv 35 \pmod 4$, $n \equiv 38 \pmod 4$ and $n \equiv 37 \pmod 4$, respectively. In conclusion,

$$s_{max} = \begin{cases} n/2; & n = 32 + 4j \\ (n-1)/2; & n = 35 + 4j \\ (n-2)/2; & n = 38 + 4j \\ (n-3)/2; & n = 37 + 4j \end{cases} \tag{5.5}$$

where $j = 0, 1, 2, 3, \ldots$. The compact expression on the right-hand side of eqn. (3) is consistent with (5).

The expression of $s_{min}(n)$ in eqn. (3) is obtainable alternatively by introducing $N = n$ and $S = s - 2$ in accord with eqn. (4.13) into (3.67), which applies for benzenoids.

When s is Given, Which are the Possible Values of n? In order to answer this question an analogous approach was used as in the above treatment. The result is

$$s + 2\lceil s/2 \rceil \leq n \leq s + 2\lfloor (1/12)(s^2 - 10s + 4) \rfloor \tag{5.6}$$

associated with the interval of 2. This equation is the special case of (3.69) for $g = 1$. In (6) the upper and lower bounds are realized for all the possible values of s, and again the intermediate values should be taken in steps by two.

Example: Which H_{19} formulas are possible for single coronoids? For $s = 19$, eqn. (6) gives $39 \leq n \leq 47$. The answer is — $C_{39}H_{19}$, $C_{41}H_{19}$, $C_{43}H_{19}$, $C_{45}H_{19}$, $C_{47}H_{19}$.

According to a direct approach for $n_{min}(s)$ one starts with

$$n = 2s + n_i \tag{5.7}$$

Clearly, $n = n_{min}(s)$ is fulfilled for $n_i = 0$, which covers all even s values. The inclusion of $n_i = 1$ takes all the odd s values into account. Hence

$$n_{min} = \begin{cases} 2s; & s = 16, 18, 20, 22, \ldots \\ 2s+1; & s = 17, 19, 21, 23, \ldots \end{cases} \tag{5.8}$$

The compact expression on the left-hand side of eqn. (6) is consistent with (8).

The expression of $n_{max}(s)$ in eqn. (6) may be obtained simply from (3.68) through the substitutions $N = n$ and $S = s - 2$.

5.3.2 Supplementary Inequalities

For the sake of completeness some additional inequalities of the types $x_{min}(n) \leq x \leq x_{max}(n)$ and $x_{min}(s) \leq x \leq x_{max}(s)$ are collected in Table 1. Specifically, $x = m$, n_i, n_b, n_3, m_i and ν (cf. Chapt. 4).

Table 5.1. Inequalities for some invariants of single coronoids in terms of n and s.*

Inequalities	Interval
$n + \lceil n/4 \rceil \leq m \leq 2n - 1 - \lceil (1/2)(n + 6^{1/2} n^{1/2}) \rceil$	1
$4 \lceil n/4 \rceil - n \leq n_i \leq 3n - 4 - 4 \lceil (1/2)(n + 6^{1/2} n^{1/2}) \rceil$	4
$4 - 2n + 4 \lceil (1/2)(n + 6^{1/2} n^{1/2}) \rceil \leq n_b \leq 2n - 4 \lceil n/4 \rceil$	4
$2 \lceil n/4 \rceil \leq n_3 \leq 2n - 2 - 2 \lceil (1/2)(n + 6^{1/2} n^{1/2}) \rceil$	2
$5 \lceil n/4 \rceil - n \leq m_i \leq 4n - 5 - 5 \lceil (1/2)(n + 6^{1/2} n^{1/2}) \rceil$	5
$3 \lceil n/4 \rceil - n \leq \nu \leq 2n - 3 - 3 \lceil (1/2)(n + 6^{1/2} n^{1/2}) \rceil$	3
$s + 3 \lceil s/2 \rceil \leq m \leq s + 3 \lfloor (1/12)(s^2 - 10s + 4) \rfloor$	3
$2 \lceil s/2 \rceil - s \leq n_i \leq 2 \lfloor (1/12)(s^2 - 10s + 4) \rfloor - s$	2
$2 \lceil s/2 \rceil \leq n_3 \leq 2 \lfloor (1/12)(s^2 - 10s + 4) \rfloor$	2
$3 \lceil s/2 \rceil - s \leq m_i \leq 3 \lfloor (1/12)(s^2 - 10s + 4) \rfloor - s$	3
$\lceil s/2 \rceil - s \leq \nu \leq \lfloor (1/12)(s^2 - 10s + 4) \rfloor - s$	1

* See also Table 4.4 and eqns. (5.3), (5.6).

5.4 Circumscribing and Excising

5.4.1 *Introduction*

Circumscribing of a coronoid is defined in the same way as for benzenoids (see below). Already Dias (1982a; 1986; 1987) was aware of the great importance of this process in the studies of benzenoid isomers. The concept of circumscribing is, in fact, a crucial one in the theory of "constant–isomer benzenoid series" (Dias 1990a–g; 1991a,c–e; Brunvoll and Cyvin SJ 1990; Cyvin SJ, Brunvoll, Cyvin and Tošić 1990; Cyvin SJ and Brunvoll 1991; Cyvin SJ, Brunvoll and Cyvin 1991d; Cyvin SJ 1991c; 1992c; Brunvoll, Cyvin BN and Cyvin 1992b; Cyvin SJ, Cyvin and Brunvoll 1993e).

Polycircumcoronenes and perforated polycircumcoronenes are special examples of circumscribed polyhexes, which were encountered in Par. 4.3.4. Furthermore, under the treatment of circular coronoids in the preceding chapter the process of circumscribing was found to be crucial; in fact, the parameter k of Par. 4.6.4 indicates the number of circumscribings; see especially Fig. 4.6. In general, if A is an extremal single coronoid generated during the spiral walk (cf. Par. 4.3.3), then circumscribed A is also bound to be generated sooner or later by the spiral walk.

Dias (1984a; 1984b; 1986; 1987; 1990f; 1990g) introduced the "excised internal structure" for pericondensed benzenoids (cf. also Brunvoll, Cyvin BN and Cyvin 1992b). A precise definition of this concept has been formulated as the subgraph spanned by the internal vertices of the benzenoid system (Gutman and Dias 1990; Gutman 1992). This definition represents the special case of excised pericondensed benzenoids in the sense of the subsequent paragraph. Here we assume, following John (1990), that the excising is applicable to any benzenoid B, but excised B is the empty graph if and only if B is catacondensed.

5.4.2 *Definitions of Circumscribing and Excising for Coronoids*

A coronoid is said to be *circumscribed* if its associated benzenoid is circumscribed. Hence the following definition may be formulated (cf. Cyvin SJ, Brunvoll and Cyvin 1991b; Brunvoll, Cyvin BN and Cyvin 1993b; also: John 1990).

Definition 5.1: A coronoid C is circumscribed when, if possible, a single chain of hexagons is added all the way around the outer perimeter of C so that each of the added hexagons shares exactly two non–incident edges with its two neighbours among the added hexagons.

In other words, the added hexagons constitute a primitive single coronoid. The circumscribed coronoid C is designated

$$\text{circum–C} \equiv c(C)$$

An *excised* coronoid is defined in connection with excising, a process opposite to circumscribing. The strict definition, which is formulated below, is an adaptation of the definition of an excised benzenoid by John (1990).

Definition 5.2: The coronoid C is excised by removing all the vertices on the outer perimeter of C and all their incident edges.

The excised coronoid C is designated

excis–C ≡ e(C)

5.4.3 *Possibilities of Circumscribing and Excising*

Excising. Any single or multiple coronoid C (as well as any benzenoid) can be excised.

Fig. 5.1. The degenerate coronoid system excis–C_2, where C_2 is the *2*-polyhex (double coronoid) of Fig. 3.1: excis–$C_{72}H_{26} = C_{36}H_{20}$.

The system e(C), however, is not always a coronoid. As an example, Fig. 1 shows the excised (double) coronoid from Fig. 3.1; it is seen to be a degenerate coronoid system. Figure 2 shows the excised polyhexes from Fig. 4.2, viz. five benzenoids and four single coronoids, the latter systems belonging to the proper topic of the present chapter. The examples are supplemented by

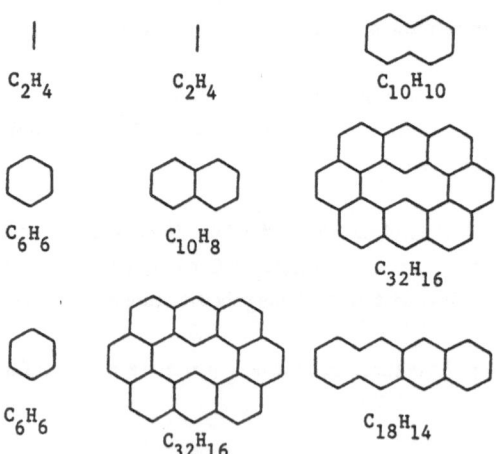

C_2H_4 C_2H_4 $C_{10}H_{10}$

C_6H_6 $C_{10}H_8$ $C_{32}H_{16}$

C_6H_6 $C_{32}H_{16}$ $C_{18}H_{14}$

Fig. 5.2. The excised *g*-polyhexes for the systems of Fig. 4.2.

Fig. 3, displaying two single coronoids of which the excised systems are disconnected.

It should not be suppressed, however, that an excised coronoid very often is another (smaller) coronoid. In particular, every time a coronoid can be circumscribed so that

$$C_1 = c(C) \tag{5.9}$$

then

$$C = e(C_1) \tag{5.10}$$

which is a coronoid.

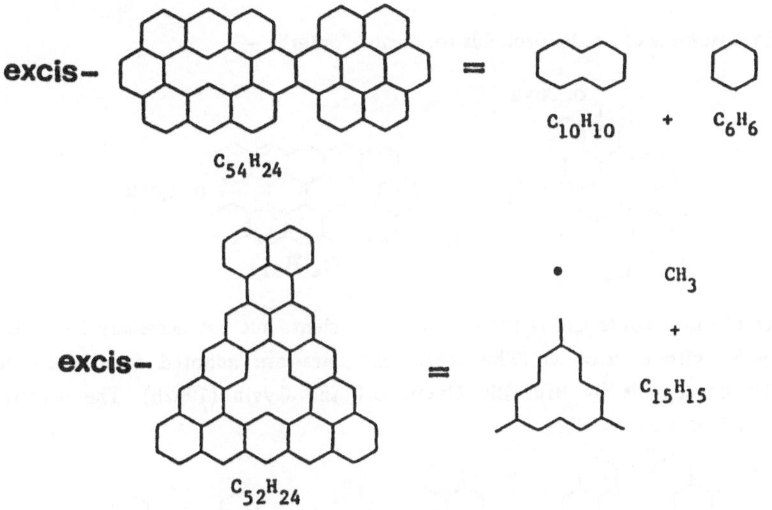

Fig. 5.3. Two single coronoids and their excised systems: excis–$C_{54}H_{24}$ = $C_{16}H_{16}$; excis–$C_{52}H_{24}$ = $C_{16}H_{18}$.

From the examples of Fig. 2, with reference to Fig. 4.2, it is seen that

$$C_{-1} = e(C) \tag{5.11}$$

does not determine unambiguously the coronoid C when C_{-1} is given. In other words, excis–C' = excis–C, where C' $\neq$ C; cf. $C_{32}H_{16} = e(C_{66}H_{22})$, $C_{32}H_{16} = e(C_{70}H_{24})$. Only the former case follows the pattern of eqns. (9) and (10) when C_1 is identified with the $C_{66}H_{22}$ coronoid of Fig. 4.2; the $C_{70}H_{24}$ coronoid of the same figure is not compatible with any circumscribed coronoid (C_1).

Circumscribing. The situation for circumscribing coronoids (as well as benzenoids) is quite different from that of excising the systems.

Assume that a coronoid C can be circumscribed as is manifested by eqn. (9). Then C_1 is determined unambiguously for the given C. Also conversely: if C_1 is given, then C may be retrieved unambiguously by virtue of eqn. (10). On the other hand, not every C can be circumscribed, depending on the formations of the outer perimeter of C. The rules are the same as for benzenoids.

For instance, it is well known (and obvious) that a benzenoid with a cove or a fjord cannot be circumscribed (Brunvoll and Cyvin 1990; Cyvin SJ, Brunvoll and Cyvin 1991b; Brunvoll, Cyvin BN and Cyvin 1992b). The corresponding rule for coronoids is formulated below.

Observation 5.1: A coronoid C with an outer cove or an outer fjord (on the outer perimeter of C) cannot be circumscribed.

Examples: Two isomers of single coronoids (o. means "outer") —

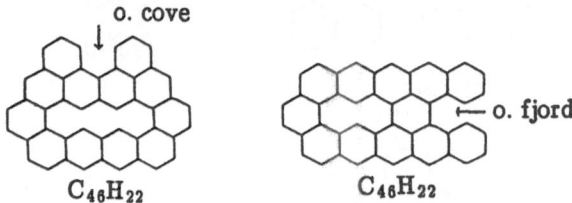

The presence of outer coves and fjords in C is a sufficient, but not necessary condition for the impossibility to circumscribe C. The below examples are adapted from a corresponding discussion for benzenoids by Brunvoll, Cyvin BN and Cyvin (1992b). The depicted single coronoids cannot be circumscribed.

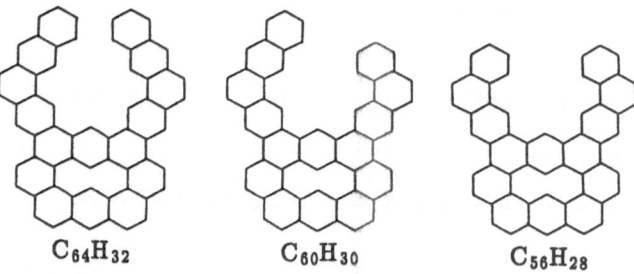

John (1990) has supplied the necessary and sufficient condition for the possibility to circumscribe a benzenoid B an unlimited number of times. When this is possible, B is called a reproducible benzenoid. The mentioned condition was re—formulated in terms of the latent row of hexagons, a possible formation on the perimeter of a benzenoid (Cyvin SJ, Cyvin, Brunvoll, Gutman and John 1993). Below we give the adaptations of this approach to coronoids.

Definition 5.3: A coronoid C is called a *reproducible coronoid* if and only if C can be circumscribed an unlimited number of times.

Definition 5.4: A *latent row* (of hexagons) in a coronoid C is a single linear chain of hexagons on the hexagonal lattice between two parallel edges of the outer perimeter of C. It should be possible to draw a line which cuts edges of all the latent row hexagons perpendicularly.

It is understood that the hexagons of the latent row are not occupied by hexagons of C. Hence the latent row is a "gap" on the outer perimeter.

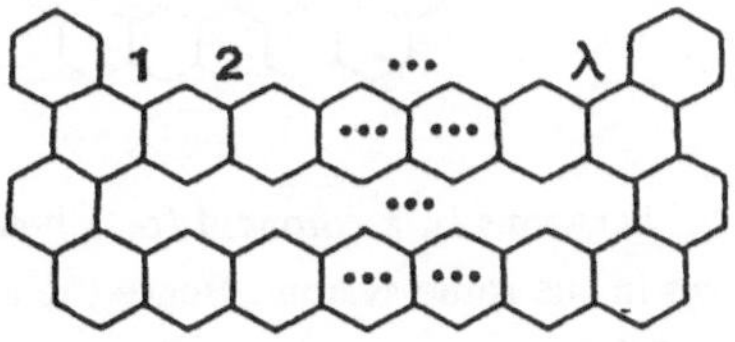

$$C_{8\lambda+24}H_{4\lambda+12}$$

Fig. 5.4. A single coronoid with a latent row of λ hexagons ($h = 2\lambda + 6$).

Figure 4 shows a single coronoid with a latent row of λ hexagons. In the last three of the above examples (viz. $C_{64}H_{32}$, $C_{60}H_{30}$ and $C_{56}H_{28}$) several latent rows with $\lambda = 2$ can be identified. Also latent rows with $\lambda = 3$ are found in these examples when it is allowed for latent rows whose hexagons are not necessarily in contact with the outer perimeter. Under this condition, which does not contradict Definition 5.4, one also finds a latent row with $\lambda = 1$ in the depicted $C_{64}H_{32}$ system. Outer coves and fjords (cf. the two examples $C_{46}H_{22}$) can also be interpreted as latent rows with $\lambda = 1$. Specifically, an outer cove is a latent "row" of one hexagon, where this hexagon is in contact with the outer perimeter. In an outer fjord there are two pairs of parallel edges, which formally define two (coalescing) latent rows corresponding to the gaps between the appropriate edges; also here the (coalescing) hexagons are in contact with the outer perimeter. In this discussion, when we speak about contact with the outer perimeter, we imagine, of course, that the latent row is filled with hexagons. This is also the case in the below definition, given as a supplement to Definition 5.4.

Definition 5.4': A latent row (of hexagons) in a coronoid C, where all the hexagons are in contact with the outer perimeter of C, is called a *proper latent row*. Otherwise the latent row is an *improper latent row* (of hexagons).

After this preparation the important condition for unlimited circumscribing can be formulated as follows.

Observation 5.2: A coronoid C is reproducible if and only if C does not possess any latent row of hexagons.

If a coronoid (or benzenoid) has only one latent row of hexagons it must obviously be a proper latent row, as is the case in Fig. 4. If there are exactly two latent rows of hexagons, either both of them are proper, or one of them is proper and the other one improper, as is exemplified below.

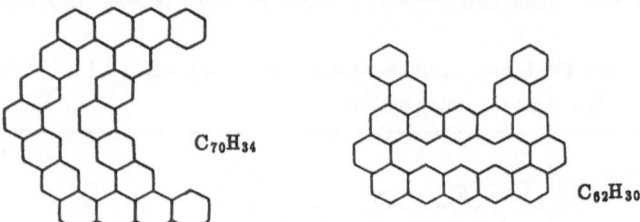

$C_{70}H_{34}$

$C_{62}H_{30}$

It is clear that an improper latent row of hexagons in a coronoid (or a benzenoid) cannot occur without the presence of proper latent rows in the same system. Hence the above observation can be given a slightly sharper formulation as follows.

Observation 5.2': A coronoid C is reproducible if and only if C does not possess any proper latent row of hexagons.

The validity of the above observations are explained by inspecting the different possibilities of the shapes of the outer perimeter of C. It is clear that the presence of a latent row hinders that C can be circumscribed without limitation. But one must also accept that any feature which hinders the unlimited circumscribing of C entails the presence of a latent row.

5.4.4 Generalizations for k–Fold Circumscribing and k–Fold Excising

Assume that a coronoid C can be circumscribed k times, and denote by C_k the k–fold circumscribed C, viz.

$$k\text{--circum--C} \equiv c_k(C)$$

Hence

$$C_k = c_k(C) \tag{5.12}$$

By definition, $c_1(C) \equiv c(C)$; consequently, eqn. (9) is seen to be the special case of (12) for $k = 1$. Under the condition of eqn. (12) we have also

$$C = e_k(C_k) \tag{5.13}$$

where $C = k\text{--excis--C}_k$ is the k–fold excised C_k. It is almost needless to mention that also $e_1(C) \equiv e(C)$.

Sometimes we shall use the notation

$$C_{-k} = e_k(C) \tag{5.14}$$

for

$$k\text{--excis--}C \equiv e_k(C)$$

if C can be excised k times. Eqn. (14) is a generalization of eqn. (11).

5.4.5 Core Coronoids

Assume that the coronoid C can be circumscribed, and define C_1 as in eqn. (9). Then also eqn. (10) is valid, and the two equations combine to

$$C_1 = c(e(C_1))$$

(5.15)

This relation indicates that C_1 is not a core coronoid. This class of coronoids is defined in analogy with the definition of a core benzenoid by John (1990).

Definition 5.5: A coronoid C_0 is a *core coronoid* if and only if $c(e(C_0)) \neq C_0$.

A coronoid which is not a core coronoid, is sometimes called a *non–core coronoid*.

In Fig. 4.2 (cf. also Fig. 2) all the systems in the top and bottom rows are core benzenoids or core simple coronoids. The two first systems in the middle row (viz. $C_{24}H_{12}$ and $C_{32}H_{14}$) are not core benzenoids, while the last system therein (viz. $C_{66}H_{22}$) is a non–core single coronoid. There exist core coronoids of different natures, as is explained below.

Firstly, if an excised coronoid C_0, viz. $e(C_0)$, is not itself a coronoid, then the circumscribing of $e(C_0)$ is not defined, and C_0 becomes automatically a core coronoid. The two last systems of the top and bottom rows of Fig. 4.2 (viz. $C_{32}H_{16}$ and $C_{48}H_{20}$) are examples. In particular, every catacondensed coronoid is a core coronoid belonging to the category under consideration. If Q is a catacondensed single coronoid without an inside feature (cf. Vol. I), then $e(Q)$ consists of the inner perimeter of Q.

Secondly, it may happen that $e(C_0)$ for a core coronoid C_0 itself is a coronoid. This kind is exemplified by $C_{70}H_{24}$, the system in the middle of the bottom row of Fig. 4.2. Here $e(C_{70}H_{24})$ is a coronoid, viz. $C_{32}H_{16}$, but $c(C_{32}H_{16}) \neq C_{70}H_{24}$; hence the condition of Definition 5.5 is fulfilled. It appears, under this category, that C_0 must have at least one special edge referred to as a fusing edge.

Definition 5.6: A *fusing edge* of a coronoid C_0 is an edge between two vertices of degree three on the outer perimeter of C_0.

The fusing edge may be characterized as a "short–cut" on the outer perimeter of C_0. If a coronoid C_0 with a fusing edge is split along this edge it separates into two polyhex fragments; if C_0 is a single coronoid the two fragments are one (smaller) single coronoid and one benzenoid. Conversely, a coronoid with a fusing edge may be interpreted as two polyhex fragments fused to each other by two free edges, one from each of the fragments.

It may happen that a core coronoid C_0 belongs to both of the above categories: $e(C_0)$ is not a coronoid and C_0 has a fusing edge. The top system ($C_{54}H_{24}$) of Fig. 3 is an example. The bottom system ($C_{52}H_{24}$) is also a core coronoid, but without a fusing edge.

If a coronoid has a fusing edge it must necessarily be a core coronoid. Notice also that all fusing edges disappear from a coronoid as soon as it is circumscribed (if possible). Finally it is noted that $e(C_0)$ very well may be a non–core coronoid even if C_0 is a core coronoid. An example is shown below.

excis–

$C_{116}H_{30}$ $C_{66}H_{22}$

The essence of the above discussion is summarized as the following observation.

Observation 5.3: A coronoid C_0 is a core coronoid if and only if (a) $e(C_0)$ is not a coronoid or (b) C_0 contains a fusing edge, or both (a) and (b) are fulfilled.

The corresponding rule applies also to benzenoids.

It is clear that C_0 is a core coronoid if the conditions of Observation 5.3 are fulfilled. But also, if a coronoid C has no fusing edge and $e(C)$ is a coronoid, it is inferred that C is restored by circumscribing $e(C)$; one only has to recall the mechanism of excising and circumscribing.

The class of *reproducible core coronoids* (analogous with the reproducible core benzenoids) is especially important. A reproducible core coronoid C_0 fulfils the condition that

$$C_{k+1} = c(C_k)$$ (5.16)

exists for every $k = 0, 1, 2, \ldots$. Here also the coronoids C_1, C_2, C_3, ... are reproducible coronoids. They are said to belong to the reproducible core coronoid C_0. In fact, the condition of eqn. (16) may be used as a definition of reproducible coronoids in analogy with John (1990) as an alternative to Definition 5.3. It is evident that any reproducible non–core coronoid C_k $(k > 0)$ belongs to a unique reproducible core coronoid. This coronoid must eventually be reached unambiguously by repeated excising of C_k: $C_{k-1} = e(C_k)$, $C_{k-2} = e(C_{k-1})$, ... According to the conditions assumed above a core coronoid appears for the first time during the repeated excising exactly as $C_0 = e(C_1)$. Then we have $C_k = k$–circum–C_0 and $C_0 = k$–excis–C_k; in consequence, eqns. (12) and (13), where C should be identified with C_0, are applicable to the case under consideration.

5.4.6 *Algebraic Treatment*

Circumscribing. Let C be a single coronoid which can be circumscribed, and define C_1 as in eqn. (9). Introduce different invariants for C and C_1 according to $C(h, n_i) = C(n; s)$ and $C_1(h_1, (n_i)_1) = C_1(n_1; s_1)$. We shall also make use of the notation $s = s' + s''$ and $s_1 = s_1' + s_1''$, where the primed and double–primed symbols pertain to the outer and inner perimeter, respectively; cf. eqns. (3.9), (3.17) and (4.3). Notice that s'' and s_1'' are identical:

$$s'' = s - s' = s_1 - s_1' \tag{5.17}$$

The number of added hexagons during circumscribing of C is s'. Hence

$$h_1 = h + s' \tag{5.18}$$

On introducing $h(n, s)$ from Table 4.1 it is obtained

$$h_1 = (1/2)(n - s) + s' = (1/2)(n + s) - s'' \tag{5.19}$$

For the number of internal vertices in $c(C) = $ circum–C one has

$$(n_i)_1 = n_i + n_b' = n_i + 2s' - 6 \tag{5.20}$$

Here $n_b' = 2s' - 6$ is the number of boundary vertices on the outer perimeter of C; cf eqn. (3.9). Now we insert $n_i(n, s)$ in accordance with Table 4.1 in order to arrive at

$$(n_i)_1 = n - 2(s - s') - 6 = n - 2s'' - 6 \tag{5.21}$$

The same result (21) is achieved more directly as:

$$(n_i)_1 = n - n_b'' = n - (2s'' + 6) \tag{5.22}$$

where $n_b'' = 2s'' + 6$ is the number of boundary vertices on the inner perimeter of either C or C_1; cf. eqn. (4.3).

We wish also to establish the connection between the formulas $(n; s)$ and $(n_1; s_1)$. The known relations between the different invariants of single coronoids (Table 4.1), combined with eqns. (19) and (21), give

$$(n_1; s_1) = (n + 2(s - s'') + 6; \ s + 6) \tag{5.23}$$

By means of eqn. (17) it is obtained from (23):

$$n_1 = n + 2s' + 6 \tag{5.24}$$

and

$$s_1' = s' + 6 \tag{5.25}$$

Another approach to the derivation of eqns. (23) − (25) invokes the benzenoids B(N; S) and $B_1(N_1; S_1)$ as associated with C and C_1, respectively. Then clearly B_1 = circum−B, and one has the formula (Brunvoll and Cyvin SJ 1990)

$$(N_1; S_1) = (N + 2S + 6; \ S + 6) \tag{5.26}$$

It is observed that $N_1 - N = n_1 - n$, and $S_1 - S = s_1 - s = s_1' - s'$. Furthermore, $S = s'$ and $S_1 = s_1'$. With the aid of these relations, eqns. (23) − (25) are easily derived when also (17) is taken into account.

Assume now that the single coronoid C can be circumscribed k times, and define $C_k(n_k; s_k)$ as k−fold circumscribed C; cf. eqn. (12). Then the generalization of eqn. (23) reads

$$(n_k; s_k) = (n + 2k(s - s'') + 6k^2; \ s + 6k) \tag{5.27}$$

Similarly, in analogy with eqns. (24) and (25) one finds

$$n_k = n + 2ks' + 6k^2 \tag{5.28}$$

and

$$s_k' = s' + 6k \tag{5.29}$$

The relations (27) − (29) are deduced most easily, in analogy with the derivation of eqns. (23) − (25) from (26). The appropriate relation for the benzenoids $B_k(N_k; S_k)$ = k−circum−B(N; S) is (Cyvin SJ 1991c; 1992c; Cyvin SJ, Brunvoll and Cyvin 1991b; Brunvoll, Cyvin BN and Cyvin 1992b; Cyvin SJ, Cyvin and Brunvoll 1993e)

$$(N_k; S_k) = (N + 2kS + 6k^2; \ S + 6k) \tag{5.30}$$

Now it should be observed that $N_k - N = n_k - n$ and $S_k - S = s_k - s = s_k' - s'$. Furthermore, $S = s'$ and $S_k = s_k'$. Finally we have, as a generalization of eqn. (17):

$$s'' = s - s' = s_k - s_k' \tag{5.31}$$

Excising. Precaution must be taken in the analytical treatment of $C_{-1}(n_{-1}; s_{-1})$ and $C(n;s)$, where C and C_{-1} obey eqn. (11). This condition is not sufficient to allow a general expression for $(n_{-1}; s_{-1})$ in terms of $(n; s)$, as is apparent from the discussion below eqn. (11). If $c(C_{-1}) = C$, then eqn. (11) is certainly valid, and also $c(e(C)) = C$; in other words, C is a non−core coronoid (cf. Definition 5.5). Under this condition eqn. (23) can give us an unambiguous answer. When the inherent equations are solved for n and s it is arrived at

$$(n;\, s) = (n_1 - 2(s_1 - s'') + 6;\ s_1 - 6) \qquad (5.32)$$

By re-naming $(n_1;\, s_1)$ and $(n;\, s)$ to $(n;\, s)$ and $(n_{-1};\, s_{-1})$, respectively, one obtains

$$(n_{-1};\, s_{-1}) = (n - 2(s - s'') + 6;\ s - 6) \qquad (5.33)$$

The first part of this relation, viz. the expression for n_{-1}, can be deduced directly from the definition of excising (Definition 5.2) as

$$n_{-1} = n - n_b' = n - 2s' + 6 = n - 2(s - s'') + 6 \qquad (5.34)$$

where $n_b' = 2s' - 6$ has been employed; cf. eqn. (3.9). The expression (34) is general for single coronoids.

A generalization of eqn. (33) is straightforward from (27), which yields

$$(n_{-k};\, s_{-k}) = (n - 2k(s - s'') + 6k^2;\ s - 6k) \qquad (5.35)$$

for $C_{-k}(n_{-k};\, s_{-k})$ and $C(n;\, s)$, where C and C_{-k} obey eqn. (14). However, eqn. (35) is not general, even if it is assumed that C can be excised k times. But if it is required that $c_k(C_{-k}) = C$, then (35) is sound. In this case C, and all C_{-j} for $j = 1, 2, 3, \ldots, k$ are non-core coronoids.

The two relations (27) and (35) can be compressed to:

$$(n_x;\, s_x) = (n + 2x(s - s'') + 6x^2;\ s + 6x) \qquad (5.36)$$

where $x = +k$ and $x = -k$ for (27) and (35), respectively ($x = \pm 1, \pm 2, \pm 3, \ldots$). For $x < 0$ eqn. (36) is sound if $c_{-x}(C_x) = C$; compare with the discussion below eqn. (35).

It remains to deduce a general expression for the formula coefficients of $C_{-1}(n_{-1};\, s_{-1})$ from $C(n;\, s)$, where C may or may not be a core coronoid. Again C and C_{-1} are supposed to obey eqn. (11). Although being somewhat tedious, we shall follow Definition 5.2 in the following deduction. Let m and m_{-1} denote the number of edges in C and C_{-1}, respectively. Then we wish to obtain m_{-1} from m. Firstly, subtract the n_b' edges of the outer perimeter of C. Secondly, subtract the t' edges of which (at least) one of the end points is a vertex of degree three on the outer perimeter of C. This last process subtracts every fusing edge (Definition 5.6) twice. Therefore, if the number of fusing edges of C is m^*, then m^* must be added. The net result is

$$m_{-1} = m - n_b' - t' + m^* \qquad (5.37)$$

or slightly re-written as

$$m - m_{-1} = n_b' + t' - m^* \tag{5.38}$$

The known relations between invariants of single coronoids (Table 4.1) yield

$$(3/2)(n - n_{-1}) - (1/2)(s - s_{-1}) = n_b' + t' - m^* \tag{5.39}$$

where $n - n_{-1} = n_b$ by virtue of eqn. (34), and $t' = s' - 6$ according to (3.10). Hence

$$n_b' = 2s' + s - s_{-1} - 2m^* \tag{5.40}$$

On inserting $n_b' = 2s' - 6$ as also was done in eqn. (34), one finally arrives at

$$s_{-1} = s - 2m^* - 6 \tag{5.41}$$

This expression, together with eqn. (34), yields

$$(n_{-1}; s_{-1}) = (n - 2(s - s'') + 6; \ s - 2m^* - 6) \tag{5.42}$$

Eqn. (42) is general for single coronoids. It is also valid when $e(C)$ is not a coronoid.

Examples

The reader is referred to Fig. 3. (a) Top row: In order to achieve $n_{-1} = 16$ and $s_{-1} = 16$ from eqn. (42), it should be inserted $n = 54$, $s = 24$, $s'' = 2$, $m^* = 1$. (b) Bottom row: In order to achieve $n_{-1} = 16$ and $s_{-1} = 18$ from (42), insert $n = 52$, $s = 24$, $s'' = 3$, $m^* = 0$.

5.4.7 Catacondensed Single Coronoids

Let Q be a catacondensed ($n_i = 0$) single coronoid with h hexagons; it has the formula $C_{4h}H_{2h}$; hence $Q \in \{C_{2s}H_s\}$. Assume that Q can be circumscribed. Then one finds easily from eqn. (23) for $Q_1(n_1; s_1) = c(Q)$, just by inserting $n = 2s$:

$$(n_1; \ s_1) = (4s - 2s'' + 6; \ s + 6) \tag{5.43}$$

Similarly from eqn. (27) one finds the generalization of (43):

$$(n_k; \ s_k) = (2s(k + 1) - 2ks'' + 6k^2; \ s + 6k) \tag{5.44}$$

which pertains to $Q_k(n_k; s_k) = c_k(Q)$, provided that Q can be circumscribed k times. Let us finally deduce the formula coefficients of $Q_{-1}(n_{-1}; \ s_{-1}) = e(Q)$ from eqn. (42). Again, just by inserting $n = 2s$, it is arrived at the remarkably simple formula

$$(n_{-1}; s_{-1}) = (2s'' + 6; \; s - 2m^* - 6) \tag{5.45}$$

It becomes still simpler on further specializations.

Assume now that Q is a catacondensed single coronoid without inside feature(s); cf. Vol. I–8.2.1. Then, as mentioned in Par. 5.4.5, $Q_{-1} = e(Q)$ consists of the inner perimeter. We shall deduce the formula of Q_{-1} in consistency with this observation. For a primitive coronoid, say $Q\dagger(2s\dagger; s\dagger)$, one has $n_b' = n_b'' + 12$; Vol. I–(4.6). On inserting $n_b' = 2s' - 6$ and $n_b'' = 2s'' + 6$ according to eqns. (3.9) and (4.3), respectively, and making use of $s' = s\dagger - s''$, it is arrived at

$$s'' = (s\dagger/2) - 6 \tag{5.46}$$

Since also $m^* = 0$ for primitive coronoids, one obtains with the aid of eqn. (45) for $Q_{-1}\dagger(n_{-1}\dagger; s_{-1}\dagger)$:

$$(n_{-1}\dagger; s_{-1}\dagger) = (2s'' + 6; \; 2s'' + 6) = (s\dagger - 6; \; s\dagger - 6) \tag{5.47}$$

It is recalled that $2s'' + 6 = n_b''$ (see above). The first parentheses on the right–hand side of (47) is valid for Q as assumed originally, viz. a catacondensed single coronoid without inside feature(s). In other words,

$$(n_{-1}; s_{-1}) = (2s'' + 6; \; 2s'' + 6) = (n_b''; n_b'') \tag{5.48}$$

This is immediately obvious because the excising will strip Q for the outside feature(s), which consist(s) of one or more catacondensed appendages. It is also possible to verify eqn. (48) in a formal way by computing m^*, the number of fusing edges, and exploit the more general equation for s_{-1} inherent in (45). Let Q and Q† be the same coronoid except for the possible outside feature(s) in Q. Then s'' is the same in Q and Q†. Let the number of hexagons in Q and Q† be h and $h\dagger$, respectively. Then

$$h = s/2 \;, \;\; h\dagger = s\dagger/2 = s'' + 6 \tag{5.49}$$

where (46) has been employed. Now the number of hexagons in the outside feature(s) of Q gives exactly the number of fusing edges; hence

$$m^* = h - h\dagger = (s/2) - s'' - 6 \tag{5.50}$$

On inserting the last expression for m^* into (45) one arrives at (48). The formula $(n_b''; n_b'')$, which appears in (48), represents clearly an $[n_b'']$annulene or the inner perimeter of Q.

Figure 5 shows three examples of $Q_{-1} = e(Q)$, to which eqn. (45) is applicable. For the top– and middle–row cases eqn. (48) is valid; in the former case (top–row) also (47). In both of these cases, $Q_{-1} = [18]$annulene. In the case of the bottom–row we must resort to eqn. (45); here

$n_{-1} = 22$, $s_{-1} = 20$ is achieved by inserting $s'' = 8$, $s = 26$, $m^* = 0$.

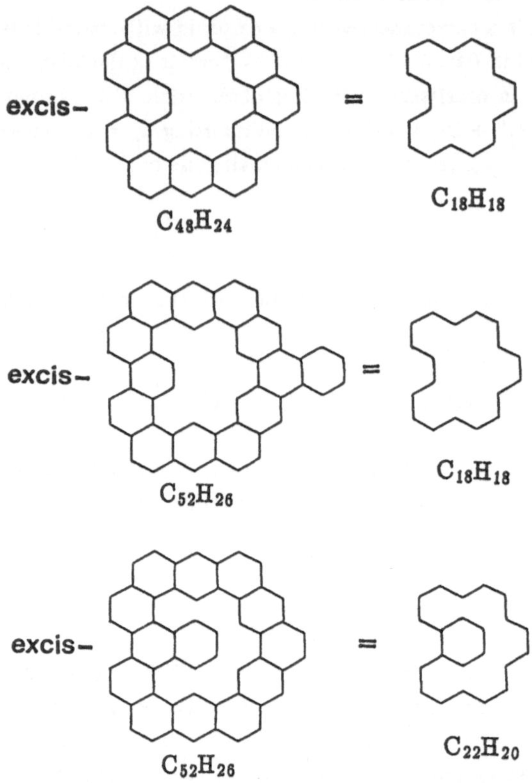

Fig. 5.5. Excising of three catacondensed single coronoids. From top: primitive coronoid; coronoid with an outside feature; coronoid with an inside feature.

5.4.8 Naphthalenic Single Coronoids

Let $C(n;s)$ be a naphthalenic single coronoid, and apply the same notation for its circumscribed and excised systems as well as their invariants as in Sect. 5.4.6. Then the transformation (4.13) applied to eqns. (26) and (30) gives straightforwardly

$$(n_1;\ s_1) = (n + 2s + 2;\ s + 6) \tag{5.51}$$

and the generalized form

$$(n_k;\ s_k) = (n + 2ks + 6k^2 - 4k;\ s + 6k) \tag{5.52}$$

respectively. The relations (51) and (52) are seen to be the special cases of (23) and (27) for $s'' = 2$, which applies to naphthalenic single coronoids (cf. Fig. 4.1). They can be tested on the numerous examples provided by Fig. 4.6. Again, as in the case of (27), eqn. (52) applies also to negative k values when the validity is assured in consistency with the discussion below eqn. (35). In the examples of Fig. 4.6 the values $k = -1$ and $k = -2$ can be used in all cases when appropriate. For instance, $e_2(C_{139}H_{31}) = C_{47}H_{19}$ (cf. the bottom row of Fig. 4.6) is reflected by eqn. (52) with $n = 139$, $s = 31$, $k = -2$.

5.5 Extremal Single Coronoids and Some of Their Subclasses

5.5.1 Extremal Single Coronoids

Formula. An extremal single coronoid, A, is treated in Par. 4.3.1. Let A have the formula $(n^a; s^a)$. Then, on inserting $n_i = (n_i)_{max}$ from eqn (4.21) into (1) it is obtained:

$$(n^a; s^a) = (2h + 5 + \lceil(12h + 21)^{1/2}\rceil; \ 5 + \lceil(12h + 21)^{1/2}\rceil) \tag{5.53}$$

where $h = 8, 9, 10, 11, ...$ indicates the number of hexagons. Eqn. (53) is easily converted to a form with $a = h - 7$ as a running index, viz.

$$(n^a; s^a) = (2a + 19 + \lceil(12a + 105)^{1/2}\rceil; \ 5 + \lceil(12a + 105)^{1/2}\rceil) \tag{5.54}$$

where $a = 1, 2, 3, 4, ...$. The first (smallest) extremal single coronoids have the formulas $C_{32}H_{16}$, $C_{35}H_{17}$, $C_{37}H_{17}$, $C_{40}H_{18}$, $C_{42}H_{18}$, $C_{45}H_{19}$, $C_{47}H_{19}$, $C_{50}H_{20}$, $C_{52}H_{20}$, $C_{54}H_{20}$, ...

The above relations for $(n^a; s^a)$ are obtained alternatively from the corresponding relations for extremal benzenoids, say $A(N^a; S^a)$, which are associated with the extremal coronoids under consideration. One has (Cyvin SJ, Brunvoll and Cyvin 1991b; Cyvin SJ 1992c; Brunvoll, Cyvin BN and Cyvin 1992b; Cyvin SJ, Cyvin and Brunvoll 1993e)

$$(N^a; S^a) = (2H + 1 + \lceil(12H - 3)^{1/2}\rceil; \ 3 + \lceil(12H - 3)^{1/2}\rceil) \tag{5.55}$$

The transformations of eqn. (4.13) yield

$$(n^a; s^a) = (2H + 1 + \lceil(12H - 3)^{1/2}\rceil; \ 5 + \lceil(12H - 3)^{1/2}\rceil) \tag{5.56}$$

where $H = 10, 11, 12, 13, ...$ are the appropriate values for the extremal single coronoids. Now eqn. (56) is converted to the forms (53) and (54) by the substitutions $H = h + 2$ and $H = a + 9$, respectively.

Properties. Many properties of extremal benzenoids, A, can immediately be adapted to extremal single coronoids, A. Relevant properties of A are treated in several places (Cyvin SJ, Cyvin, Brunvoll, Gutman and John 1993; Cyvin SJ, Cyvin and Brunvoll 1993d; 1993e) with more or less rigorous proofs therein. Here we shall not conduct special proofs for the adaptations of the different properties to the A coronoids. We are particularly interested in properties of A which pertain to formations (or absence of formations) on the perimeter of A. By the adaptation of these properties to A one correlates the outer perimeter of an A system with the (only) perimeter of an A.

It has been mentioned (cf. e.g., Par. 4.3.1) that an A coronoid is naphthalenic and perfect extremal; it is obtained from an A benzenoid by perforating it with a naphthalene hole.

Property 5.1: A coronoid A has never an outer cove or an outer fjord.

Property 5.2: A coronoid A has never a latent row of hexagons.

The following corollary emerges from Observation 5.2 and Property 5.2.

Corollary 5.1: Any A is reproducible.

In other words, any extremal single coronoid (A) can be circumscribed without limitation. However, there also exist nonextremal single coronoids which are reproducible and therefore can be circumscribed without limitation. Kekulene is an obvious example.

Property 5.3: A coronoid A has never a fusing edge.

The following corollary emerges from Observation 5.3 and Property 5.3.

Corollary 5.2: If A_0 is an extremal core single coronoid, then $e(A_0)$ is not a coronoid.

Property 5.4: For any A, an extremal single coronoid, also $c(A)$ is an extremal single coronoid.

Corollary 5.1 assures that circum—A $\equiv c(A)$ exists for every A.

5.5.2 *Circumextremal Single Coronoids*

A *circumextremal* (single) *coronoid* is defined in connection with a circumextremal benzenoid (Cyvin SJ, Zhang, Cyvin and Guo 1993), previously characterized as a pericondensed protrusive benzenoid (for references, see below). Let it be identified by $P(N^p; S^p)$. By definition, a circumextremal benzenoid is a circumscribed extremal benzenoid.

Definition 5.7: A single coronoid (P) is circumextremal if its associated benzenoid, P, is a circumextremal benzenoid, and excis–P is the associated benzenoid to a (smaller) extremal single coronoid.

Some explanations should be attached to this somewhat complicated definition. It is clear that a circumscribed extremal (single) coronoid is circumextremal; write circum–A = P. Then it is also clear that excis–P = A. Assume now that the naphthalene hole of P is moved as near as possible to the outer perimeter, and call the new coronoid P'. Then P and P' are isomers, but P' cannot be excised to give another coronoid; excis–P' becomes a degenerate system. Nevertheless, P' falls under the definition of circumextremal coronoids. An illustration is given below.

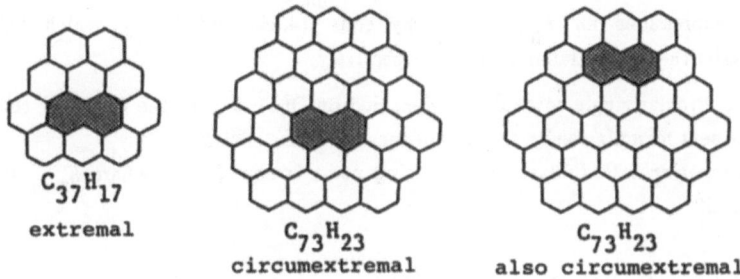

C$_{37}$H$_{17}$
extremal

C$_{73}$H$_{23}$
circumextremal

C$_{73}$H$_{23}$
also circumextremal

In simple terms: all the circumextremal single coronoids are obtained as circumscribed extremal single coronoids and their isomers. Once more, in other words: a circumextremal single coronoid has the same formula as a circumscribed extremal single coronoid.

Let a circumextremal single coronoid be identified by P(n^P; s^P). Then it is obtained from eqns. (51) and (53):

$$(n^P; s^P) = (2h + 17 + 3\lceil(12h+21)^{1/2}\rceil;\ 11 + \lceil(12h+21)^{1/2}\rceil) \tag{5.57}$$

where $h = 8, 9, 10, 11, \ldots$. Similarly from eqns. (51) and (54):

$$(n^P; s^P) = (2a + 31 + 3\lceil(12a+105)^{1/2}\rceil;\ 11 + \lceil(12a+105)^{1/2}\rceil) \tag{5.58}$$

where $a = 1, 2, 3, 4, \ldots$. The first (smallest) circumextremal single coronoids have the formulas C$_{66}$H$_{22}$, C$_{71}$H$_{23}$, C$_{73}$H$_{23}$, C$_{78}$H$_{24}$, C$_{80}$H$_{24}$, C$_{85}$H$_{25}$, C$_{87}$H$_{25}$, C$_{92}$H$_{26}$, C$_{94}$H$_{26}$, C$_{96}$H$_{26}$, \ldots .

A coronoid P is a naphthalenic perforated P. Hence the relations for (n^P; s^P) may be deduced alternatively from (Cyvin SJ, Brunvoll and Cyvin 1991a; Cyvin SJ 1992c; Brunvoll, Cyvin BN and Cyvin 1992b)

$$(N^P; S^P) = (2H' + 13 + 3\lceil(12H'-3)^{1/2}\rceil;\ 9 + \lceil(12H'-3)^{1/2}\rceil) \tag{5.59}$$

In analogy with the derivation of eqn. (56) from (55) it is obtained

$$(n^{\text{p}};\ s^{\text{p}}) = (2H' + 13 + 3\lceil(12H'-3)^{1/2}\rceil;\ \ 11 + \lceil(12H'-3)^{1/2}\rceil) \tag{5.60}$$

where $H' = 10, 11, 12, 13, ...$ are the appropriate values for circumextremal single coronoids. Eqn. (60) is converted to (57) and (58) by the substitutions $H' = h + 2$ and $H' = a + 9$, respectively.

The class of circumextremal single coronoids (P) is a subclass of the extremal single coronoids (A); cf. Property 5.4.

5.5.3 Circular Single Coronoids

A circular coronoid has by definition $h = h_{\text{max}}(s)$; cf. Definition 4.1. Here we recall the important expressions for h_{max} given by eqns. (4.44) and (4.50), which belong to the Harary–Harborth– and Balaban picture, respectively.

Let a circular single coronoid be identified by $O(n^{\text{o}};\ s^{\text{o}})$. In the following the expressions for the pertinent formula coefficients are deduced in both "pictures".

Harary–Harborth Picture. The starting point is eqn. (4.44), where h_{max} is substituted with the result

$$(1/2)(n^{\text{o}} - s) = \lfloor(1/12)(s^2 - 10s + 4)\rfloor \tag{5.61}$$

When solved for n^{o} this equation yields

$$(n^{\text{o}};\ s^{\text{o}}) = (s + 2\lfloor(1/12)(s^2 - 10s + 4)\rfloor;\ s) \tag{5.62}$$

where $s\ (= s^{\text{o}}) = 16, 17, 18, 19, ...$. It may be checked that this relation reproduces all the formulas of Fig. 4.6 (including those for the degenerate coronoids with $s = 14$ and $s = 15$). In Table 2 a larger number of the formulas in question are collected. An alternative derivation of (62) is based on the fact that a coronoid O is a naphthalenic perforated circular benzenoid O (cf. Par. 4.6.4). One has (Cyvin SJ 1992c; Brunvoll, Cyvin BN and Cyvin 1992b; Cyvin SJ, Cyvin and Brunvoll 1993e)

$$(N^{\text{o}};\ S^{\text{o}}) = (S + 2\lfloor(1/12)(S^2 - 6S)\rfloor;\ S) \tag{5.63}$$

Herefrom:

$$(n^{\text{o}};\ s^{\text{o}}) = (S + 2\lfloor(1/12)(S^2 - 6S)\rfloor;\ S + 2) \tag{5.64}$$

where $S = 14, 15, 16, 17, ...$ are the S values of interest. On inserting $S = s - 2$ in (64) this relation will indeed be transformed to eqn. (62).

Table 5.2. Formulas for circular single coronoids.*

k	ϵ					
	0	1	2	3	4	5
0	$(C_{24}H_{14})$	$(C_{27}H_{15})$	$C_{32}H_{16}$	$C_{37}H_{17}$	$C_{42}H_{18}$	$C_{47}H_{19}$
1	$C_{54}H_{20}$	$C_{59}H_{21}$	$C_{66}H_{22}$	$C_{73}H_{23}$	$C_{80}H_{24}$	$C_{87}H_{25}$
2	$C_{96}H_{26}$	$C_{103}H_{27}$	$C_{112}H_{28}$	$C_{121}H_{29}$	$C_{130}H_{30}$	$C_{139}H_{31}$
3	$C_{150}H_{32}$	$C_{159}H_{33}$	$C_{170}H_{34}$	$C_{181}H_{35}$	$C_{192}H_{36}$	$C_{203}H_{37}$
4	$C_{216}H_{38}$	$C_{227}H_{39}$	$C_{240}H_{40}$	$C_{253}H_{41}$	$C_{266}H_{42}$	$C_{279}H_{43}$
5	$C_{294}H_{44}$	$C_{307}H_{45}$	$C_{322}H_{46}$	$C_{337}H_{47}$	$C_{352}H_{48}$	$C_{367}H_{49}$
6	$C_{384}H_{50}$	$C_{399}H_{51}$	$C_{416}H_{52}$	$C_{433}H_{53}$	$C_{450}H_{54}$	$C_{467}H_{55}$

* Formulas for degenerate coronoids in parentheses.

Balaban Picture. Starting from eqn. (4.50) and following the same procedure as above one obtains

$$(1/2)(n^O - 6k - 14 - \epsilon) = 3k^2 + 9k + (k+2)\epsilon - \lceil \epsilon/6 \rceil \tag{5.65}$$

where

$$6k + 14 + \epsilon = s \tag{5.66}$$

Herefrom:

$$(n^O; s^O) = (6k^2 + 24(k+1) + (2k+5)\epsilon - 2\lceil \epsilon/6 \rceil; \ 6k + 14 + \epsilon) \tag{5.67}$$

where (when the two degenerate coronoids $C_{24}H_{14}$ and $C_{27}H_{15}$ are included) $k = 0, 1, 2, 3, \ldots,$ and $\epsilon = 0, 1, 2, 3, 4, 5$ (cf. Par. 4.6.4). Again an alternative derivation is possible. For $O(N^O; S^O)$ one has (Cyvin SJ 1992c; Cyvin SJ, Cyvin and Brunvoll 1993e)

$$(N^O; S^O) = (6K^2 + 12K + (2K+3)\epsilon + 6 - 2\lceil \epsilon/6 \rceil; \ 6K + 6 + \epsilon) \tag{5.68}$$

from which one obtains

$$(n^O; o^O) = (6K^2 + 12K + (2K+3)\epsilon + 0 - 2\lceil \epsilon/6 \rceil; \ 6K + 8 + \epsilon) \tag{5.69}$$

Now, $K = k + 1$ renders the relation into the form (67).

New Picture. Cyvin BN, Cyvin and Brunvoll (1993) presented for the first time a generating function which reproduces a sequence of chemical formulas. Their work was inspired by an algorithm of Dias (1993), who considered differences in n values. Here we apply the same methods for the $C_n H_s$ formulas of circular single coronoids. Introduce n^o as a function of s and label it n_s^o; $n_{16}^o = 32$, $n_{17}^o = 37$, etc. We are looking for the generating function defined by

$$n^o(x) = \sum_{s=16}^{\infty} n_s^o x^s = 32x^{16} + 37x^{17} + 42x^{18} + 47x^{19} + 54x^{20} + \dots \tag{5.70}$$

Consider the differences $\Delta_s = n_s^o - n_{s-1}^o$ $(s > 16)$, while $\Delta_{16} = 5$ $(s = 16)$ by definition. Then, with the notation from the Balaban picture,

$$\Delta_s = 2k + 3 \ (\epsilon = 1), \quad \Delta_s = 2k + 5 \ (\epsilon \neq 1) \tag{5.71}$$

In other words, $\Delta_s = 5, 5, 5, 5, 7, 5, 7, 7, 7, 7, 9, 7, 9, 9, 9, 9, 11, 9, \dots$ for $s = 16, 17, 18, \dots$, 33. The relations (71) are easily verified by means of (67). The elementary result

$$p(x) = (5 - 3x)(1 - x)^{-2} = \sum_{p=0}^{\infty} (2p + 5)x^p = 5 + 7x + 9x^2 + 11x^3 + \dots \tag{5.72}$$

gives a clue to the generating function for Δ_s, say $\Delta(x)$. The next elementary step is

$$x^{14} p(x^6) = 5x^{14} + 7x^{20} + 9x^{26} + 11x^{32} + \dots \tag{5.73}$$

In consequence,

$$\Delta(x) = \sum_{s=16}^{\infty} \Delta_s x^s = (1 + x^2 + x^3 + x^4 + x^5 + x^7)x^{14}p(x^6) - 5x^{14}$$
$$= 5x^{16} + 5x^{17} + 5x^{18} + 5x^{19} + 7x^{20} + 5x^{21} + 7x^{22} + 7x^{23}$$
$$+ 7x^{24} + 7x^{25} + 9x^{26} + 7x^{27} + 9x^{28} + \dots \tag{5.74}$$

Finally,

$$n^o(x) = (1 - x)^{-1}[\Delta(x) + 27x^{16}] \tag{5.75}$$

On combining (72) − (75) it was arrived at the explicit forms:

$$n^o(x) = x^{14}(1 - x)^{-1}[27x^2 - 5 + (1 + x^2 + x^3 + x^4 + x^5 + x^7)(5 - 3x^6)(1 - x^6)^{-2}]$$

$$= \frac{x^{16}(32 + 5x + 5x^2 + 5x^3 + 7x^4 + 5x^5 - 57x^6 - 3x^7 - 3x^8 - 3x^9 - 5x^{10} - 3x^{11} + 27x^{12})}{(1 - x)(1 - x^6)^2} \tag{5.76}$$

The generating function (76) reproduces the formulas of circular single coronoids (cf. Table 2) as is indicated in (70). The degenerate systems (formulas for $s < 16$) are excluded.

The circular single coronoids form a subclass of the extremal single coronoids (cf. Par. 4.6.2).

5.6 Extreme Single Coronoids

5.6.1 *Definition*

An *extreme coronoid* (not to be confused with an extremal coronoid) is analogous with an extreme benzenoid (with a few exceptions specified below); cf. Cyvin SJ (1992c), which has also been called an extreme–left benzenoid (Cyvin SJ 1992c; Cyvin SJ, Cyvin and Brunvoll 1993e).

Definition 5.8: An extreme coronoid is defined by having the minimum number of hexagons (h) for a given number of internal vertices (n_i): $h = h_{\min}(n_i)$.

Assume that E is an extreme single coronoid. Every A, an extremal single coronoid, is an E coronoid; cf. Par. 4.3.2. In other words, the extremal single coronoids form a subclass of the extreme single coronoids. Figure 4.3 is especially elucidating. Here all asterisks (with and without circles) correspond to E coronoids. The nonextremal extreme single coronoids (corresponding to asterisks without circles in Fig. 4.3) constitute an important class to be treated in a subsequent paragraph.

5.6.2 *Formula*

There exists a formula which pertains to an extreme single coronoid E, say $(n^e; \ s^e)$, for every $n_i = 0, 1, 2, 3, \ldots$ .The pair of invariants of $E(h, n_i)$ is given by eqn. (4.23). Herefrom one obtains the expressions:

$$(n^e; \ s^e) = (16 - n_i + 4\lceil (n_i/2) + (1/2)(6n_i + 60)^{1/2}\rceil;$$
$$8 - n_i + 2\lceil (n_i/2) + (1/2)(6n_i + 60)^{1/2}\rceil) \tag{5.77}$$

The first (smallest) extreme single coronoids have the formulas $C_{32}H_{16}$, $C_{35}H_{17}$, $C_{38}H_{18}$, $C_{37}H_{17}$, $C_{40}H_{18}$, $C_{43}H_{19}$, $C_{42}H_{18}$, $C_{45}H_{19}$, $C_{48}H_{20}$, $C_{47}H_{19}$, ...

A coronoid E is not necessarily naphthalenic. However, for every formula $(n^e; \ s^e)$ a naphthalenic perforated $E(N^e; \ S^e)$ is found, where E is the associated benzenoid to E and an extreme benzenoid. Here we are not able to derive eqn. (77) on this basis since the corresponding expressions for $(N^e; \ S^e)$ have not been given before. However, we can derive these expressions from (77). By means of eqn. (4.13) and $n_i = N_i - 10$ from (4.12) one obtains

$$(N^e;\ S^e) = (6 - N_i + 4\lceil(1/2)(N_i + 6^{1/2}N_i^{1/2})\rceil;$$
$$6 - N_i + 2\lceil(1/2)(N_i + 6^{1/2}N_i^{1/2})\rceil) \tag{5.78}$$

For $N_i = 0, 1, 2, 3, \ldots$ these expressions correspond to the benzenoids defined by $H = H_{min}(N_i)$ for their numbers of hexagons. These are the extreme benzenoids except for $N_i = 0$; in this case eqn. (78) gives C_6H_6, while the extreme (or extreme–left) benzenoid for $N_i = 0$ by definition is $C_{10}H_8$.

5.6.3 Nonextremal Extreme Single Coronoids

Notation and Preliminary Treatment. Let an extreme single coronoid which is not extremal be identified by $X(n^x;\ s^x)$. These formulas occur for certain values of $n_i = n_i^x$, but not for $n_i = n_i^a$, where n_i^a pertains to the extremal single coronoids. From eqns. (53) and (54) one obtains

$$n_i^a = 2h - 5 - \lceil(12h+21)^{1/2}\rceil = 2a + 9 - \lceil(12a+105)^{1/2}\rceil \tag{5.79}$$

where $h = 8, 9, 10, 11, \ldots$, and $a = 1, 2, 3, 4, \ldots$. The first (lowest) values of n_i^a are $n_i^a = 0, 1, 3, 4, 6, 7, 9, 10, 12, 14, \ldots$. An obvious strategy for detecting successively the formulas $(n^x;\ s^x)$ for nonextremal extreme single coronoids would be: use eqn. (77) for $n_i = 0, 1, 2, 3, \ldots$, but skip every n_i value which coincides with n_i^a.

Instead of this elaborate specification of the $(n^x;\ s^x)$ formulas we are able to derive explicit expressions for the coefficients of these formulas. Before this derivation some additional properties of the X coronoids are to be treated as a preparation.

Depictions. The X coronoids are not generated by the spiral walk (Par. 4.3.3), but this process can be used, with a slight modification, to produce an X coronoid for every h value which is realized in this class of coronoids. Consider a three–contact addition (Vol. I – Fig. 2.2) of a hexagon X during the spiral walk; it is accompanied by $\Delta n_i = 2$. This occurs for the first time by the addition of hexagon No. 10 (cf. Fig. 4.4). Now, instead of adding the hexagon X, another hexagon X' can always be added by a two–contact addition (cf. Fig. 4.4), for instance by going in the opposite direction of the spiral walk. For the smallest example this modification is illustrated by the addition of hexagon No. 10'; cf. the top–left depiction of Fig. 6. The two–contact addition (of the hexagon X') is accompanied by $\Delta n_i = 1$.

Figure 6 shows the ten smallest X coronoids generated in the way described above. The chemical formulas are indicated, but only one of the several existing isomers for each formula is displayed. Some of the additional isomers are found in Fig 7; this figure demonstrates at the same time the existence of X systems which are not naphthalenic.

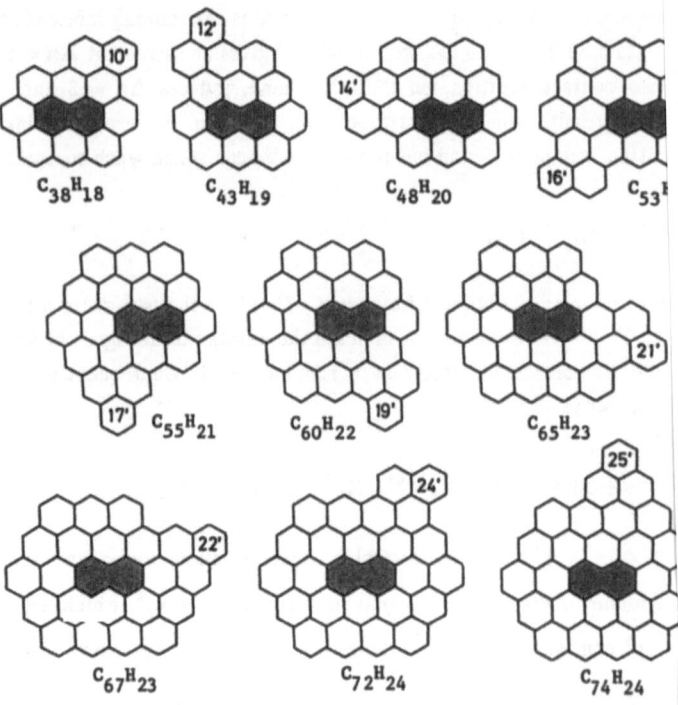

Fig. 5.6. One isomer each of the smallest nonextremal extreme single coronoids. numerals indicate the number of last added hexagon during the modified spiral wa

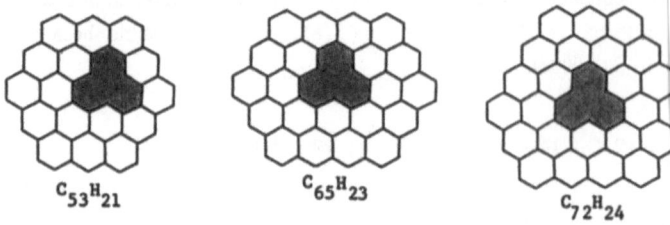

Fig. 5.7. Examples of nonextremal extreme single coronoids which are not naphtl

Formula. In accord with the above description every nonextremal extreme
(X) corresponds to an extremal single coronoid, A(h), for which A($h-1$) → A(h) i
by $\Delta n_i = 2$. The formulas of these special A coronoids are obtained from eqn. (57
the running index from $h = 1$; i.e. $h = 1, 2, 3, 4, \ldots$. These A coronoids are associa

circ
(60
0:
+C
thr
sys

Th
val
int
not

wh
con
dif
fou

wh
26

ex

wl
si
wl

B

imextremal benzenoids starting from $H' = 3$, where H' is the running index of eqns. (59) and . The three—contact addition of a hexagon to A(h–1) gives an increment $\Delta n = 2$, while $\Delta s =$ e., $+C_2$. A two—contact addition, on the other hand, induces $\Delta n = 3$ and $\Delta s = 1$: i.e., $_3$H. As the net result, when the two—contact addition is executed instead of the e—contact addition, one more C and one more H have been added when compared with the A em. Hence

$$(n^X; s^X) = (n^P + 1; s^P + 1) \tag{5.80}$$

final expressions are readily obtained from eqn. (57), where the parameter h should have the es $h = 1, 2, 3, 4, \dots$. Here h does not indicate the number of hexagons of the X system of rest, but should be considered merely as a running index. Consequently, we change the ation to $h = x$ and write

$$(n^X; s^X) = (2x + 18 + 3\lceil (12x+21)^{1/2}\rceil; \ 12 + \lceil (12x+21)^{1/2}\rceil) \tag{5.81}$$

re $x = 1, 2, 3, 4, \dots$. The first ten formulas for the smallest nonextremal extreme simple onoids (X) are found in Fig. 6. The X systems occur at $n_i = n_i^X$, which are the n_i values erent from n_i^a of eqn. (79). These n_i^X values can be specified by means of eqn. (81). It is nd:

$$n_i^X = 2x - 6 + \lceil (12x+21)^{1/2}\rceil \tag{5.82}$$

re $x = 1, 2, 3, 4, \dots$. The first (lowest) values of n_i^X are $n_i^X = 2, 5, 8, 11, 13, 16, 19, 21, 24,$ ..... . The same sequence is characterized in a different way at the end of Par. 4.3.3.

Here an alternative derivation of eqn. (81) is again possible because of the known ressions (Cyvin SJ 1992c)

$$(N^X; S^X) = (2H' + 14 + 3\lceil (12H'-3)^{1/2}\rceil; \ 10 + \lceil (12H'-3)^{1/2}\rceil) \tag{5.83}$$

ich pertain to nonextremal extreme benzenoids, say $X(N^X; S^X)$. Now, a nonextremal extreme gle coronoid (X) for any formula $(n^X; s^X)$ can be obtained as a naphthalenic perforated X, in ich case $H' = 3, 4, 5, 6, \dots$ are the H' values of interest. Consequently:

$$(n^X; s^X) = (2H' + 14 + 3\lceil (12H'-3)^{1/2}\rceil; \ 12 + \lceil (12H'-3)^{1/2}\rceil) \tag{5.84}$$

refrom eqn. (81) emerges readily on substituting H' by $x + 2$.

5.7 Table of Formulas

The possible C_nH_s formulas for single coronoids are listed conveniently in an (h, n_i) coordinate system as shown in Table 3. A part of this table is included as a subset of Table 3.5.

It follows straightforwardly from eqn. (1) that the table of formulas for single coronoids can be built up recursively by the scheme:

$$C_nH_s(h, n_i) \rightarrow C_{n-1}H_{s-1}(h, n_i+1)$$
$$\downarrow$$
$$C_{n+4}H_{s+2}(h+1, n_i)$$

The initial formula (at the upper–left corner), along with its coordinates, is $C_{32}H_{16}$ (8, 0).

In Table 3 "everything is perfect". The formulas at the extreme right of each row pertain to perfect extremal single coronoids, and therefore the right–hand boundary is a perfect staircase; cf. Par. 3.7.3.

A formula C_nH_s is called an *even–carbon* formula when n and s (which necessarily have the same parity) are even integers, while C_nH_s is an *odd–carbon* formula when n and s are odd. In Table 3 (as well as in Table 3.5) it is found that the even–carbon– and odd–carbon formulas occur in alternating columns. They are characterized by $n_i = 0, 2, 4, 6, ...$ and $n_i = 1, 3, 5, 7, ...$, respectively.

The staircase boundary of the table of formulas can obviously consist of steps with exactly either one or two formulas of the horizontal part of the step. We shall presently refer to these steps as a *short step* and a *long step*, respectively. The occurrence of a short (resp. long) step depends on the increment Δn_i for $A(h) \rightarrow A(h+1)$, whether it is 1 (resp. 2); cf. Par. 4.3.3.

Examples:

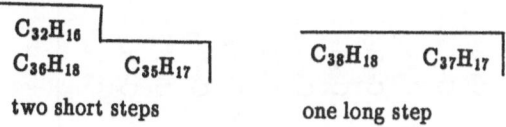

two short steps   one long step

Figure 8 shows a mapping of the C_nH_s formulas for single coronoids in terms of a dot diagram, which may be compared with Fig. 3.5. In Fig. 8 the formulas for some classes are indicated. Those for the extremal single coronoids (A) are found at the extreme right of each row. The formulas for the circular single coronoids (O), which also are extremal, are found on the rim above each short step; with the exception of the smallest coronoid ($C_{32}H_{16}$, $h = 8$) they are on the rim of a long step. Finally, the formulas for nonextremal extreme single coronoids (X) are situated on the long steps, next to the rim. The A and X coronoids taken together are the extreme single coronoids (E); their formulas are seen to be found at the top of each column.

Table 5.3. Formulas for single coronoids.

h	n_i 0	1	2	3	4	5	6	7	8	9
8	$C_{32}H_{16}$									
9	$C_{36}H_{18}$	$C_{35}H_{17}$								
10	$C_{40}H_{20}$	$C_{39}H_{19}$	$C_{38}H_{18}$	$C_{37}H_{17}$						
11	$C_{44}H_{22}$	$C_{43}H_{21}$	$C_{42}H_{20}$	$C_{41}H_{19}$	$C_{40}H_{18}$					
12	$C_{48}H_{24}$	$C_{47}H_{23}$	$C_{46}H_{22}$	$C_{45}H_{21}$	$C_{44}H_{20}$	$C_{43}H_{19}$	$C_{42}H_{18}$			
13	$C_{52}H_{26}$	$C_{51}H_{25}$	$C_{50}H_{24}$	$C_{49}H_{23}$	$C_{48}H_{22}$	$C_{47}H_{21}$	$C_{46}H_{20}$	$C_{45}H_{19}$		
14	$C_{56}H_{28}$	$C_{55}H_{27}$	$C_{54}H_{26}$	$C_{53}H_{25}$	$C_{52}H_{24}$	$C_{51}H_{23}$	$C_{50}H_{22}$	$C_{49}H_{21}$	$C_{48}H_{20}$	$C_{47}H_{19}$
15	$C_{60}H_{30}$	$C_{59}H_{29}$	$C_{58}H_{28}$	$C_{57}H_{27}$	$C_{56}H_{26}$	$C_{55}H_{25}$	$C_{54}H_{24}$	$C_{53}H_{23}$	$C_{52}H_{22}$	$C_{51}H_{21}$
16	$C_{64}H_{32}$	$C_{63}H_{31}$	$C_{62}H_{30}$	$C_{61}H_{29}$	$C_{60}H_{28}$	$C_{59}H_{27}$	$C_{58}H_{26}$	$C_{57}H_{25}$	$C_{56}H_{24}$	$C_{55}H_{23}$
17	$C_{68}H_{34}$	$C_{67}H_{33}$	$C_{66}H_{32}$	$C_{65}H_{31}$	$C_{64}H_{30}$	$C_{63}H_{29}$	$C_{62}H_{28}$	$C_{61}H_{27}$	$C_{60}H_{26}$	$C_{59}H_{25}$
18	$C_{72}H_{36}$	$C_{71}H_{35}$	$C_{70}H_{34}$	$C_{69}H_{33}$	$C_{68}H_{32}$	$C_{67}H_{31}$	$C_{66}H_{30}$	$C_{65}H_{29}$	$C_{64}H_{28}$	$C_{63}H_{27}$
19	$C_{76}H_{38}$	$C_{75}H_{37}$	$C_{74}H_{36}$	$C_{73}H_{35}$	$C_{72}H_{34}$	$C_{71}H_{33}$	$C_{70}H_{32}$	$C_{69}H_{31}$	$C_{68}H_{30}$	$C_{67}H_{29}$
20	$C_{80}H_{40}$	$C_{79}H_{39}$	$C_{78}H_{38}$	$C_{77}H_{37}$	$C_{76}H_{36}$	$C_{75}H_{35}$	$C_{74}H_{34}$	$C_{73}H_{33}$	$C_{72}H_{32}$	$C_{71}H_{31}$
21	$C_{84}H_{42}$	$C_{83}H_{41}$	$C_{82}H_{40}$	$C_{81}H_{39}$	$C_{80}H_{38}$	$C_{79}H_{37}$	$C_{78}H_{36}$	$C_{77}H_{35}$	$C_{76}H_{34}$	$C_{75}H_{33}$
22	$C_{88}H_{44}$	$C_{87}H_{43}$	$C_{86}H_{42}$	$C_{85}H_{41}$	$C_{84}H_{40}$	$C_{83}H_{39}$	$C_{82}H_{38}$	$C_{81}H_{37}$	$C_{80}H_{36}$	$C_{79}H_{35}$
23	$C_{92}H_{46}$	$C_{91}H_{45}$	$C_{90}H_{44}$	$C_{89}H_{43}$	$C_{88}H_{42}$	$C_{87}H_{41}$	$C_{86}H_{40}$	$C_{85}H_{39}$	$C_{84}H_{38}$	$C_{83}H_{37}$

(cont.)

Table 5.3 (continued)

h	\multicolumn{10}{c}{n_i}
	10	11	12	13	14	15	16	17	18	19
15	$C_{50}H_{20}$									
16	$C_{54}H_{22}$	$C_{53}H_{21}$	$C_{52}H_{20}$							
17	$C_{58}H_{24}$	$C_{57}H_{23}$	$C_{56}H_{22}$	$C_{55}H_{21}$	$C_{54}H_{20}$					
18	$C_{62}H_{26}$	$C_{61}H_{25}$	$C_{60}H_{24}$	$C_{59}H_{23}$	$C_{58}H_{22}$	$C_{57}H_{21}$				
19	$C_{66}H_{28}$	$C_{65}H_{27}$	$C_{64}H_{26}$	$C_{63}H_{25}$	$C_{62}H_{24}$	$C_{61}H_{23}$	$C_{60}H_{22}$	$C_{59}H_{21}$		
20	$C_{70}H_{30}$	$C_{69}H_{29}$	$C_{68}H_{28}$	$C_{67}H_{27}$	$C_{66}H_{26}$	$C_{65}H_{25}$	$C_{64}H_{24}$	$C_{63}H_{23}$	$C_{62}H_{22}$	
21	$C_{74}H_{32}$	$C_{73}H_{31}$	$C_{72}H_{30}$	$C_{71}H_{29}$	$C_{70}H_{28}$	$C_{69}H_{27}$	$C_{68}H_{26}$	$C_{67}H_{25}$	$C_{66}H_{24}$	$C_{65}H_{23}$
22	$C_{78}H_{34}$	$C_{77}H_{33}$	$C_{76}H_{32}$	$C_{75}H_{31}$	$C_{74}H_{30}$	$C_{73}H_{29}$	$C_{72}H_{28}$	$C_{71}H_{27}$	$C_{70}H_{26}$	$C_{69}H_{25}$
23	$C_{82}H_{36}$	$C_{81}H_{35}$	$C_{80}H_{34}$	$C_{79}H_{33}$	$C_{78}H_{32}$	$C_{77}H_{31}$	$C_{76}H_{30}$	$C_{75}H_{29}$	$C_{74}H_{28}$	$C_{73}H_{27}$

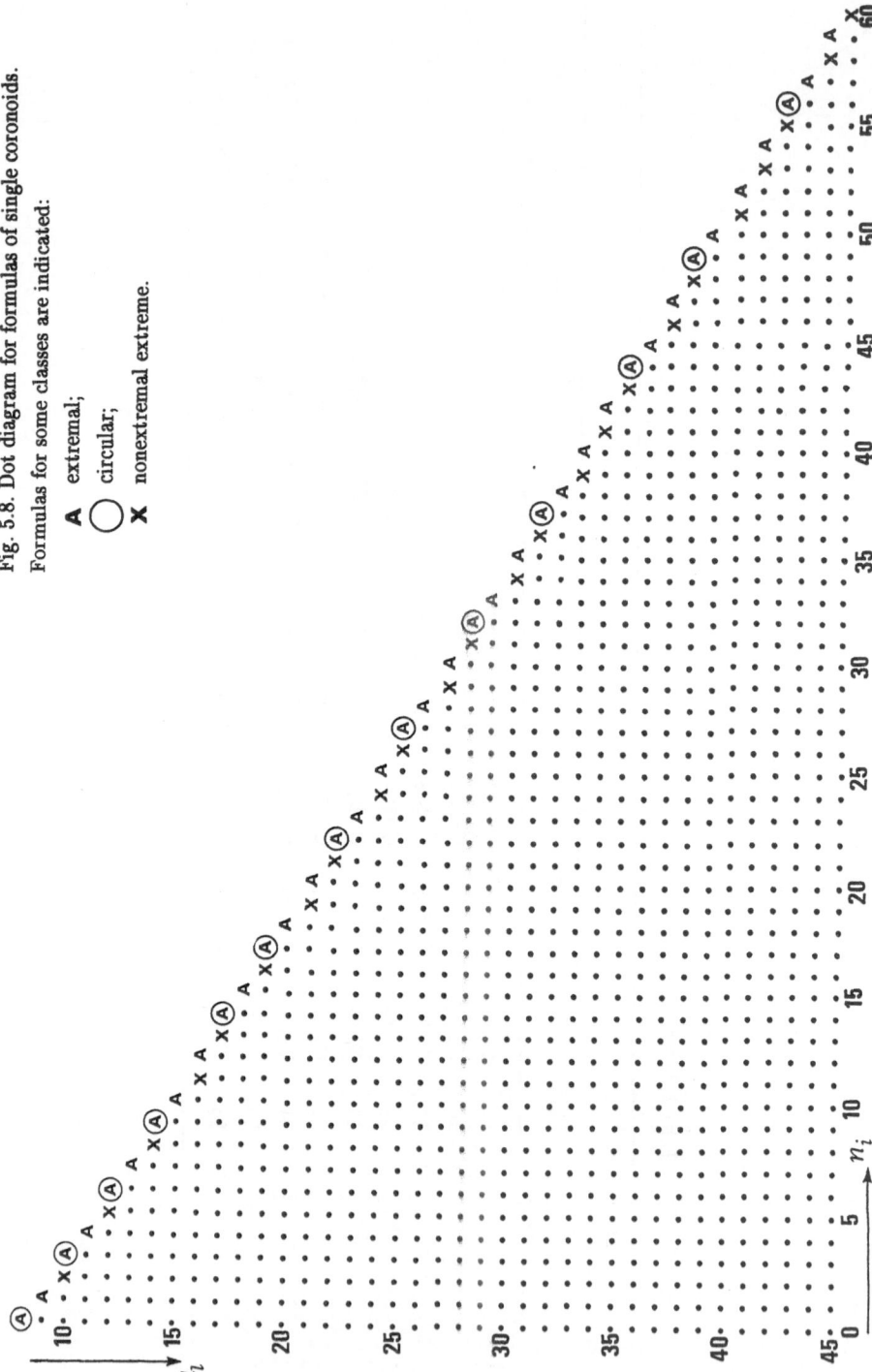

Fig. 5.8. Dot diagram for formulas of single coronoids.
Formulas for some classes are indicated:

A extremal;
◯ circular;
X nonextremal extreme.

5.8 Single Coronoid Isomers and Number of Edges

A single coronoid isomer $C \in \{C_n H_s\}$ has m edges, and the function $m(n, s)$ is found in Table 4.1. In general, m does not determine uniquely a formula $C_n H_s$ for C. However, it was found that this happens exactly for twelve m values. They are listed, together with the corresponding (h, n_i) invariants and the $C_n H_s$ formulas in Table 4. This feature can be demonstrated by means of the upper and lower bounds of n and s as functions of m.

Table 5.4. Single coronoid formulas which are uniquely determined by the number of edges.

m	h	n_i	Formula
40	8	0	$C_{32}H_{16}$
44	9	1	$C_{35}H_{17}$
45	9	0	$C_{36}H_{18}$
47	10	3	$C_{37}H_{17}$
48	10	2	$C_{38}H_{18}$
49	10	1	$C_{39}H_{19}$
50	10	0	$C_{40}H_{20}$
51	11	4	$C_{40}H_{18}$
52	11	3	$C_{41}H_{19}$
53	11	2	$C_{42}H_{20}$
56	12	4	$C_{44}H_{20}$
57	12	3	$C_{45}H_{21}$

Start with the second formula of Table 4.4. Insert successively

$$h = m - n \tag{5.85}$$

and

$$h = (1/3)(m - s) \tag{5.86}$$

These relations are readily obtained from Table 4.1. Herefrom the following inequalities are deduced.

$$1 + \lceil (2m/3) + (1/3)(4m+5)^{1/2} \rceil \leq n \leq m - \lceil m/5 \rceil \tag{5.87}$$

$$3 - 2m + 3\lceil (2m/3) + (1/3)(4m+5)^{1/2} \rceil \leq s \leq m - 3\lceil m/5 \rceil \tag{5.88}$$

In the case of eqn. (87) the interval (in the sense of Tables 4.3 and 4.4) is 1, while it is 3 in the case of (88). The allowed m values are $m = 40, 44, 45, 47, 48, 49, 50, ...$ (cf. Table 4.2). Now it may happen that the upper and lower bounds in (87) and (88) coincide. It happens actually for the m values listed in Table 4.

Examples

1. From eqn. (87): (a) $m = 40$ gives $32 \leq n \leq 32$; hence $n = 32$. (b) $m = 57$ gives $45 \leq n \leq 45$; hence $n = 45$.

2. From eqn. (88): (a) $m = 40$ gives $16 \leq s \leq 16$; hence $s = 16$. (b) $m = 57$ gives $21 \leq s \leq 21$; hence $s = 16$.

In the above Example 1 the unique $C_n H_s$ formulas are determined by means of another version of the connections between the different invariants, viz.

$$s = 3n - 2m \qquad (5.89)$$

It gives (a) $s = 16$ and (b) $s = 21$. Similarly, in the case of Example 2, the relation of interest is

$$n = (1/3)(2m + s) \qquad (5.90)$$

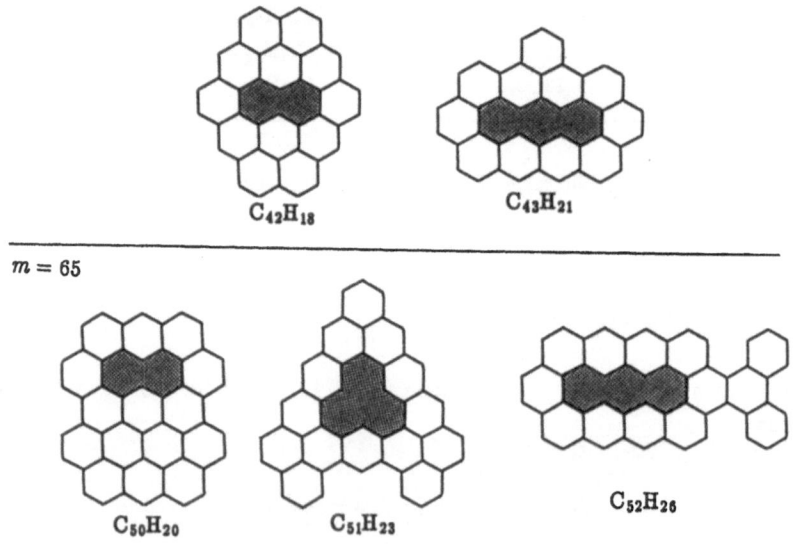

$m = 54$

$C_{42}H_{18}$ $C_{43}H_{21}$

$m = 65$

$C_{50}H_{20}$ $C_{51}H_{23}$ $C_{52}H_{26}$

Fig. 5.9. Representatives of different single coronoid isomers with same number of edges (m).

It gives (a) $n = 42$ and (b) $n = 43$. In conclusion, the numbers of edges (a) $m = 40$ and (b) $m = 57$ are compatible with the unique formulas (a) $C_{42}H_{18}$ and (b) $C_{43}H_{21}$, respectively, for single coronoids.

The first ambiguity occurs for $m = 54$. In this case eqn. (87) gives $42 \leq n \leq 43$; hence $n = 42, 43$. Furthermore, (88) gives $18 \leq s \leq 21$; hence, when the interval 3 is taken into account, $s = 18, 23$. In conclusion, $m = 54$ is compatible with the two formulas $C_{42}H_{18}$ and $C_{43}H_{21}$. Figure 9 shows one representative for each of these formulas. By the way, $C_{42}H_{18}$ pertains to circular single coronoids, of which the other one of the two isomers is depicted in Fig. 4.6. Altogether there are eleven m values, viz. $m = 54, 55, 58, 59, 60, 61, 62, 63, 64, 66, 67$, which are compatible with two formulas each. Each pair of these formulas, like $C_{42}H_{18}/C_{43}H_{21}$, consists of one even–carbon– and one odd–carbon formula.

Three formulas for single coronoids are possible when $m = 65$ as the smallest m value. Two of them are even–carbon formulas, while one is an odd–carbon formula: $C_{50}H_{20}$, $C_{51}H_{23}$, $C_{52}H_{26}$. Here $C_{50}H_{20}$ pertains to extremal single coronoids, while the $C_{52}H_{26}$ isomers are catacondensed. Figure 9 includes one representative for each of these three formulas.

Chapter 6

FURTHER STUDIES OF THE CHEMICAL FORMULAS

OF SINGLE CORONOIDS

6.1 Introduction

The focussing on C_nH_s formulas for single coronoids is continued in this chapter. It starts with a treatment of sequences of associated formulas for extremal single coronoids. This study leads to another subdivision of the extremal coronoids: ground forms and higher members. The pertinent sections can be considered as a preparation to the enumeration of coronoid isomers, which is treated in the subsequent chapter. Also the process of building–up (Sect. 6.5) is highly relevant to the enumeration of isomers.

6.2 Ground Forms and Higher Members

6.2.1 *Introduction*

Among the extremal benzenoids one distinguishes between the ground forms and higher members, concepts which were introduced (Brunvoll and Cyvin SJ 1990; Cyvin SJ and Brunvoll 1991; Cyvin SJ, Brunvoll and Cyvin 1991d; Cyvin SJ 1991c; 1992c; Cyvin SJ, Cyvin and Brunvoll 1993e) in connection with the constant–isomer benzenoid series. For a first survey of references, see Par. 5.4.1. Additional works on this topic are available: Dias (1992a, 1993); Cyvin BN, Cyvin and Brunvoll (1993).

A higher member (of benzenoids) is a circumextremal benzenoid (and itself an extremal benzenoid). A ground form is an extremal benzenoid which is not circumextremal. Assume that $(N_0; S_0)$ is a formula for a ground form benzenoid G. Then the formulas $(N_0; S_0)$, $(N_1; S_1)$, ..., $(N_k; S_k)$, ... are said to form a constant–isomer series when $(N_k; S_k)$ pertains to k–fold circumscribed G. All these formulas are known to have the same number of benzenoid isomers.

For extremal single coronoids we shall define ground forms and higher members in connection with a sequence of formulas $(n_0; s_0)$, $(n_1; s_1)$, ..., $(n_k; s_k)$, ... in analogy with the corresponding classes for benzenoids. However, the formulas of such a sequence have not the same cardinalities, i.e. they do not pertain to constant numbers of isomers of single coronoids.

6.2.2. Definitions and Preliminary Treatment

Definition 6.1: A *ground form* (of single coronoids), G, is an extremal single coronoid which is not circumextremal.

Definition 6.2: A *higher member* (of single coronoids) is synonymous with a circumextremal single coronoid.

Assume that G has the formula $(n_0; s_0)$: G $\in \{n_0; s_0\}$. Furthermore, $c_k(G) \in \{n_k; s_k\}$ ($k > 0$). Then a pair of formulas $(n_i; s_i)$ and $(n_j; s_j)$, where $i, j = 1, 2, 3, ..., k$ and $i \neq j$, are said to be associated with each other, and the sequence $(n_0; s_0), (n_1; s_1), ..., (n_k; s_k), ...$ is a *sequence of associated formulas* (for single coronoids). We shall also speak about ground forms and higher members as belonging to a sequence of associated formulas; here $(n_0; s_0)$ pertains to a set of ground forms, and $(n_k; s_k)$ for $k > 0$ to sets of higher members.

The higher members occur with certain numbers of hexagons, say h^p. From eqns. (5.57) and (5.58) one obtains

$$h^p = h + 3 + \lceil (12h + 21)^{1/2} \rceil = a + 10 + \lceil (12a + 105)^{1/2} \rceil \tag{6.1}$$

where $h = 8, 9, 10, 11, ...$, and $a = 1, 2, 3, 4, ...$. Herefrom the first (lowest) values of h^p are found to be $h^p = 22, 24, 25, 27, 28, 30, 31, 33, 34, 35, ...$. One of the possible strategies to derive all formulas $(n_0; s_0)$ for G is: use eqn. (5.53) for $h = 8, 9, 10, 11, ...$, but skip every h value which coincides with h^p.

Instead of this elaborate specification of the $(n_0; s_0)$ formulas we shall be able to give a full account of them in terms of comprehensive specifications and explicit expressions. All the three "pictures" are going to be invoked: the Harary–Harborth–, the Balaban– and the new picture. But first some additional properties of the systems under consideration are needed. In this connection we shall introduce two subclasses of the ground forms and higher members: perfect and imperfect.

6.2.3 Perfect and Imperfect Ground Forms

Consider the spiral walk (cf. Par. 4.3.3, and especially Fig. 4.4), and focus the attention on the event when a hexagon is added to a circular coronoid. This is a case when $\Delta n_i = 1$, a two–contact addition (cf. Vol I–2.1.1, especially I–Fig. 2.2). In this way a ground form of a single coronoid is generated since the system is extremal, but obviously not circumextremal. This feature has an analogy among benzenoids: a ground form benzenoid is generated by a two–contact addition of a hexagon to a circular benzenoid.

Let $O(n^o; s^o)$ denote a circular single coronoid as in Chapt. 5. Furthermore, introduce $G^P(n_0{}^{pg}; s_0{}^{pg})$ as a ground form obtained by a two–contact addition of a hexagon to O. Then the formula coefficients obey

$$(n_0{}^{pg}; s_0{}^{pg}) = (n^o + 3; s^o + 1) \qquad (6.2)$$

Definitions 6.3: (a) A *perfect ground form* (of single coronoids), G^P, is a single coronoid with the formula $(n^o + 3; s^o + 1)$. (b) An *imperfect ground form*, G^i, is a ground form single coronoid which is not perfect.

The spiral walk generates exactly one G system $(G^P$ or $G^i)$ for every possible formula $(n_0; s_0)$. It should be observed, however, that there may be several nonisomorphic systems $G \in \{n_0; s_0\}$ for a given formula, and it may happen that some of them have not the structure of a hexagon added to a circular coronoid.

The associated benzenoid to G^P is a ground form benzenoid, while the associated benzenoid to G^i is a higher member (circumscribed ground form). These properties might serve as alternative definitions of perfect and imperfect coronoid ground forms. In other words: among the extremal single coronoids which are not circumextremal, (a) a perfect ground form is a (naphthalenic) perforated ground form benzenoid, while (b) an imperfect ground form is a perforated higher member benzenoid. The origin of imperfect ground forms is, in simple terms, that the corresponding ground form benzenoids are too small to be perforable by a naphthalene hole.

Table 1 gives a survey of $(n_0; s_0)$ formulas for the G systems up to $s_0 = 25$. There is exactly one G^P for every $s_0 \geq 17$. One G^i system occurs for $s_0 = 16$; otherwise the G^i formulas duplicate s_0 values from G^P. In one case $(s_0 = 20)$ the s_0 value is even triplicated (cf. Table 1). The table includes information on the benzenoids (A) which are associated with the coronoid ground forms. Here H is the number of hexagons of A. The symbol $\{j, \delta, k\}$ refers to a coding system (Cyvin SJ and Brunvoll 1991; Cyvin SJ 1991c; Cyvin SJ, Cyvin and Brunvoll 1993e) for benzenoid ground forms and higher members. In this three–parameter code: $j = 0, 1, 2, \ldots.$; $\delta = 1, 2, 3, 4, 5, 6$; $k = 0, 1, 2, \ldots.$. In particular, k indicates the number of circumscribings of a ground form; consequently, $k = 0$ pertains to ground forms and $k > 0$ to higher members. It is seen in Table 1 that, indeed, the A benzenoids which are associated with G^P (marked by "pg") have $k = 0$, while those associated with G^i ("ig") have $k > 0$. Usually $k = 0$ or $k = 1$; only in the case of three times $s_0 = 20$, one finds $k = 0, 1, 2$.

The definitions of perfect and imperfect coronoid higher members follow.

Table 6.1. Formula for the first (smallest) ground forms of single coronoids (G) and their associated benzenoids (A)

G			A		
h	Formula	Type*	H	Formula	{j,δ,k}
8	$C_{32}H_{16}$	o,ig	10	$C_{32}H_{14}$	{0,1,1}
9	$C_{35}H_{17}$	pg	11	$C_{35}H_{15}$	{1,2,0}
10	$C_{37}H_{17}$	o,ig	12	$C_{37}H_{15}$	{0,2,1}
11	$C_{40}H_{18}$	pg	13	$C_{40}H_{16}$	{1,3,0}
12	$C_{42}H_{18}$	o,ig	14	$C_{42}H_{16}$	{0,3,1}
13	$C_{45}H_{19}$	pg	15	$C_{45}H_{17}$	{1,4,0}
14	$C_{47}H_{19}$	o,ig	16	$C_{47}H_{17}$	{0,4,1}
15	$C_{50}H_{20}$	pg	17	$C_{50}H_{18}$	{1,5,0}
16	$C_{52}H_{20}$	ig	18	$C_{52}H_{18}$	{0,5,1}
17	$C_{54}H_{20}$	o,ig	19	$C_{54}H_{18}$	{−1,5,2}
18	$C_{57}H_{21}$	pg	20	$C_{57}H_{19}$	{1,6,0}
19	$C_{59}H_{21}$	o,ig	21	$C_{59}H_{19}$	{0,6,1}
20	$C_{62}H_{22}$	pg	22	$C_{62}H_{20}$	{2,1,0}
21	$C_{64}H_{22}$	ig	23	$C_{64}H_{20}$	{1,1,1}
23	$C_{69}H_{23}$	pg	25	$C_{69}H_{21}$	{2,2,0}
26	$C_{76}H_{24}$	pg	28	$C_{76}H_{22}$	{2,3,0}
29	$C_{83}H_{25}$	pg	31	$C_{83}H_{23}$	{2,4,0}

* Abbreviations: ig imperfect (ground form); o circular; pg perfect (ground form).

Definition 6.4: (a) A *perfect higher member* (of single coronoids) is defined by having the same formula as a k–fold ($k \geq 1$) circumscribed perfect ground form. (b) An *imperfect higher member* has the same formula as a k–fold circumscribed imperfect ground form.

More elaborate definitions or characterizations of the systems in Definition 6.4 could be based on the corresponding associated benzenoids in the style of Definition 5.7; cf. also the explanations below the latter definition (Par. 5.5.2).

6.2.4 *Formulas for Perfect Ground Forms and for Higher Members*

General formulations for the formulas of the title classes have been worked out.
Harary–Harborth Picture. From eqns. (5.62) and (2) it is obtained for the formulas of the

G^P systems:

$$(n_0{}^{pg};\, s_0{}^{pg}) = (4 - s + 2\lfloor(1/12)(s^2 + 3)\rfloor;\; s) \tag{6.3}$$

where $s\ (= s_0{}^{pg}) = 17, 18, 19, \ldots$. It may be checked that this relation reproduces the formulas of Table 1 which pertain to the perfect ground forms (marked "pg"). A more extensive list is given in Table 2. Also the formulas for the degenerate coronoids therein are reproduced by (3) if the values of s are extended appropriately. An alternative derivation of eqn. (3) is possible on exploiting the fact that the G^P systems are naphthalenic perforated ground form benzenoids, $G(N_0;\, S_0)$. It is known (Cyvin and Brunvoll 1991; Cyvin SJ 1991c; 1992c; Cyvin SJ, Cyvin, Brunvoll, Zhang and Guo 1992; Cyvin SJ, Cyvin and Brunvoll 1993e):

$$(N_0;\, S_0) = (S + 2\lfloor(1/12)(S^2 - 8S + 19)\rfloor;\; S) \tag{6.4}$$

Herefrom,

$$(n_0{}^{pg};\, s_0{}^{pg}) = (S + 2\lfloor(1/12)(S^2 - 8S + 19)\rfloor;\; S + 2) \tag{6.5}$$

where $S = 15, 16, 17, \ldots$ are the S values of interest. On inserting $S = s - 2$ in (5) one indeed arrives at eqn. (3).

Let $(n_k{}^{pg};\, s_k{}^{pg})$ symbolize the formula for a k-fold circumscribed G^P system. Then this formula (for $k > 0$) pertains to the perfect higher members of single coronoids. From (3) and (5.52) one obtains the general formulation:

Table 6.2. Formulas for perfect ground forms of single coronoids.*

| j | ϵ | | | | | |
	0	1	2	3	4	5
0	$(C_{27}H_{15})$	$(C_{30}H_{16})$	$C_{35}H_{17}$	$C_{40}H_{18}$	$C_{45}H_{19}$	$C_{50}H_{20}$
1	$C_{57}H_{21}$	$C_{62}H_{22}$	$C_{69}H_{23}$	$C_{76}H_{24}$	$C_{83}H_{25}$	$C_{90}H_{26}$
2	$C_{99}H_{27}$	$C_{106}H_{28}$	$C_{115}H_{29}$	$C_{124}H_{30}$	$C_{133}H_{31}$	$C_{142}H_{32}$
3	$C_{153}H_{33}$	$C_{162}H_{34}$	$C_{173}H_{35}$	$C_{184}H_{36}$	$C_{195}H_{37}$	$C_{206}H_{38}$
4	$C_{219}H_{39}$	$C_{230}H_{40}$	$C_{243}H_{41}$	$C_{256}H_{42}$	$C_{269}H_{43}$	$C_{282}H_{44}$
5	$C_{297}H_{45}$	$C_{310}H_{46}$	$C_{325}H_{47}$	$C_{340}H_{48}$	$C_{355}H_{49}$	$C_{370}H_{50}$
6	$C_{387}H_{51}$	$C_{402}H_{52}$	$C_{419}H_{53}$	$C_{436}H_{54}$	$C_{453}H_{55}$	$C_{470}H_{56}$

* Formulas for degenerate coronoids in parentheses.

$$(n_k^{pg};\ s_k^{pg}) = (6k^2 + (2k{-}1)s - 4(k{-}1) + 2\lfloor(1/12)(s^2{+}3)\rfloor;\ \ s + 6k) \tag{6.6}$$

Eqn. (3) is the special case of (6) for $k = 0$.

Balaban Picture. From eqns. (5.67) and (2) one obtains a general formulation for the G^p systems as an alternative to eqn. (3):

$$(n_0^{pg};\ s_0^{pg}) = (6j^2 + 24j + (2j{+}5)\epsilon + 27 - 2\lceil\epsilon/6\rceil;\ \ 6j + 15 + \epsilon) \tag{6.7}$$

where $j = 0, 1, 2, \ldots$, and $\epsilon = 0, 1, 2, 3, 4, 5$. The parameter ϵ is the same as in the formula of a circular single coronoid (O) which is connected with the formula of G^p through eqn. (2). The parameters j and ϵ may be retrieved from the coefficient $s = s_0^{pg}$ of a G^p system in the following way.

$$j = \lfloor(1/6)(s - 15)\rfloor \tag{6.8}$$

$$\epsilon = s - 6j - 15 = s - 3 - 6\lfloor(1/6)(s - 3)\rfloor \tag{6.9}$$

Eqn. (7) reproduces the formulas of Table 2, including the two formulas of the degenerate coronoids.

As an alternative to eqn. (6) we also give the formulas for perfect higher members of circular coronoids, as obtained from eqns. (7) and (5.52).

$$(n_k^{pg};\ s_k^{pg}) = ((6j{+}k)^2 + 2(j{+}k)\epsilon + 24j + 26k + 27 + 5\epsilon - 2\lceil\epsilon/6\rceil;\ \ 6(j{+}k) + 15 + \epsilon) \tag{6.10}$$

New Picture. Define in analogy with eqn. (5.70):

$$n^{pg}(x) = \sum_{s=17}^{\infty} n_s^{pg} x^s = 35x^{17} + 40x^{18} + 45x^{19} + 50x^{20} + 57x^{21} + \ldots \tag{6.11}$$

Then, on translating eqn. (2) into generating functions one obtains readily

$$n^{pg}(x) = xn^{O}(x) + 3x^{17}(1 - x)^{-1} \tag{6.12}$$

and in explicit form:

$$n^{pg}(x) = \frac{x^{17}(35 + 5x + 5x^2 + 5x^3 + 7x^4 + 5x^5 - 63x^6 - 3x^7 - 3x^8 - 3x^9 - 5x^{10} - 3x^{11} + 30x^{12})}{(1 - x)(1 - x^6)^2} \tag{6.13}$$

157

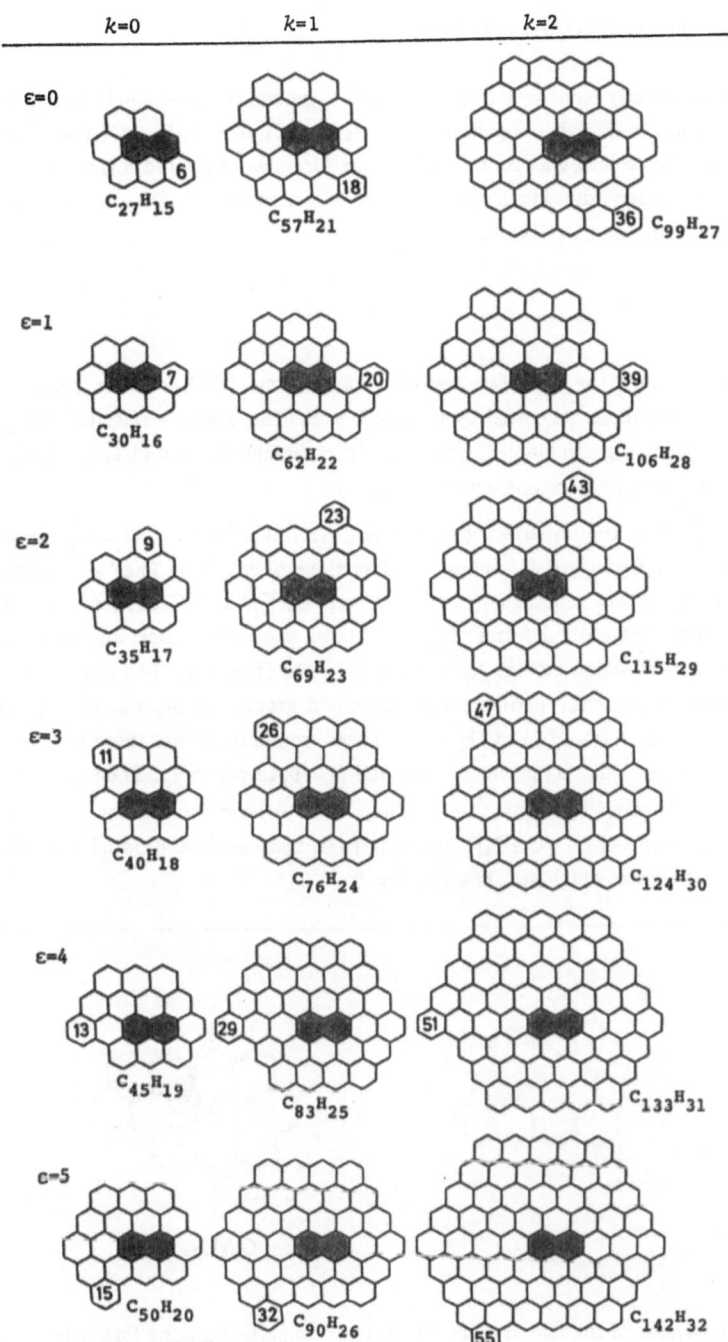

Fig. 6.1. The smallest perfect ground forms of single coronoids, augmented by two degenerate coronoids. The inscribed numerals indicate the number of the last added hexagon during the spiral walk.

6.2.5 *Formulas for Imperfect Ground Forms*

There are exactly eight formulas for imperfect ground forms of single coronoids, viz. those which are listed under the label "ig" in Table 1: $C_{32}H_{16}$, $C_{37}H_{17}$, $C_{42}H_{18}$, $C_{47}H_{19}$, $C_{52}H_{20}$, $C_{54}H_{20}$, $C_{59}H_{21}$, $C_{64}H_{22}$. These formulas contain all the six formulas of ground forms which at the same time are circular coronoids. There are only two formulas of non–circular ground form single coronoids, viz. $C_{52}H_{20}$ and $C_{64}H_{22}$.

6.2.6 *Depictions*

Figure 1 shows the smallest perfect ground forms (G^p), one for each formula. The depicted isomers are those generated by the spiral walk. The reader is referred to Fig. 4.6, which shows the corresponding circular coronoids, O, from which the ground forms are generated by two–contact additions of hexagons, one to each O.

With regard to the imperfect ground forms (G^i), one finds six of them ($C_{32}H_{16}$, $C_{37}H_{17}$, $C_{42}H_{18}$, $C_{47}H_{19}$, $C_{54}H_{20}$, $C_{59}H_{21}$) among the depictions of Fig. 4.6. They are at the same time circular coronoids. There remain two formulas, viz. $C_{52}H_{20}$ and $C_{64}H_{22}$, which pertain to the non–circular imperfect ground forms. Figure 2 shows one isomer each for these two formulas. They are the isomers which emerge from the spiral walk (Fig. 4.4). The heavy–line contours in these depictions indicate the perimeters of benzenoid ground forms, viz. $G \in \{C_{22}H_{12}\}$ at the left–hand side and $G' \in \{C_{30}H_{14}\}$ at the right. These ground forms are those which are generated by the spiral walk for benzenoids when $H = 6$ and $H = 9$, respectively. The G and G' systems are too small for perforating by corona holes. They have to be circumscribed first; (circum–G) $\in$ $\{C_{52}H_{18}\}$ and (circum–G') $\in \{C_{64}H_{20}\}$ are indeed the benzenoids associated with the isomers of imperfect ground forms which are shown in Fig. 2.

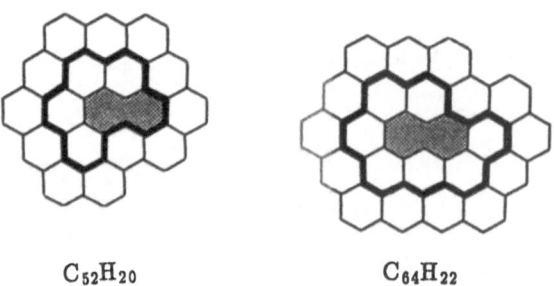

$C_{52}H_{20}$ $C_{64}H_{22}$

Fig. 6.2. Two isomers of the non–circular G^i systems: imperfect ground form single coronoids.

6.2.7 *Mapping of Formulas*

Figure 3 shows a dot diagram for the C_nH_s formulas of single coronoids as in Fig. 5.8. However, more detailed information about the extremal (A) systems is included: formulas for ground forms and higher members, both perfect and imperfect, are indicated.

6.3 Formula Index

A formula index (x) has been defined for a benzenoid $B(H, N_i) \in \{N; S\}$; cf. Cyvin SJ, Brunvoll and Cyvin (1991b); Brunvoll, Cyvin BN and Cyvin (1993b). It is given by the expressions

$$x = \lfloor (1/12)[(2H - N_i)^2 - 12H + 15] \rfloor = \lfloor (1/12)[(S-4)^2 - 6(N-S) + 3] \rfloor \qquad (6.14)$$

The value of x gives the following information about B. (i) $x > 0$; B is a nonextremal benzenoid. If it can be circumscribed x times, then x–circum–B is a ground form benzenoid (which is extremal), and there always exists (at least) one isomer $B \in \{N; S\}$ which can be circumscribed arbitrarily many times. (ii) $x = 0$; B is a ground form benzenoid. (iii) $x < 0$; B is a higher member (circumextremal benzenoid), and specifically an $|x|$–fold circumscribed ground form.

The coefficients of $(N; S)$ were transferred to those of a naphthalenic single coronoid, say $B(h, n_i) \in \{n; s\}$, whose associate benzenoid is $B(N; S)$. Then the expressions (14) read:

$$x = \lfloor (1/12)[(2h - n_i - 6)^2 - 12h - 9] \rfloor = \lfloor (1/12)[(s-6)^2 - 6(n-s) - 9] \rfloor \qquad (6.15)$$

From the information on B, the associated benzenoid to B, we deduce the following properties of the naphthalenic coronoid $B(n; s)$.

(i) $x > 0$; B is a nonextremal coronoid. If it is naphthalenic and can be circumscribed x times, then x–circum–B is a perfect ground form (which is extremal). The formula of this ground form is given by

$$(n_x; s_x) = (n + 2xs + 6x^2 - 4x, \; s + 6x) \qquad (6.16)$$

as obtained from eqn. (5.52) simply by replacing k with x.

(ii) $x = 0$; B is a perfect ground form.

(iii) $x < 0$; B is either an imperfect ground form (which is extremal) or a higher member (circumextremal coronoid), perfect or imperfect.

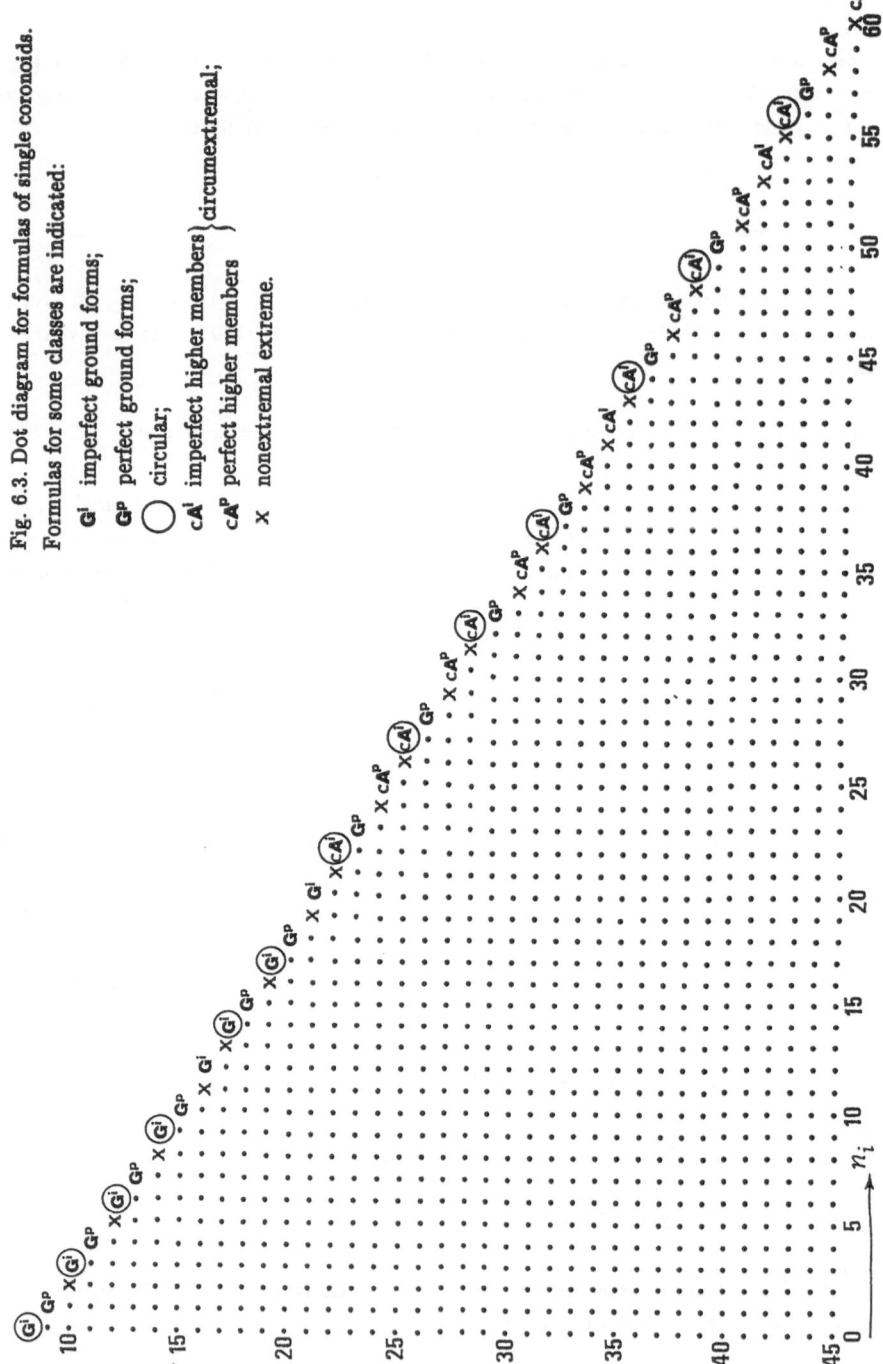

Fig. 6.3. Dot diagram for formulas of single coronoids.
Formulas for some classes are indicated:

G^I imperfect ground forms;

G^P perfect ground forms;

◯ circular;

cA^I imperfect higher members ⎱ circumextremal;
cA^P perfect higher members ⎰

× nonextremal extreme.

In order to distinguish between the different possibilities within (iii), consider the coefficients obtained by inserting the pertinent negative x value into (16). A useful special case of (16) is (for $x = -1$):

$$(n_{-1}; s_{-1}) = (n - 2s + 10; \ s - 6) \tag{6.17}$$

(a) Suppose that x, when inserted into (16), gives the formula C_6H_8.

Then $x = -2$ indicates that B is an imperfect ground form with the formula $C_{54}H_{20}$ (viz. perforated circumcoronene; cf. Fig. 4.6). If $x < -2$, B is an imperfect higher member, and an isomer of B can be produced by $(|x| - 2)$–fold circumscribing of $C_{54}H_{20}$, perforated circumcoronene (cf. Fig. 4.6). Here excis–$C_{54}H_{20}$ = $C_{24}H_{14}$ (cf. Fig. 4.6), while C_6H_8 represents the acyclic system hexatriene:

It corresponds formally to "naphtha– C_6H_8

lenic perforated benzene (C_6H_6)".

(b) Suppose that x, when inserted into (16), gives one of the seven formulas $C_{10}H_{10}$, $C_{13}H_{11}$, $C_{16}H_{12}$, $C_{19}H_{13}$, $C_{22}H_{14}$, $C_{27}H_{15}$, $C_{30}H_{16}$.

Then $x = -1$ indicates that B is an imperfect ground form with one of the formulas $C_{32}H_{16}$, $C_{37}H_{17}$, $C_{42}H_{18}$, $C_{47}H_{19}$, $C_{52}H_{20}$, $C_{59}H_{21}$ or $C_{64}H_{22}$, respectively (cf. Figs. 4.6 and 2). Here excis–$C_{32}H_{16}$ = $C_{10}H_{10}$, excis–$C_{37}H_{17}$ = $C_{13}H_{11}$, etc. These excised coronoids are degenerate systems of which one isomer each is depicted below (cf. also the diagrams of Par. 4.6.4, Fig. 4.6 and Fig. 2).

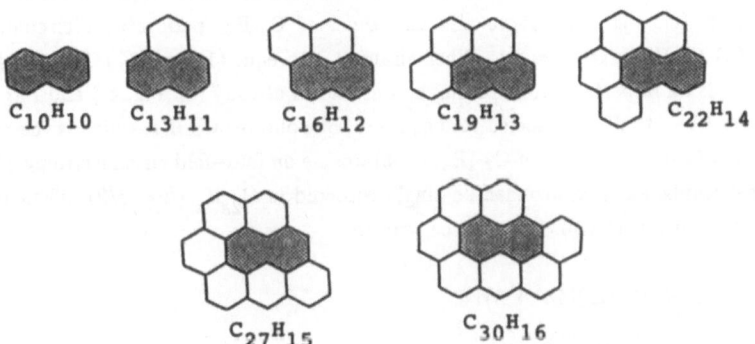

If $x < -1$, then B is an imperfect higher member, and an isomer of B can be produced by $(|x|-1)$–fold circumscribing of the respective ground form. The formula of this ground form is retrieved as $(n_{x+1}; s_{x+1})$ from eqn. (16).

(c) Suppose that x, when inserted into (16), gives a formula different from the eight formulas specified under (a) and (b). Then B is a perfect higher member, and an isomer of B can be produced by $|x|$—fold circumscribing of a perfect ground form with the formula $(n_x;\ s_x)$ according to eqn. (16).

Examples

(i) $C_{44}H_{20}$ (cf. the below diagram); $x = 3$. The right–hand system in the diagram is a perfect ground form. It may be checked that its formula, viz. $C_{206}H_{38}$, fits eqn. (3).

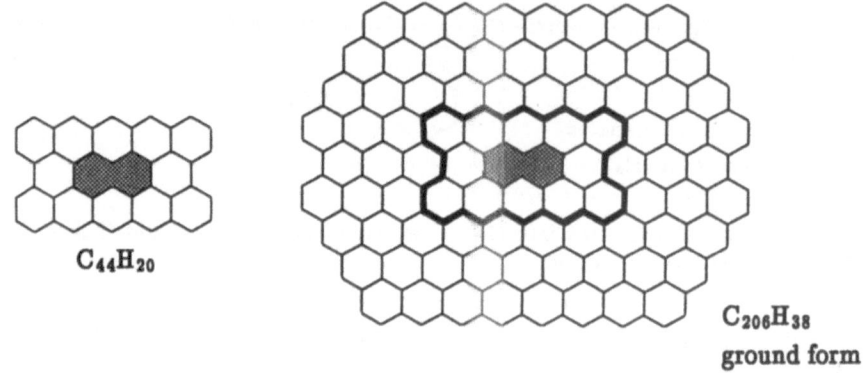

$$C_{44}H_{20}$$

$$C_{206}H_{38}$$
ground form

(ii) All the formulas in Table 2 exhibit $x = 0$.

(iii)(a) $C_{96}H_{26}$ gives $x = -3$. When inserted into eqn. (16) — together with $n = 96$, $s = 26$ — one obtains C_6H_8. Hence an isomer of $C_{96}H_{26}$ is circum–$C_{54}H_{20}$, where $C_{54}H_{20}$ indicates perforated circumcoronene (cf. Fig. 4.6, where also an isomer of $C_{96}H_{20}$ perforated dicircumcoronene is depicted). (b) $C_{32}H_{16}$ gives $x = -1$. When inserted into eqn. (16) or (17), one obtains $C_{10}H_{10}$. Hence $C_{32}H_{16}$ is an imperfect ground form, as was known already (cf. Fig. 4.6 and Par. 6.2.5). (c) $C_{264}H_{24}$ gives $x = -4$. When inserted into eqn. (16) one obtains $C_{40}H_{18}$, which is the formula of a perfect ground form. An isomer of $C_{264}H_{42}$ is obtainable on four–fold circumscribing $C_{40}H_{18}$.

The formula for a catacondensed single coronoid is $C_{2s}H_s$ ($h = s/2$). Then the formula index, in accordance with eqn. (15), can be written

$$x = 2 - 3h + \lfloor (1/12)(4h^2 + 3) \rfloor \tag{6.18}$$

Similar expressions could be obtained for the systems with $n_i = 1$, $n_i = 2$, etc. Numerical values of x for all the formulas of Table 5.3 are collected in Table 3.

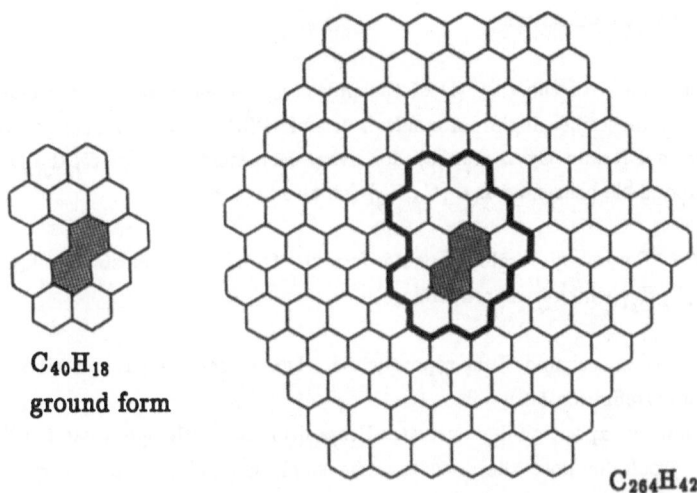

$C_{40}H_{18}$
ground form

$C_{264}H_{42}$

Table 6.3. Formula indices (x) for single coronoids. For the listing of the formulas, see Table 5.3.

h	n_i 0	1	2	3	4	5	6	7	8	9
8	−1									
9	2	0								
10	5	3	1	−1						
11	9	7	4	2	0					
12	14	11	8	6	3	1	−1			
13	19	16	13	10	7	5	2	0		
14	25	22	18	15	12	9	6	4	1	−1
15	32	28	24	21	17	14	11	8	5	3
16	39	35	31	27	23	20	16	13	10	7
17	47	43	38	34	30	26	22	19	15	12
18	56	51	46	42	37	33	29	25	21	18
19	65	60	55	50	45	41	36	32	28	24
20	75	70	64	59	54	49	44	40	35	31
21	86	80	74	69	63	58	53	48	43	39
22	97	91	85	79	73	68	62	57	52	47
23	109	103	96	90	84	78	72	67	61	56

h	n_i 10	11	12	13	14	15	16	17	18	19
15	0									
16	4	2	−1							
17	9	6	3	1	−2					
18	14	11	8	5	2	0				
19	20	17	13	10	7	4	1	−1		
20	27	23	19	16	12	9	6	3	0	
21	34	30	26	22	18	15	11	8	5	2
22	42	38	33	29	25	21	17	14	10	7
23	51	46	41	37	32	28	24	20	16	13

6.4 A Property of Circumscribing

In connection with the case (i) of the preceding section a certain statement about the possibility to circumscribe single coronoids is useful. With regard to benzenoids it is known (Brunvoll, Cyvin BN and Cyvin 1993b): for every benzenoid formula $C_N H_S$ there exists (at least) one reproducible benzenoid $B \in \{C_N H_S\}$ (cf. also Sect. 6.3).

Proposition 6.1: For every single coronoid formula $C_n H_s$ there exists (at least) one single coronoid $C \in \{C_n H_s\}$ which is reproducible.

It is recalled that a coronoid C is reproducible when C can be circumscribed an unlimited number of times (Definition 5.3).

The following exposition demonstrates Proposition 6.1, although it can hardly be accepted as a rigorous proof. Suppose that one extreme single coronoid, E, for every relevant $C_n H_s$ formula according to the spiral walk (cf. Par. 5.6.3, especially Fig 5.6) has been generated. These E systems, which clearly are reproducible, represent one isomer of each formula at the top of the columns in the table of formulas as displayed in Table 5.3. Representatives down the columns are obtained by annelating catacondensed fragments to the E systems. Suppose that a single linear chain of a hexagons is annelated to each E at a free edge sufficiently far from the last added hexagon (during the spiral— or modified spiral walk) so as to avoid the formation of a latent row of hexagons (Definition 5.4). Then the resulting coronoid system also becomes reproducible, and by varying a all the $C_n H_s$ formulas are covered. Below we show examples for two columns of formulas, one starting with a formula for extremal single coronoids ($C_{40} H_{18}$), the other with a formula for nonextremal extreme single coronoids ($C_{43} H_{19}$).

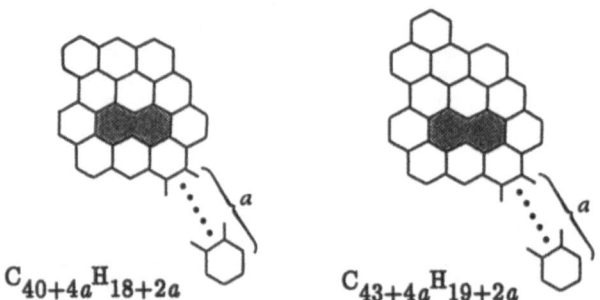

$$C_{40+4a} H_{18+2a} \qquad\qquad C_{43+4a} H_{19+2a}$$

From the above exposition we can give a statement slightly sharper than Proposition 6.1:

Proposition 6.1': For every single coronoid formula $C_n H_s$ there exists (at least) one naphthalenic single coronoid $C \in \{C_n H_s\}$ which is reproducible.

6.5 Building–Up

6.5.1 *Introduction and Basic Concepts*

It is well known (and obvious) that all single coronoids ($h > 8$) can be generated successively by additions of one hexagon each time in all possible positions when starting from the primitive single coronoids (Vol. I–2.1.1). Under this process, referred to as the building–up, it must be allowed for additions to the inner perimeter (if possible) as well as to the outer perimeter. If one wishes to generate all single coronoids only by additions of hexagons to the outer perimeter, one must start from the basic single coronoids (Vol. I–8.2.2, 8.3), which include all the primitive coronoids, but also some non–primitive (with inside features). The corresponding principle for benzenoids is simpler since a benzenoid has only one perimeter. It has been stated in the form (Cyvin SJ, Cyvin and Brunvoll 1993e): "It is well known (and obvious) that all benzenoids with $h + 1$ hexagons can be generated by additions of one hexagon each time in all the possible positions to all benzenoids with h hexagons." This formulation is adopted to single coronoids in the following principle.

Principle 6.1: All single coronoids with $h + 1$ hexagons are obtained by (a) adding one hexagon each time in all the possible positions to all single coronoids with h hexagons and (b) including all the primitive single coronoids with $h + 1$ hexagons.

Altogether there are five types of addition of hexagons (cf., e.g., Vol. I–2.1.1, especially I–Fig.2.2). A survey is given in Table 4. The designations for modes of hexagons and formations of the perimeters (free edge, fissure, etc.) are explained elsewhere (Cyvin SJ and Gutman 1988; Vol. I–3.2.1, especially I–Fig.3.1).

Table 6.4. Characterization of the additions of hexagons.

	Designation	Description*	Acquired mode	Δn_i	$C_{\Delta n} H_{\Delta s}$
(i)	one–contact	fusion (or annelation) to a free edge	L_1	0	$C_4 H_2$
(ii)	two–contact	filling a fissure	P_2	1	$C_3 H_1$
(iii)	three–contact	embedding in a bay	L_3	2	$C_2 H_0$
(iv)	four–contact	covering a cove	P_4	3	$C_1 H_{-1}$
(v)	five–contact	immersing in a fjord (fiord)	L_5	4	$C_0 H_{-2}$

* Cyvin SJ, Cyvin BN, Brunvoll J (1993). Topics in Current Chemistry 166: 65

166

6.5.2 *Fundamental Building–Up Principle*

A principle for the building–up of single coronoids, which is stronger than Principle 6.1, shall presently be referred to as the *fundamental building–up principle*. It is formulated in the following.

Principle 6.2: All single coronoids with $h + 1$ hexagons are obtained by (a) executing a one–, two– or three–contact addition of one hexagon each time in all positions to all single coronoids with h hexagons and (b) including all the primitive single coronoids with $h + 1$ hexagons.

The corresponding strengthening of the building–up principle for benzenoids quoted above (Par. 6.5.1) has been formulated elsewhere as a theorem (Cyvin SJ, Cyvin and Brunvoll 1993e).

In order to demonstrate Principle 6.2 it should firstly be ascertained that every single coronoid has a free edge. In fact, it has several free edges, at least six on the outer perimeter, but this is not crucial for our demonstration. Altogether a free edge occurs in the hexagon modes L_1, A_2, P_2 and L_3. A non–primitive coronoid cannot have all its free edges in A_2–mode hexagons. Therefore a free edge in one of the modes L_1, P_2 or L_3 must be present, and these modes are only found in outside and inside features (Vol.I–8.2.1); recall that a primitive coronoid consists exclusively of L_2– and A_2–mode hexagons. Consequently, an L_1–, P_2– or L_3 hexagon can be removed from a non–primitive single coronoid without opening the corona hole, leaving behind another (smaller) single coronoid. In this way it must be possible to tear down successively any single non–primitive coronoid to a primitive coronoid. It is implied that one L_1–, P_2– or L_3–mode hexagon is removed every time in this tearing–down process, which is opposite to the

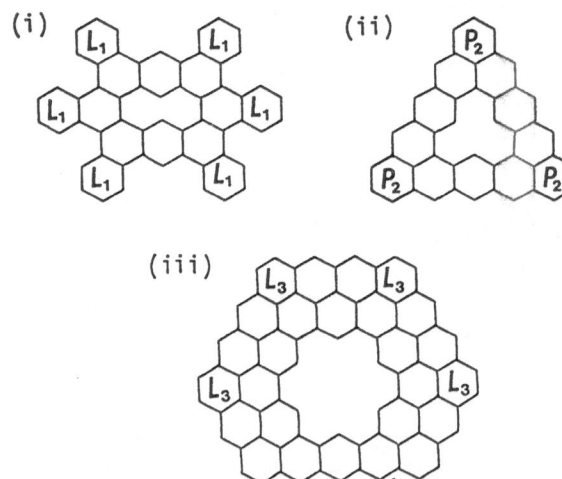

(i) (ii) (iii)

Fig. 6.4. Examples of
(i) L_1–added–
(ii) P_2–added– and
(iii) L_3–added single coronids.

building–up. Therefore it is also possible to build up a single non–primitive coronoid by adding successively an L_1–, P_2– or L_3 hexagon every time. But such additions are exactly those which have been called one–, two– and three–contact additions, respectively (cf. Table 4).

Principle 6.2 states the sufficiency of one–, two– and three–contact additions in the building–up process. It is easy to prove that these three types also are necessary in general. That is for instance proved by examples where all free edges occur in (i) L_1–, (ii) P_2– or (iii) L_3–mode hexagons exclusively (see Fig. 4). It is clear that the last step in the building–up of the above systems had to employ a (i) one–contact–, (ii) two–contact– and (iii) three–contact addition.

Let the above types of systems be termed L_1–added, P_2–added and L_3–added for (i), (ii) and (iii), respectively. A precise definition is given below.

Definition 6.5: A single coronoid C is L_1–added, P_2–added or L_3–added when all the free edges of C occur in L_1–mode–, P_2–mode– or L_3–mode hexagons, respectively, apart from those free edges which occasionally may be found in A_2.

For instance, any circumscribed single coronoid without internal features is an L_3–added system. It has obviously all its free edges in L_3–mode hexagons. The three classes under consideration do not represent a subdivision of all single coronoids; it is not difficult to construct systems where the last addition is optional between two of the modes L_1, P_2 and L_3, or between all of them. The latter case is illustrated below. The small (thick) arrows indicate sites of addition.

6.5.3 *Addition Units*

The additions of hexagons were described by Dias (1982a; 1983; 1986; 1987; 1989b; 1990c; 1990f) as attachments of certain units, e.g. C_4H_2, C_3H and C_2, as is illustrated in the above diagram. These *addition units*, as we shall call them, are given as $C_{\Delta n}H_{\Delta s}$ in the last column of Table 4. Dias (1989b; 1990c) used the term "aufbau units" in connection with the "aufbau" of benzenoids. This terminology, although having been adopted elsewhere (Cyvin SJ, Cyvin and Brunvoll 1993e) shall not be adopted here. The three units mentioned above, which correspond to the hexagon additions (i), (ii) and (iii) according to the designations in Table 4, were called "elementary aufbau units" by Dias (1989b; 1990c). It is recalled that these additions are crucial in the fundamental building–up principle (Principle 6.2). Dias (1984b; 1989a) considered also the "pyrolytic ring closure concomitant with H_2 loss" (Dias 1989a), which corresponds to our five–contact addition (v) and implies entirely a negative "addition" unit (C_0H_{-2}). Also in the case of the four–contact addition (iv), the "addition" unit (viz. C_1H_{-1}) should be taken in a symbolic sense.

6.5.4 *Application of the Fundamental Building–Up Principle to Isomers*

Principle 6.2 (Par. 6.5.2) is readily applicable to the generation of C_nH_s isomers of single coronoids; cf. Cyvin SJ (1992c), and Cyvin SJ, Cyvin and Brunvoll (1993e) on the analogous rules for benzenoids.

Principle 6.3: All pericondensed ($n_i > 0$) single coronoid C_nH_s isomers are generated by attaching (i) one C_4H_2 unit to all $C_{n-4}H_{s-2}$ isomers, (ii) one C_3H unit to all $C_{n-3}H_{s-1}$ isomers and (iii) one C_2 unit to all $C_{n-2}H_s$ isomers in all possible positions.

It should be understood that if one or two of the classes $C_{n-4}H_{s-2}$, $C_{n-3}H_{s-1}$ and $C_{n-2}H_s$ do not exist, then the remaining one or two classes are sufficient for the generation.

According to Principle 6.2 the (i) one–, (ii) two– and (iii) three–contact additions should be applied to those isomers which have one hexagon less than in C_nH_s, and (i) the same number of internal vertices, (ii) one internal vertex less and (iii) two internal vertices less. These classes are exactly those of the $C_{n-4}H_{s-2}$, $C_{n-3}H_{s-1}$ and $C_{n-2}H_s$ isomers, respectively. It is recalled that the attachment of a (i) C_4H_2–, (ii) C_3H– or (iii) C_2 unit is equivalent to a (i) one–, (ii) two– or (iii) three–contact addition, respectively.

Principle 6.3 is illustrated in Fig. 5.

6.5.5 *Simplified Building–Up Principles*

A single coronoid formula C_nH_s may be situated in such a place of Table 5.3 (and its extension) that one or two of the formulas $C_{n-4}H_{s-2}$, $C_{n-3}H_{s-1}$ and $C_{n-2}H_s$ do not exist. These special cases lead to simplifications of the building–up process (cf. also the comment below

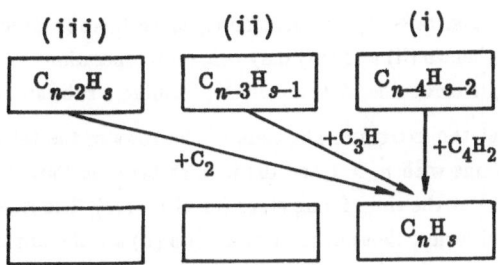

Fig. 6.5. Illustration of the fundamental building—up principle applied to isomers (Principle 6.3). The formulas are arranged as in Table 5.3.

Principle 6.3). Corresponding simplifications have been discussed for benzenoids by Dias (for references, see Par. 6.5.3) and, more systematically, elsewhere (Cyvin SJ, Cyvin and Brunvoll 1993e).

   Catacondensed Systems. The catacondensed single coronoids fall outside the scope of Principle 6.3. Instead, we have the following rules.

Principle 6.4: All catacondensed ($n_i = 0$) single coronoid $C_n H_s$ isomers are obtained by (a) attaching one $C_4 H_2$ unit to all $C_{n-4} H_{s-2}$ in all possible positions and (b) including all the primitive single coronoid $C_n H_s$ isomers.

This principle implies that the one—contact additions (i) are sufficient for generating all the non—primitive catacondensed single coronoids with $h + 1$ hexagons from the catacondensed single coronoids with h hexagons. This property is clearly sound since the number of internal vertices (n_i) should not be allowed to increase during the additions. The same property may also be inferred from the positions of the pertinent $C_n H_s$ formulas in Table 5.3 (first formula column). The only thinkable scheme of addition is clearly $C_{n-4} H_{s-2} \rightarrow C_n H_s$, which corresponds to the last column in Fig. 5. All the $h + 1$ systems under consideration (non—primitive catacondensed single coronoids) are L_1—added.

   Systems with One Internal Vertex. The $C_n H_s$ formulas for the single coronoids with $n_i = 1$ are found in the second formula column of Table 5.3. Here $C_{35} H_{17}$ has a unique position. Its isomers can only be generated by the attachments of $C_3 H$ according to $C_{32} H_{16} \rightarrow C_{35} H_{17}$. The $C_{35} H_{17}$ single coronoids are P_2—added.

   Otherwise ($h > 9$) the two schemes listed as (i) and (ii) are sufficient in order to generate the single coronoids with $n_i = 1$: $C_{n-4} H_{s-2} \rightarrow C_n H_s$; $C_{n-3} H_{s-1} \rightarrow C_n H_s$ (cf. Fig. 5).

   Extreme Systems. The extreme single coronoids, E (Sect. 5.6), have formulas (n^e; s^e) situated at the tops of the columns of the table of formulas (cf. Fig. 5.8). Then it is obvious that an E coronoid cannot be generated by the scheme (i) since there is no formula right above (n^e; s^e). Therefore only the schemes (ii) and (iii) may come into operation.

The extremal single coronoids, A (Par. 5.5.1), forming a subclass of the extreme systems, have formulas (n^a; s^a) at the extreme right ends of the rows of the table of formulas. Here we shall consider the A systems with $n_i > 1$ and distinguish between those with formulas (a) at the rim of short steps and (b) at the rim of long steps (cf. Sect. 5.7). The isomers of the former class (a) are the perfect ground forms; those of the latter class (b) are the imperfect ground forms and circumextremal systems, perfect or imperfect. Figure 3 provides a comprehensive illustration of these features. From the positions of the formulas for the class (b) coronoids it is inferred that these isomers are generated exclusively by the scheme (iii). There are no formulas in the places which would be required for additions according to the schemes (i) and (ii):

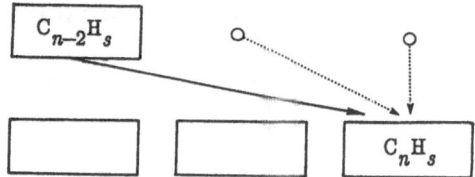

6.5.6 Formations Available for Addition

Here the formations free edge, fissure, bay, cove and fjord of single coronoids C are considered.

Definition 6.6: A free edge, fissure, bay, cove or fjord of C is said to be *available for addition* (of a hexagon) if a hexagon can be fused to the free edge, made to fill the fissure, embedded in the bay, made to cover the cove or immersed in the fjord of C, respectively, so that a new single coronoid is generated.

Every single coronoid (C) has free edges available for addition (viz. fusion), a fact in consistency with the discussion of Par. 6.5.2. An analogous rule does not exist among the other formations in question.

On the other hand, a free edge of C is not always available for addition of a hexagon. The left–hand drawing in the below diagram shows (by heavy lines) one inner free edge and two outer free edges which are not available for addition. Similarly, the existence of fissures (inner and outer) not available for addition is demonstrated by the middle drawing of the below diagram. Finally, the right–hand drawing exemplifies one inner and one outer bay not available for addition.

The situation for coves is different: every cove (inner or outer) of C is evidently available for addition of a hexagon. A fjord of C is always available for addition except for the inner fjords of a naphthalene hole; it is not allowed to immerse a hexagon into a fjord of a naphthalene hole of C when a new coronoid is supposed to be generated.

6.5.7 Further Deductions

The following rule should be clear in view of the discussion of Par. 6.5.5.

Observation 6.1: If a $C_n H_s$ single coronoid C has a hexagon in one of the addition modes, viz. L_1, P_2, L_3, P_4 or L_5, then there must exist (at least) one single coronoid isomer C' with the formula $C_{n-4}H_{s-2}$, $C_{n-3}H_{s-1}$, $C_{n-2}H_s$, $C_{n-1}H_{s+1}$ or $C_n H_{s+2}$, respectively.

This rule is explained by the fact that C is generated by the appropriate addition of one hexagon to C'. The addition modes are specified and characterized in Table 4. The following illustration, actually an extension of Fig. 5, is relevant.

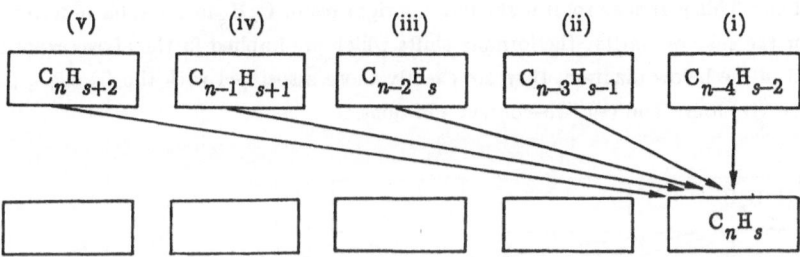

We can say more about C' with the formulas given in Observation 6.1. There must exist such an isomer which possesses a fissure, bay, cove or fjord available for addition in the cases $C_{n-3}H_{s-1}$, $C_{n-2}H_s$, $C_{n-1}H_{s+1}$ and $C_n H_{s+2}$, respectively. With regard to a coronoid C' which has the formula $C_{n-4}H_{s-2}$ (also encountered in Observation 6.1) one can only say that such an isomer must exist with a free edge available for fusion, but this is a property common to all single coronoids, as is stated in Par. 6.5.6.

A corollary of Observation 6.1 is formulated below.

Property 6.1: An extreme single coronoid E has never an L_1—mode hexagon.

If E $\in$ $\{C_nH_s\}$ had an L_1—mode hexagon, then there would exist a single coronoid with the formula $C_{n-4}H_{s-2}$, right above C_nH_s in the table of formulas — a contradiction. Property 6.1 is in particular possessed by the extremal single coronoids A, which form a subclass of the E coronoids. In Par. 6.5.5 it was distinguished between A systems (a) at the rim of short steps and (b) at the rim of long steps. For the latter category we can formulate another corollary of Observation 6.1 as below.

Property 6.2: An imperfect ground form or a circumextremal single coronoid, $A^{(b)}$, has neither an L_1— or P_2—mode hexagon.

If $A^{(b)}$ $\in$ $\{C_nH_s\}$ had an L_1— or P_2—mode hexagon, then there would exist a single coronoid with the formula $C_{n-4}H_{s-2}$ or $C_{n-3}H_{s-1}$, respectively, but the corresponding sites in the table of formulas are vacant — again a contradiction.

Observation 6.2: If a C_nH_s single coronoid C has a fissure, bay, cove or fjord available for addition, then there must exist (at least) one single coronoid isomer C' with the formula $C_{n+3}H_{s+1}$, $C_{n+2}H_s$, $C_{n+1}H_{s-1}$ or C_nH_{s-2}, respectively.

Here an appropriate C' isomer is obtained from C by filling the fissure, embedding a hexagon in the bay, covering the cove, or immersing a hexagon into the fjord, respectively.

Since every single coronoid (C) has a free edge available for fusion it is inferred (in addition to Observation 6.2) that for every C_nH_s formula there exists a $C_{n+4}H_{s+2}$ formula for these systems. This is indeed so; it is the formula right below C_nH_s in the table of formulas.

For the sake of clarity, the formula shifts which are implied in the above exposition, are illustrated in the below diagram; they are exactly those associated with the (i) one—, (ii) two—, (iii) three—, (iv) four— and (v) five—contact additions.

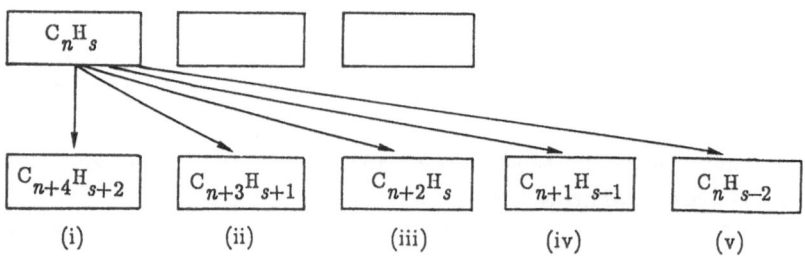

Also with the aid of Observation 6.2 several corollaries may be deduced. Some of them are given below and may be explained in the same way as Property 6.1 and Property 6.2.

Property 6.3: An extremal single coronoid has never a cove and never a fjord in addition to the inner fjords of its naphthalene hole.

This property is sharper than Property 5.1 of Par. 5.5.1.

Property 6.4: A nonextremal extreme single coronoid X has never a fjord except for the inner fjords of its naphthalene hole if the system is naphthalenic.

If X has a formula in a (long) step which is immediately followed by a short step of the staircase boundary, then X is also void of coves. This situation occurs for the constellations

$$\begin{matrix} \text{X} \text{Ⓐ} \\ \bullet \;\; \bullet \;\; \text{A} \end{matrix}$$

in the symbolism of Fig. 5.8. In order to have a possibility for an X coronoid with a cove one must look for two consecutive long steps like:

$$\begin{matrix} \text{X A} \\ \bullet \;\; \bullet \;\; \text{X} \text{Ⓐ} \end{matrix} \qquad \text{or} \qquad \begin{matrix} \text{X A} \\ \bullet \;\; \bullet \;\; \text{X A} \end{matrix}$$

It is seen (Table 5.3 and Fig. 5.8) that $X \in \{C_{53}H_{21}\}$ are the smallest ($h = 16$) X isomers where coves possibly may occur. Then comes $X \in \{C_{65}H_{23}\}$ ($h = 21$) and $X \in \{C_{72}H_{24}\}$ ($h = 24$). These coronoid isomers are at the same time those where a phenalene hole, which implies inner coves, cannot be excluded. Indeed, such isomers with phenalene holes can be constructed, as is exemplified by Fig. 5.7. Naphthalenic single coronoid isomers with the same formulas and outer coves are also possible:

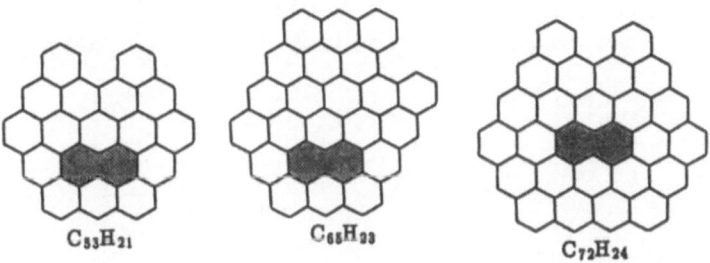

$$C_{53}H_{21} \qquad\qquad C_{65}H_{23} \qquad\qquad C_{72}H_{24}$$

In Sect. 5.7 the shape of the staircase boundary of the table of formulas was discussed. It was inferred that it consists exclusively of short (one–formula) steps and long (two–formula) steps. Here we shall give alternative demonstrations for the fact that there cannot be any "extra long steps" or "high steps". Suppose there was an extra long (three–formula) step as illustrated below.

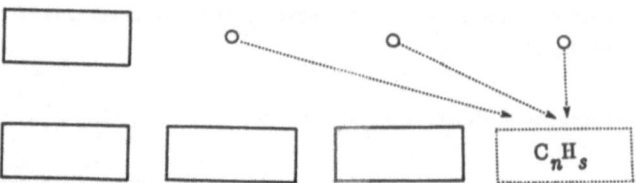

How could an isomer with the formula in the dotted frame possibly be generated? All the positions for formulas of which at least one would be required, are vacant. Suppose now there was a high step as:

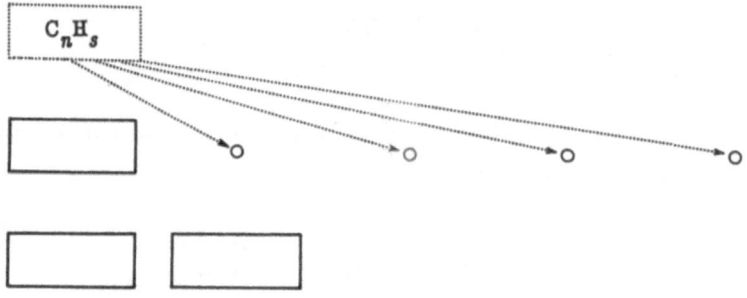

Then any isomer with the formula C_nH_s in the dotted frame would be void of fissures, bays, coves and fjords available for addition. If one of these formations were present, then there would exist a formula in one of the sites which actually are empty as indicated. But a single coronoid (or a benzenoid) as characterized above, is not thinkable.

SUPPLEMENT

Introductory Remarks

In this supplement, some of the most important formulas and definitions of classes of single coronoids from Chapters 4–6 are summarized. In a conclusion, this material is placed in a broader context.

Summary of Formulas and Definitions

A single coronoid has the *chemical formula* $C_n H_s \equiv (n;\ s)$ given by

$$(5.1) \qquad (n;\ s) = (4h - n_i;\ 2h - n_i)$$

in terms of the h number of hexagons and n_i number of internal vertices. Conversely,

$$(5.2) \qquad (h,\ n_i) = (\tfrac{1}{2}(n{-}s),\ n{-}2s)$$

For a given n, the possible values of s obey

$$(5.3) \qquad 2 - n + 2\lceil (1/2)(n + 6^{1/2} n^{1/2}) \rceil \leq s \leq n - 2\lceil n/4 \rceil$$

where $n = 32, 35, 36, 37, 38, \dots$, and the coefficients $(n,\ s)$ should have the same parity. For a given s, the possible values of n obey

$$(5.6) \qquad s + 2\lceil s/2 \rceil \leq n \leq s + 2\lfloor (1/12)(s^2 - 10s + 4) \rfloor$$

where $s = 16, 17, 18, 19, \dots$, and the coefficients $(n,\ s)$ should have the same parity.

An *extremal single coronoid* $A(n^a;\ s^a)$ has by definition $n_i = (n_i)_{max}$ for a given h. One has

$$(4.21) \qquad (n_i)_{max} = 2h - 5 - \lceil (12h + 21)^{1/2} \rceil$$

where $h = 8, 9, 10, 11, \dots$. The formula of A is given by

$$(5.53) \qquad (n^a;\ s^a) = (2h + 5 + \lceil (12h + 21)^{1/2} \rceil;\ 5 + \lceil (12h + 21)^{1/2} \rceil)$$

A *circular single coronoid* $O(n^o; s^o)$ has by definition $h = h_{max}$ for a given s. One has

(4.44) $h_{max} = \lfloor (1/12)(s^2 - 10s + 4) \rfloor$

where $s = 16, 17, 18, 19, \ldots$. The formula of O is given by: (i) in the *Harary–Harborth picture*

(5.62) $(n^o; s^o) = (s + 2\lfloor (1/12)(s^2 - 10s + 4)\rfloor; s)$

where $s\ (= s^o) = 16, 17, 18, 19, \ldots$; (ii) in the *Balaban picture*

(5.67) $(n^o; s^o) = (6k^2 + 24(k+1) + (2k+5)\epsilon - 2\lceil \epsilon/6 \rceil; 6k + 14 + \epsilon)$

where (when two degenerate coronoids $C_{24}H_{14}$ and $C_{27}H_{15}$ are included) $k = 0, 1, 2, 3, \ldots$, and $\epsilon = 0, 1, 2, 3, 4, 5$; (iii) in the *new picture*, $n^o(x) = \sum\limits_{s=16}^{\infty} n_s^o x^s$, where n^o as a function of s is labeled n_s^o. It was found

(5.76) $n^o(x) = x^{14}(1-x)^{-1}[27x^2 - 5 + (1 + x^2 + x^3 + x^4 + x^5 + x^7)(5 - 3x^6)(1 - x^6)^{-2}]$

$$= \frac{x^{16}(32 + 5x + 5x^2 + 5x^3 + 7x^4 + 5x^5 - 57x^6 - 3x^7 - 3x^8 - 3x^9 - 5x^{10} - 3x^{11} + 27x^{12})}{(1-x)(1-x^6)^2}$$

All extremal single coronoids A are naphthalenic. The circular single coronoids O are also extremal; they form a subclass of the A systems. Some expressions for circumscribing naphthalenic single coronoids, C, are given below.

Assume that $C(n;s)$ is a naphthalenic single coronoid. If C can be circumscribed and $C_1(n_1; s_1) = \text{circum-}C(n;s)$, then

(5.51) $(n_1; s_1) = (n + 2s + 2; s + 6)$

where $s = 16, 17, 18, 19, \ldots$. In generalized form, assume that C can be k–fold circumscribed and $C_k(n_k; s_k) = k\text{-circum-}C(n;s)$. Then

(5.52) $(n_k; s_k) = (n + 2ks + 6k^2 - 4k; s + 6k)$

where $s = 16, 17, 18, 19, \ldots$.

A *circumextremal single coronoid* $P(n^P;s^P)$ is a circumscribed extremal single coronoid, viz. circum–A, or a C_nH_s isomer of circum–A. An expression for $(n^P; s^P)$ is given in eqn. (5.57).

A *ground form* $G(n_0;s_0)$ of single coronoids is an extremal single coronoid which is not circumextremal. The formula $(n_0; s_0)$ is the first term in a *sequence of associated formulas*, $(n_0; s_0)$, $(n_1; s_1)$, ... , $(n_k; s_k)$, ... Here any formula $(n_k; s_k)$ is obtained from eqn. (5.52) given above, where $s = s_0$. The formulas $(n_k; s_k)$ for $k > 0$ pertain to *higher members* $G_k(n_k;s_k)$ of single coronoids, which coincide exactly with the circumextremal single coronoids P. A ground form G may be: (a) *perfect ground form* G^P of single coronoids, which is a naphthalenic perforated ground form benzenoid; (b) *imperfect ground form* G^i of single coronoids, which is a naphthalenic perforated higher member benzenoid.

(a) A perfect ground form $G^P(n_0^{pq};s_0^{pq})$ of single coronoids has the formula

$$(6.3) \qquad (n_0^{pq}; s_0^{pq}) = (4 - s + 2\lfloor(1/12)(s^2 + 3)\rfloor; \; s)$$

where s $(= s_0^{pq})$ = 17, 18, 19, 20, ... This corresponds to the Harary–Harborth picture. Expressions for $(n_0^{pq}; s_0^{pq})$ in the Balaban picture and in the new picture have also been deduced and are given in eqn. (6.7) and (6.13), respectively.

(b) There are exactly eight formulas for imperfect ground forms G^i of single coronoids, viz.: $C_{32}H_{16}$, $C_{37}H_{17}$, $C_{42}H_{18}$, $C_{47}H_{19}$, $C_{52}H_{20}$, $C_{54}H_{20}$, $C_{59}H_{21}$, $C_{64}H_{22}$.

Conclusion

The treatment of single coronoid which is summarized above, follows a pattern which was developed for benzenoids (Cyvin SJ and Brunvoll 1991; Cyvin SJ 1991c; 1992c; Brunvoll, Cyvin BN and Cyvin 1992b; Cyvin SJ, Cyvin and Brunvoll 1993e). The same pattern was also adapted to: all–benzenoids (Cyvin SJ 1991a; Cyvin BN, Brunvoll and Cyvin 1992a); fluoranthenoids/ fluorenoids (Cyvin SJ 1992b; Cyvin SJ, Cyvin and Brunvoll 1993b); biphenylenoids (Cyvin SJ, Cyvin and Brunvoll 1993a); helicenes (Cyvin BN, Guo et al. 1992; Cyvin SJ, Zhang, Cyvin and Guo 1993; Cyvin SJ and Cyvin 1993). Some general formulations for mono–q–polyhexes with $q \leq$ 6 are available (Cyvin SJ, Cyvin and Brunvoll 1993a; Cyvin BN, Cyvin and Brunvoll 1993). It is reasonable to expect that similar approaches could be applied to further classes of polygonal systems. Finally we wish to mention that Hansen and Zheng (1993a) recently have published a relevant work on best possible bounds of graph invariants under a rather general scope.

Chapter 7

GENERATION AND ENUMERATION

OF SINGLE CORONOID ISOMERS

7.1 Tabulation of Complete Data

7.1.1 Introduction and Survey

The present chapter deals with the numbers of C_nH_s single coronoid isomers (cardinalities of genus 1 of the C_nH_s formulas, $|C_nH_s|_1$; cf. Sect. 3.8). Complete data of such isomers are known for $8 \leq h \leq 14$. This is to understand that the $|C_nH_s|_1$ values are known for all C_nH_s formulas which are compatible with the given h values. These numbers of isomers, classified according to the Kekuléan and non–Kekuléan systems, have been given for $8 \leq h \leq 11$ by Cyvin SJ and Brunvoll (1989), and for $12 \leq h \leq 14$ by Cyvin SJ, Brunvoll and Cyvin (1991a). They are summarized in Table 1. The last reference (Cyvin SJ, Brunvoll and Cyvin 1991a) contains also $|C_{50}H_{20}|_1 = 51$, $|C_{52}H_{20}|_1 = 12$ and $|C_{54}H_{20}|_1 = 2$, but these numbers represent incomplete data for $h = 15$, 16 and 17, respectively.

7.1.2 Classifications According to neo and the Color Excess

Table 2 shows the numbers of single coronoid isomers according to the neo classification (n normal; e essentially disconnected; o non–Kekuléan). Furthermore, the non–Kekuléan systems, o, are classified according to their color excess (Δ). The table is arranged in a way which was found suitable for benzenoid isomers (Brunvoll, Cyvin BN and Cyvin 1992b). All the numbers in this table, although they have not been given before explicitly, can be deduced from the data of Volume I, Tables I–9.1 to I–9.8.

7.1.3 Classification According to the Corona Holes

The pair of invariants (h^o, n_i^o) pertain to the corona hole (considered as a benzenoid) of a single coronoid C, while (h, n_i) are the corresponding invariants of C. For a given h value, the range of h^o values was found to be – Vol. I–eqn. (9.8)

$$2 \leq h^o \leq \lfloor (1/12)(h^2 - 6h) \rfloor + 1 \tag{7.1}$$

Table 7.1. Numbers of single coronoid isomers: complete data for $8 \leq h \leq 14$.

h	n_i	Formula	Kekuléan	non—Kekuléan	Total
8	0	$C_{32}H_{16}$	1^a	0	1^a
9	0	$C_{36}H_{18}$	3^a	0	3^a
	1	$C_{35}H_{17}$	0	2^b	2^b
10	0	$C_{40}H_{20}$	$15^{c,d}$	0	$15^{c,d}$
	1	$C_{39}H_{19}$	0	15^b	15^b
	2	$C_{38}H_{18}$	9^b	3^b	12^b
	3	$C_{37}H_{17}$	0	1^b	1^b
11	0	$C_{44}H_{22}$	$62^{c,d}$	0	$62^{c,d}$
	1	$C_{43}H_{21}$	0	96^b	96^b
	2	$C_{42}H_{20}$	61^b	21^b	82^b
	3	$C_{41}H_{19}$	0	36^b	36^b
	4	$C_{40}H_{18}$	5^b	2^b	7^b
12	0	$C_{48}H_{24}$	312^d	0	312^d
	1	$C_{47}H_{23}$	0	552^e	552^e
	2	$C_{46}H_{22}$	436^e	144^e	580^e
	3	$C_{45}H_{21}$	0	329^e	329^e
	4	$C_{44}H_{20}$	104^e	38^e	142^e
	5	$C_{43}H_{19}$	0	37^e	37^e
	6	$C_{42}H_{18}$	2^e	0	2^e
13	0	$C_{52}H_{26}$	1435^d	0	1435^d
	1	$C_{51}H_{25}$	0	2986^e	2986^e
	2	$C_{50}H_{24}$	2593^e	900^e	3493^e
	3	$C_{49}H_{23}$	0	2487^e	2487^e
	4	$C_{48}H_{22}$	912^e	354^e	1266^e
	5	$C_{47}H_{21}$	0	521^e	521^e
	6	$C_{46}H_{20}$	110^e	48^e	158^e
	7	$C_{45}H_{19}$	0	17^e	17^e

Table 7.1 (continued).

h	n_i	Formula	Kekuléan	non–Kekuléan	Total
14	0	$C_{56}H_{28}$	6785^f	0	6785^f
	1	$C_{55}H_{27}$	0	15500^e	15500^e
	2	$C_{54}H_{26}$	14779^e	5318^e	20097^e
	3	$C_{53}H_{25}$	0	16469^e	16469^e
	4	$C_{52}H_{24}$	7050^e	2928^e	9978^e
	5	$C_{51}H_{23}$	0	4777^e	4777^e
	6	$C_{50}H_{22}$	1359^e	610^e	1969^e
	7	$C_{49}H_{21}$	0	590^e	590^e
	8	$C_{48}H_{20}$	81^e	33^e	114^e
	9	$C_{47}H_{19}$	0	4^e	4^e

[a] Dias JR (1982). J Chem Inf Comput Sci 22:15; wrong numbers therein are omitted.
[b] Cyvin SJ, Brunvoll J (1989). Chem Phys Letters 164: 635
[c] Brunvoll J, Cyvin BN, Cyvin SJ (1987). J Chem Inf Comput Sci 27: 14
[d] Balaban AT, Brunvoll J, Cioslowski J, Cyvin BN, Cyvin SJ, Gutman I, He WC, He WJ, Knop JV, Kovačević M, Müller WR, Szymanski K, Tošić R, Trinajstić N (1987). Z Naturforsch 42a: 863
[e] Cyvin SJ, Brunvoll J, Cyvin BN (1991). J Math Chem 8: 63
[f] He WJ, He WC, Wang QX, Brunvoll J, Cyvin SJ (1988). Z Naturforsch 43a: 693

However, all benzenoids with the h^o values prescribed above need not be inspected (for $h > 8$). Take for instance $h = 12$, for which the maximum h^o value is 7 in accordance with eqn. (1). Here only one out of the 331 benzenoids with seven hexagons is of interest, viz. coronene. The corresponding corona hole ($h^o = 7$, $n_i^o = 6$) occurs in kekulene ($h = 12$), while all single coronoids with other $h^o = 7$ corona holes are larger ($h > 12$).

In order to shed some light on the above problem, introduce

$$h^* = 2h^o - n_i^o + 4 \tag{7.2}$$

Then, according to the last equation in Volume I, I–(9.9), a primitive single coronoid around the hole (h^o, n_i^o) has $h = h^*$ (and $n_i = 0$). This of course presupposes that the corona hole benzenoid, which is characterized by (h^o, n_i^o), can be circumscribed. The following restrictions are valid for primitive single coronoids.

$$h^o \geq h - 2 - \lfloor h/2 \rfloor \quad , \quad n_i^o \geq h - 2\lfloor h/2 \rfloor \tag{7.3}$$

Table 7.2. Numbers of single coronoid isomers, classified according to *neo* and Δ values: complete data.*

h	n_i	Formula	$\Delta = 0$		o (non–Kekuléan)			
			n	e	$\Delta=1$	2	3	4
8	0	$C_{32}H_{16}$	1					
9	0	$C_{36}H_{18}$	3					
	1	$C_{35}H_{17}$	0		2			
10	0	$C_{40}H_{20}$	15					
	1	$C_{39}H_{19}$	0		15			
	2	$C_{38}H_{18}$	9		0	3		
	3	$C_{37}H_{17}$	0		1	0		
11	0	$C_{44}H_{22}$	62					
	1	$C_{43}H_{21}$	0		96			
	2	$C_{42}H_{20}$	59	2	0	21		
	3	$C_{41}H_{19}$	0	0	35	0	1	
	4	$C_{40}H_{18}$	5	0	0	2	0	
12	0	$C_{48}H_{24}$	312					
	1	$C_{47}H_{23}$	0		552			
	2	$C_{46}H_{22}$	404	32	0	144		
	3	$C_{45}H_{21}$	0	0	319	0	10	
	4	$C_{44}H_{20}$	97	7	0	38	0	
	5	$C_{43}H_{19}$	0	0	35	0	2	
	6	$C_{42}H_{18}$	2	0	0	0	0	
13	0	$C_{52}H_{26}$	1435					
	1	$C_{51}H_{25}$	0		2986			
	2	$C_{50}H_{24}$	2290	303	0	900		
	3	$C_{49}H_{23}$	0	0	2410	0	77	
	4	$C_{48}H_{22}$	804	108	0	354	0	
	5	$C_{47}H_{21}$	0	0	500	0	21	
	6	$C_{46}H_{20}$	102	8	0	47	0	1
	7	$C_{45}H_{19}$	0	0	17	0	0	0

Table 7.2 (continued)

h	n_i	Formula	$\Delta = 0$		o (non–Kekuléan)			
			n	e	$\Delta=1$	2	3	4
14	0	$C_{56}H_{28}$	6785					
	1	$C_{55}H_{27}$	0		15500			
	2	$C_{54}H_{26}$	12456	2323	0	5318		
	3	$C_{53}H_{25}$	0	0	15894	0	575	
	4	$C_{52}H_{24}$	5728	1322	0	2920	0	8
	5	$C_{51}H_{23}$	0	0	4593	0	184	0
	6	$C_{50}H_{22}$	1212	147	0	604	0	6
	7	$C_{49}H_{21}$	0	0	564	0	26	0
	8	$C_{48}H_{20}$	78	3	0	33	0	0
	9	$C_{47}H_{19}$	0	0	4	0	0	·0

* Abbreviations: e essentially disconnected; n normal; o non–Kekuléan.

Here the lower bound of n_i^o is 0 when h is even and 1 when h is odd. Now let h^o, for the fixed value of h, increase by steps of unity. Simultaneously n_i^o will increase by steps of two. This process is supposed to be continued so that all (h^o, n_i^o) combinations should be included which are compatible with the restriction of eqn. (3.43) or (4.22) with the substitutions $H = h^o$, $N_i = n_i^o$. Example: the $h = 12$ primitive single coronoids have $h^o = 4, 5, 6, 7$ and $n_i^o = 0, 2, 4, 6$, respectively; notice that both combinations (6, 4) and (7, 6) represent extremal benzenoids.

It is also needed to consider corona hole benzenoids which cannot be circumscribed. Such corona holes are associated with the non–primitive basic single coronoids (Vol. I–8, especially I–8.3). The well known smallest non–primitive basic coronoid (I–8.2.2 and I–Fig. 8.2) has $h = 12$. Its corona hole (benzo[c]phenanthrene) has $h^o = 4$, $n_i^o = 0$. Here again $h = h^*$, where h^* is defined in (2), but $n_i = 1$. The next–smallest non–primitive basic coronoid, which has $h = 13$, is also well known (references as above). Its corona hole (pentahelicene) has $h^o = 5$, $n_i^o = 0$. In this case $h = h^* - 1$ and $n_i = 0$. The two smallest non–primitive basic coronoids are depicted in the following (cf. also the top row of I–Fig.8.1).

	h^o	n_i^o	h	Holes of primitive coronoids	Holes of non–primitive (basic) coronoids
①	2	0	8		
②	3	1	9		
③	3	0	10		
④	4	2	10		
⑤	4	1	11		
⑥	5	3	11		
⑦	4	0	12		
⑧	5	2	12		
⑨	6	4	12		
⑩	7	6	12		
⑪	5	0	13		

(cont.)

h^o	n_i^o	h	Holes of primitive coronoids	Holes of non–primitive (basic) coronoids
⑫ 5	1	13		
⑬ 6	3	13		
⑭ 7	5	13		
⑮ 8	7	13		
⑯ 5	0	14		
⑰ 6	1	14		
⑱ 6	2	14		

(cont.)

h^o	n_i^o	h	Holes of primitive coronoids	Holes of non–primitive (basic) coronoids
⑲ 7	4	14	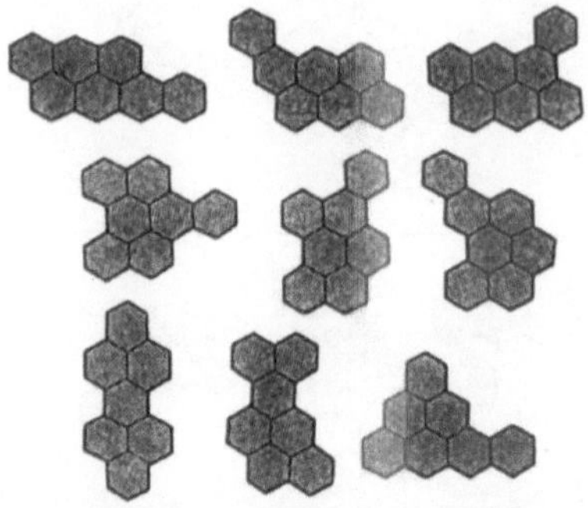	
⑳ 8	6	14	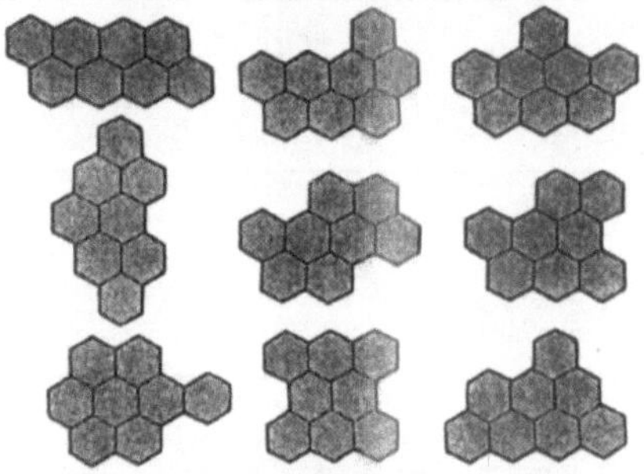	
㉑ 9	8	14	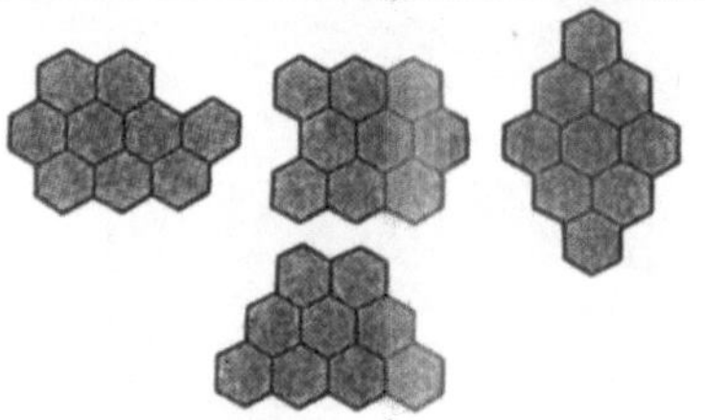	
㉒ 10	10	14	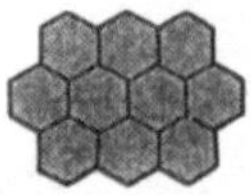	

Fig. 7.1. Corona holes (represented as benzenoids) of basic single coronoids with $h \leq 14$. All these benzenoid forms are found in Brunvoll, Cyvin BN and Cyvin (1992b). Correction: for $C_{28}H_{14}$ (corresponding to $h^o = 8$, $n_i^o = 6$), two forms in the cited reference should be interchanged in order to give the classification (n/o_2) correctly.

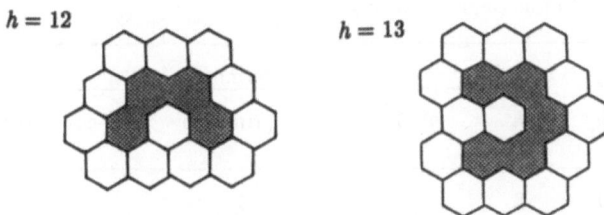

$h = 12$ $h = 13$

These two cases exemplify corona hole benzenoids with a cove and a fjord, respectively. They are representative for all the non–primitive basic single coronoids with $h = 12$, 13 and 14 (I–Fig. 8.2), which are those of the prime interest in this section. For the sake of clarity we repeat the relevant rules in strict formulations. For $12 \leq h \leq 14$ the non–primitive basic single coronoids are of two kinds. (a) The corona hole benzenoid possesses exactly one cove; then $h = h^*$, $n_i = 1$. (b) The corona hole possesses exactly one fjord; then $h = h^* - 1$, $n_i = 0$.

The corona holes, represented as benzenoids, of all basic single coronoids for $8 \leq h \leq 14$ (primitive and non–primitive) are depicted in Fig. 1. The encircled numerals identify the different classes with characteristic combinations of h^o, n_i^o and h. Table 3 includes a listing of the corona holes in question or the appropriate basic single coronoids, which amounts to the same.

The situation is not so simple for the non–primitive basic single coronoids with $h \geq 15$. Already for $h = 15$ there is one instance not covered by the cases (a) and (b) above. It is depicted below (left–hand drawing).

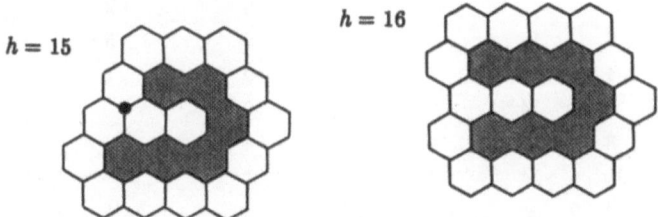

$h = 15$ $h = 16$

Here $h^o = 6$, $n_i^o = 0$. The pertinent corona hole benzenoid has one fjord, but when a hexagon is immersed a cove is created. As a result, one obtains $h = h^* - 1$, $n_i = 1$ (the internal vertex is indicated by a black dot). All the other non–primitive basic single coronoids with $h = 15$ (I–Fig.8.2) follow the schemes (a) and (b). The listing for $h = 15$ is included in Table 3. Another anomalous case is depicted in the right–hand drawing of the above diagram, in which $h^o = 7$, $n_i^o = 0$. The corona hole benzenoid has one fjord, but when a hexagon is immersed another fjord is created. The result is $h = h^* - 2$, $n_i = 0$. These examples show that the usefulness of eqn. (2) is somewhat limited. Still more bizarre cases exist, even with $h > h^*$. This feature is exemplified by the left–hand bottom system of I–Fig.8.1:

Table 7.3. Numbers of basic single coronoids.

Corona hole(s)			h		Number	
$(X)^a$	h^o	n_i^o	h^*	h^*-1	primitive	non-primitive
①	2	0	8		1	
②	3	1	9		1	
③	3	0	10		2	
④	4	2	10		1	
⑤	4	1	11		1	
⑥	5	3	11		1	
⑦	4	0	12		4	1
⑧	5	2	12		3	
⑨	6	4	12		3	
⑩	7	6	12		1	
⑪	5	0		13	0	1
⑫	5	1	13		4	2
⑬	6	3	13		4	
⑭	7	5	13		3	
⑮	8	7	13		1	
⑯	5	0	14		8	3
⑰	6	1		14	0	2
⑱	6	2	14		9	5
⑲	7	4	14		9	1
⑳	8	6	14		9	
㉑	9	8	14		4	
㉒	10	10	14		1	
	6	0		15	0	4
	6	1	15		12	10
	7	2		15	0	8
	7	3	15		14	11
	8	4		15	0	2
	8	5	15		16	5
	9	7	15		14	1
	10	9	15		9	
	11	11	15		2	
	12	13	15		1	

^{a} See Fig. 7.1.

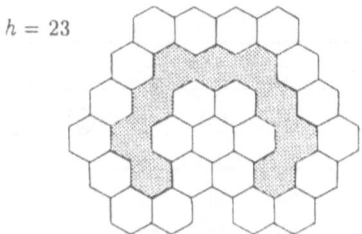

$h = 23$

Here $h^\circ = 9$, $n_i^\circ = 0$. Furthermore, $h = h^* + 1$ (and $n_i = 6$).

The occurrence of corona holes is cumulative in the following sense. All the holes which occur in the basic (primitive and non–primitive) single coronoids with h hexagons each, are also found in single coronoids with more than h hexagons. A detailed account on the numbers of single coronoid isomers with the different classes of holes (identified by encircled numerals), is furnished by Table 4.

Table 7.4. Numbers of single coronoid isomers, classified according to corona holes: complete data.

h	n_i	Formula	Corona hole(s)* ①	②	③	④	⑤	⑥
8	0	$C_{32}H_{16}$	1					
9	0	$C_{36}H_{18}$	2	1				
	1	$C_{35}H_{17}$	2					
10	0	$C_{40}H_{20}$	11	1	2	1		
	1	$C_{39}H_{19}$	13	2				
	2	$C_{38}H_{18}$	12					
	3	$C_{37}H_{17}$	1					
11	0	$C_{44}H_{22}$	45	7	6	2	1	1
	1	$C_{43}H_{21}$	76	10	7	3		
	2	$C_{42}H_{20}$	72	9	1			
	3	$C_{41}H_{19}$	35	1				
	4	$C_{40}H_{18}$	7					

(cont.)

Table 7.4 (continued)

h	n_i	Formula	Corona hole(s)* ①	②	③	④	⑤	⑥	⑦	⑧	⑨	⑩	⑪
12	0	$C_{48}H_{24}$	217	29	34	11	7	3	4	3	3	1	
	1	$C_{47}H_{23}$	412	60	47	17	9	6	1				
	2	$C_{46}H_{22}$	462	58	42	17	1						
	3	$C_{45}H_{21}$	289	32	7	1							
	4	$C_{44}H_{20}$	135	7									
	5	$C_{43}H_{19}$	37										
	6	$C_{42}H_{18}$	2										
13	0	$C_{52}H_{26}$	975	143	161	45	42	18	15	14	8	1	1
	1	$C_{51}H_{25}$	2109	320	294	102	69	38	21	16	14	1	0
	2	$C_{50}H_{24}$	2625	379	285	107	51	33	9	3	1		
	3	$C_{49}H_{23}$	2003	247	164	56	14	3					
	4	$C_{48}H_{22}$	1085	116	52	13							
	5	$C_{47}H_{21}$	480	38	3								
	6	$C_{46}H_{20}$	156	2									
	7	$C_{45}H_{19}$	17										
14	0	$C_{56}H_{28}$	4458	647	797	218	227	85	94	87	48	5	4
	1	$C_{55}H_{27}$	10469	1643	1648	548	444	229	159	129	95	7	5
	2	$C_{54}H_{26}$	14274	2175	1885	703	418	249	161	103	87	9	1
	3	$C_{53}H_{25}$	12426	1738	1334	466	261	135	63	33	12	1	
	4	$C_{52}H_{24}$	7920	965	729	231	93	32	6	2			
	5	$C_{51}H_{23}$	3998	459	237	76	6	1					
	6	$C_{50}H_{22}$	1771	155	38	5							
	7	$C_{49}H_{21}$	569	21									
	8	$C_{48}H_{20}$	114										
	9	$C_{47}H_{19}$	4										

(cont.)

Table 7.4 (continued)

h	n_i	Formula	Corona hole(s)*										
			⑫	⑬	⑭	⑮	⑯	⑰	⑱	⑲	⑳	㉑	㉒
13	0	$C_{52}H_{26}$	4	4	3	1							
	1	$C_{51}H_{25}$	2										
14	0	$C_{56}H_{28}$	27	26	17	3	8	2	9	9	9	4	1
	1	$C_{55}H_{27}$	43	36	29	7	3	0	5	1			
	2	$C_{54}H_{26}$	25	5	2								

* See Fig. 7.1.

7.2 Forms

The actual forms of a number of single coronoid isomers are shown in Fig. 2. All these forms for $h \leq 14$ are found to be consistent with the relevant numbers and classifications of Table 2 and of Table 4. The depictions go somewhat beyond $h = 14$. All the depicted forms for $h \geq 13$ are extremal coronoids.

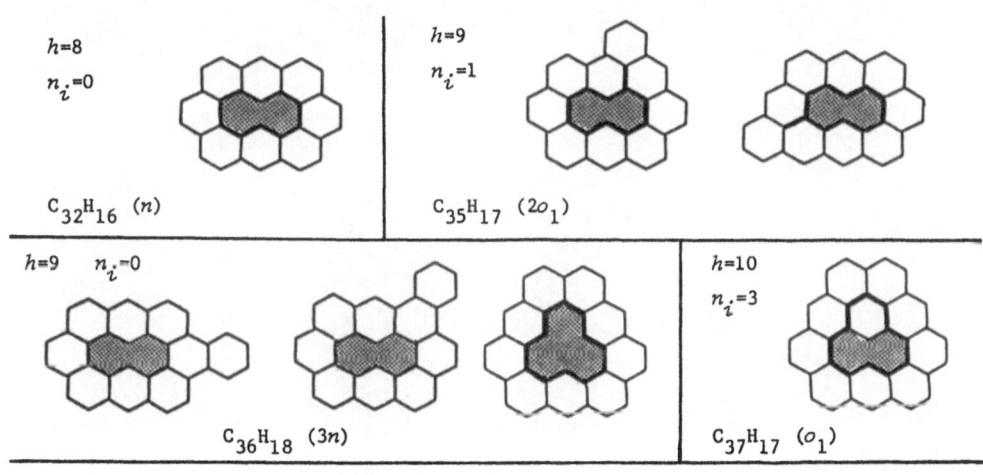

$h=8$ $n_i=0$ $C_{32}H_{16}$ (n)

$h=9$ $n_i=1$ $C_{35}H_{17}$ $(2o_1)$

$h=9$ $n_i=0$ $C_{36}H_{18}$ $(3n)$

$h=10$ $n_i=3$ $C_{37}H_{17}$ (o_1)

(cont.)

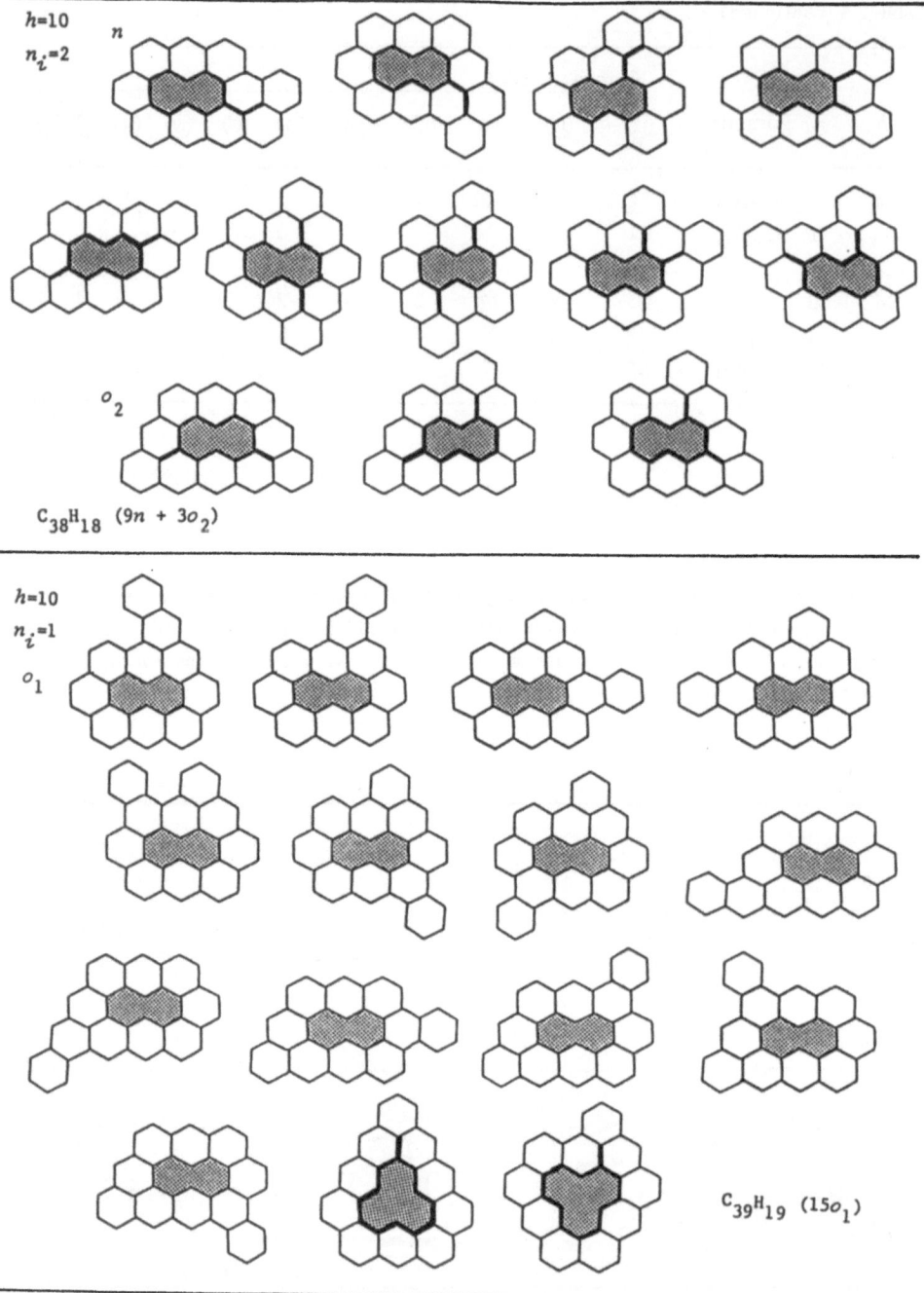

$h=10$
$n_i=2$
n

o_2

$C_{38}H_{18}$ $(9n + 3o_2)$

$h=10$
$n_i=1$
o_1

$C_{39}H_{19}$ $(15o_1)$

(cont.)

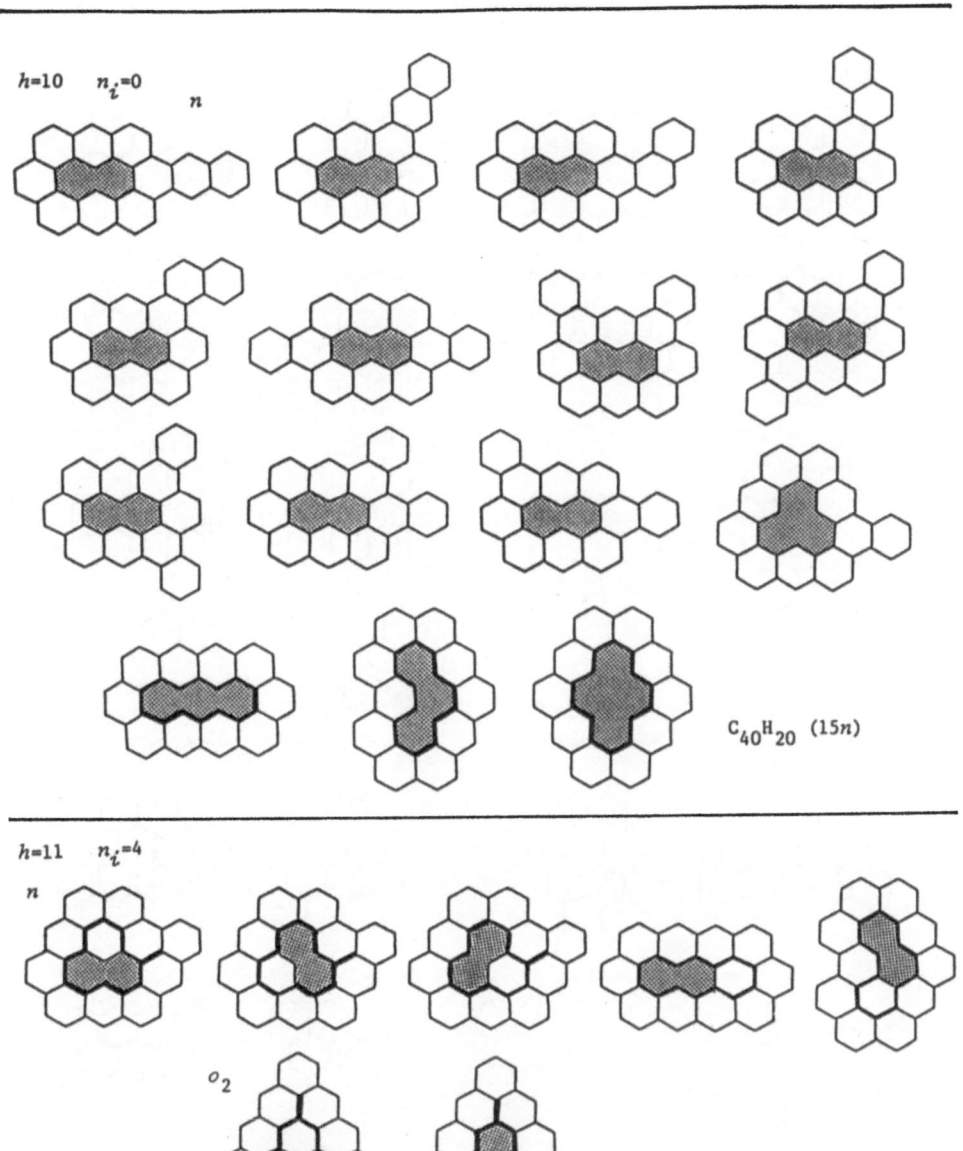

$h=10 \quad n_i=0$

n

$C_{40}H_{20} \quad (15n)$

$h=11 \quad n_i=4$

n

o_2

$C_{40}H_{18} \quad (5n + 2o_2)$

(cont.)

$h=11$ $n_i=3$

o_1

(cont.)

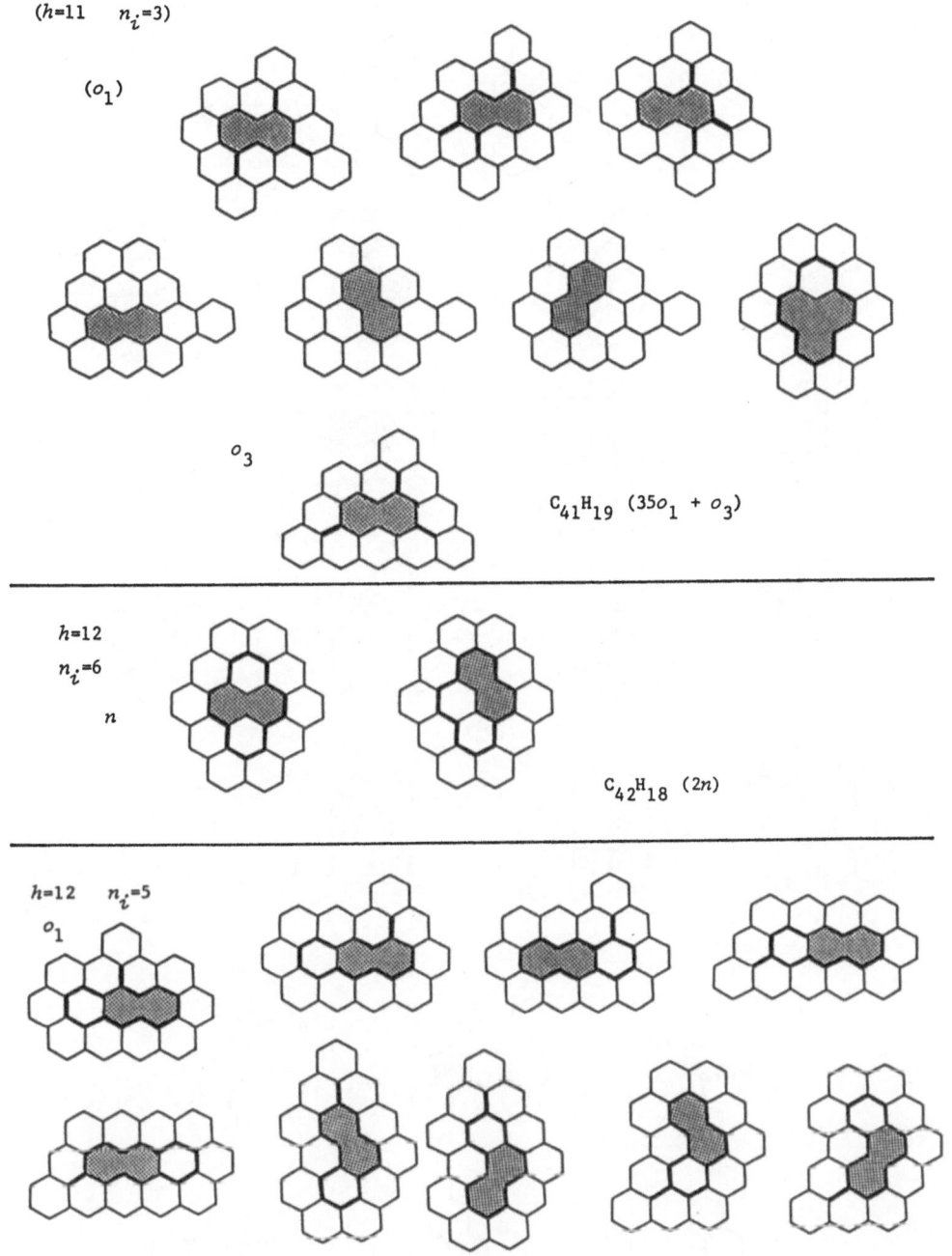

$(h=11 \quad n_i=3)$

(o_1)

o_3

$C_{41}H_{19} \quad (35o_1 + o_3)$

$h=12$
$n_i=6$

n

$C_{42}H_{18} \quad (2n)$

$h=12 \quad n_i=5$
o_1

(cont.)

(h=12)

(n_i=5)

(o_1)

(cont.)

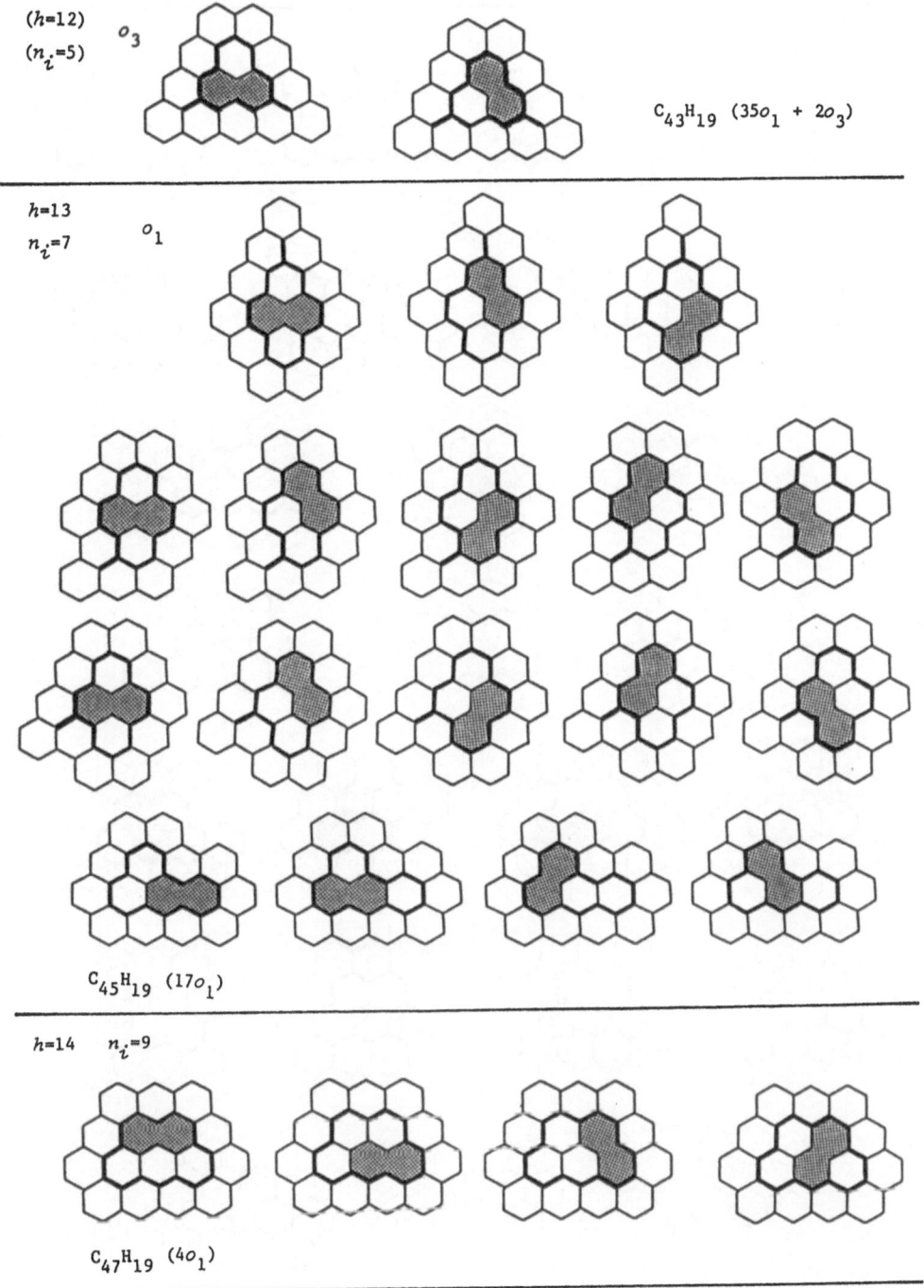

$(h=12)$ o_3
$(n_i=5)$

$C_{43}H_{19}$ $(35o_1 + 2o_3)$

$h=13$ o_1
$n_i=7$

$C_{45}H_{19}$ $(17o_1)$

$h=14$ $n_i=9$

$C_{47}H_{19}$ $(4o_1)$

(cont.)

$h=15$ $n_i=10$ n

(cont.)

$(h=15$
$n_i=10)$

(n)

o_2

$C_{50}H_{20}$ $(37n + 14o_2)$

(cont.)

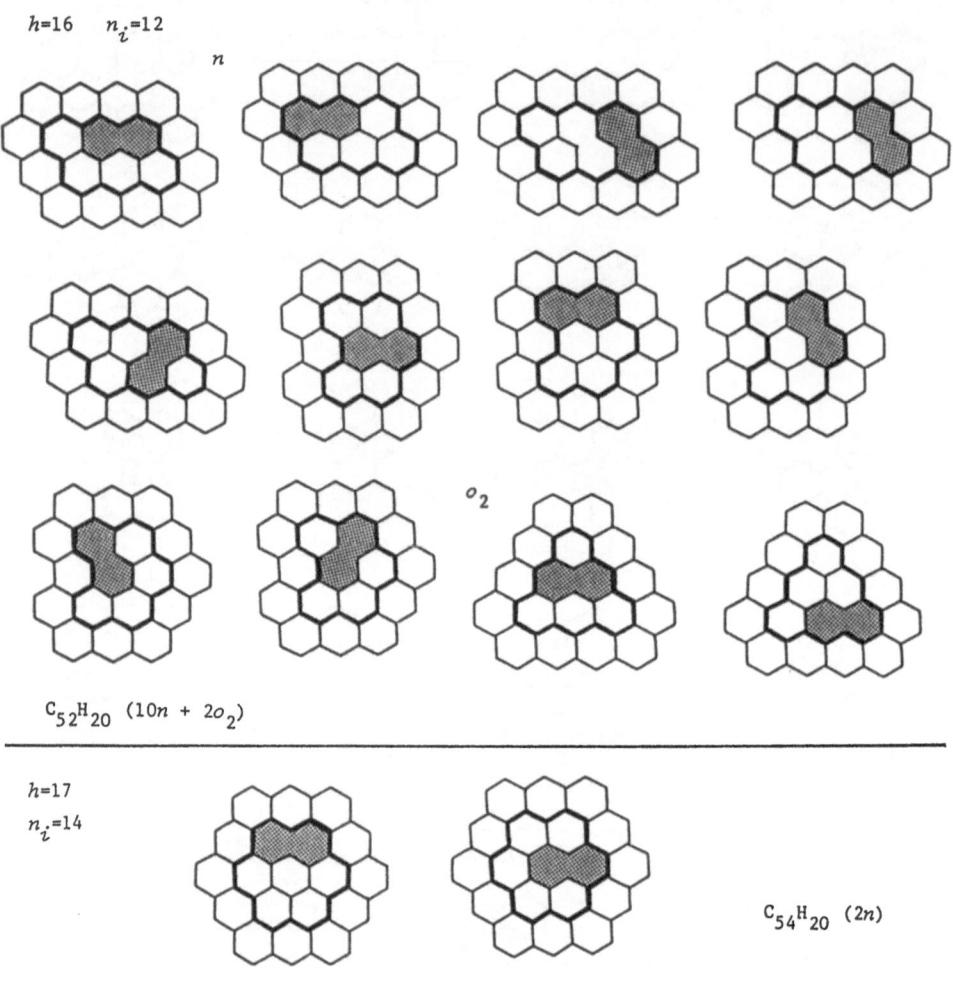

Fig. 7.2. Forms of single coronoid isomers. Abbreviations: n normal; o non–Kekuléan, with the Δ values indicated as subscript $- o_\Delta$. When the associated benzenoid is strictly pericondensed, the contour of its excised internal structure is given by heavy lines.

7.3 Methods of Generation

7.3.1 *Introduction*

The numbers of Tables 2 and 4 are based on computerized generations and enumerations (Vol.I–9.3). The forms of Fig. 2, on the other hand, were generated without computer aid. In the following, two methods for generating single coronoid isomers are demonstrated.

7.3.2 *Application of the Fundamental Building–Up Principle*

Let us demonstrate the generation of all $C_{42}H_{20}$ ($h = 11$, $n_i = 2$) single coronoid isomers by means of the fundamental building–up principle (cf. Par. 6.5.4):

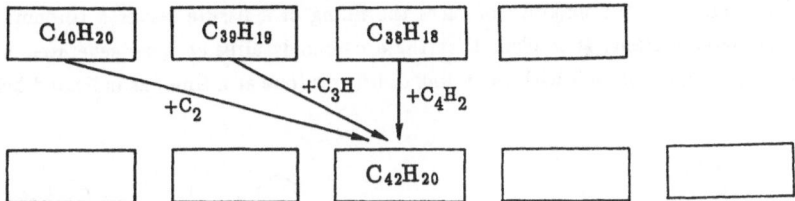

(i) Start with one–contact additions to the 12 ($9n+3o_2$) $C_{38}H_{18}$ isomers, which are depicted in Fig. 2. The sites of annelations (to be executed one at a time) are indicated by heavy strokes in the below diagram. Symmetry was taken into account in order to avoid isomorphic systems.

$7n$ $7n$ $8n$ $4n$

$3n$ $5n$ $4n$ $8n$ $7n$

(cont.)

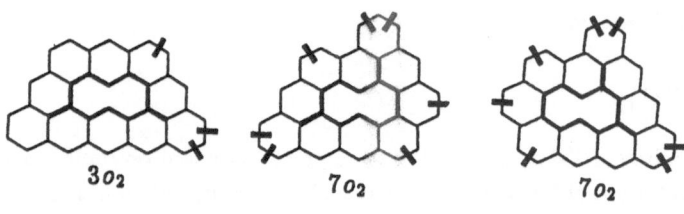

$3o_2$ $\qquad$ $7o_2$ $\qquad$ $7o_2$

$$C_{38}H_{18} + C_4H_2$$

This analysis was particularly simple. Since the initial coronoids do not possess any L_1–mode hexagons, it is ascertained that isomorphic systems cannot be generated from two different initial coronoids. Annelations do not affect the Δ values. Furthermore, an annelated normal coronoid is also normal. In total, we have so far generated 70 ($53n+17o_2$) $C_{42}H_{20}$ isomers.

(ii) The next step are the two–contact additions to the $15o_1$ $C_{39}H_{19}$ isomers, which also are found in Fig. 2. In the majority of cases the filling of a fissure leaves a coronoid with an L_1–mode hexagon present. It is clear that these coronoids already were generated under the annelations (i). One is only left with the filling of fissures (one at a time) as indicated below.

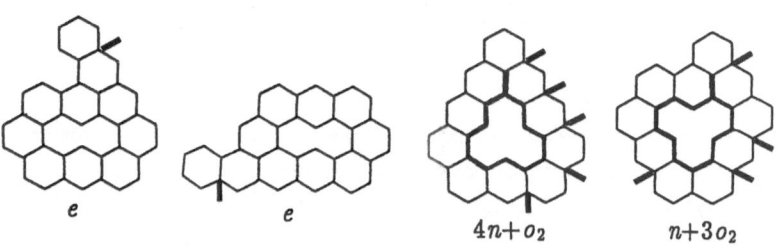

e $\qquad$ e $\qquad$ $4n+o_2$ $\qquad$ $n+3o_2$

$$C_{39}H_{19} + C_3H$$

Here again precaution was taken in order to avoid isomorphic systems. In particular, the two essentially disconnected (e) systems can each be generated from two initial coronoids. Under this point the number of generated $C_{42}H_{20}$ isomers is 11 ($5n+2e+4o_2$).

(iii) It remains to inspect the three–contact additions to the $15n$ $C_{40}H_{20}$ isomers (see Fig. 2). Now it is of interest to look for an embedding of a hexagon into a bay so that the resulting system has neither an L_1– or a P_2–mode hexagon. Systems with L_1– and /or P_2–mode hexagons were obviously generated already under the points (i) or (ii). In other words we are looking for an L_3–added coronoid. There is only one possibility: one normal (n) system is generated by embedding a hexagon into the bay, which is marked by a heavy stroke; see below.

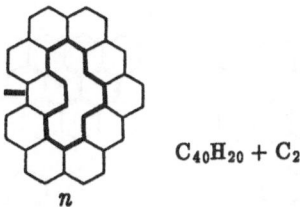

$C_{40}H_{20} + C_2$

Altogether we have generated all the nonisomorphic $C_{42}H_{20}$ single coronoid isomers: 82 $(59n+2e+21o_2)$ in consistency with the entries in Table 2. Also the distribution of corona holes is consistent with Table 4: 72 naphthalene $(h^o = 2,\ n_i^o = 0)$ holes, 9 phenalene $(h^o = 3,\ n_i^o = 1)$ holes, and 1 phenanthrene $(h^o = 3,\ n_i^o = 0)$ hole.

7.3.3 Perforating Benzenoids

Introductory Remarks. There is so much information available on the forms of benzenoid isomers that the perforating of benzenoids lends itself as a method for generating single coronoid isomers. The systematic listings of the forms of coronoid isomers in Fig. 2 are based on this method. All the forms of benzenoids which were needed for this purpose, were taken from Cyvin SJ, Brunvoll and Cyvin (1991d) and from Brunvoll, Cyvin BN and Cyvin (1992b). These two publications are very useful as standard sources of reference. The attack of Dias (1991d) on one of them (Cyvin SJ, Brunvoll and Cyvin 1991d) is clearly unjustified. The last table in his communication (Dias 1991d) still contains erroneous numbers for the $C_{58}H_{20}$ benzenoid isomers (also given in Dias 1990g); they have been corrected around the same time by Cyvin SJ, Brunvoll and Cyvin (1991a).

Example. The method of perforating benzenoids shall presently be exemplified by the $C_{46}H_{20}$ ($h = 13$, $n_i = 6$) single coronoid isomers. The application of the fundamental building–up principle would be somewhat laborious in this case; it would imply the appropriate additions to (i) 2 $C_{42}H_{18}$, (ii) 37 $C_{43}H_{19}$ and (iii) 142 $C_{44}H_{20}$ coronoid isomers.

The majority of the $C_{46}H_{20}$ isomers, which are to be generated, are naphthalenic. By virtue of eqn. (4.12) they emerge as the benzenoids with $H = 15$, $N_i = 16$ perforated by one naphthalene hole each; they are the 70 $C_{46}H_{18}$ benzenoid isomers, distributed according to $47n + 2e + 20o_2 + o_4$ and depicted by Cyvin SJ, Brunvoll and Cyvin (1991d). The pertinent forms are reproduced below with inscribed numerals indicating the number of sites for the naphthalene hole, chosen so as to be compatible with nonisomorphic systems. The abbreviations n (normal), e (essentially disconnected), o_Δ (non–Kekuléan) in the below diagram pertain to the $C_{46}H_{18}$ benzenoids. The naphthalenic coronoids remain to be of the same nature after perforating the e, o_2 and o_4 benzenoids. Also most of the perforated n benzenoids remain normal.

n

(cont.)

(n)

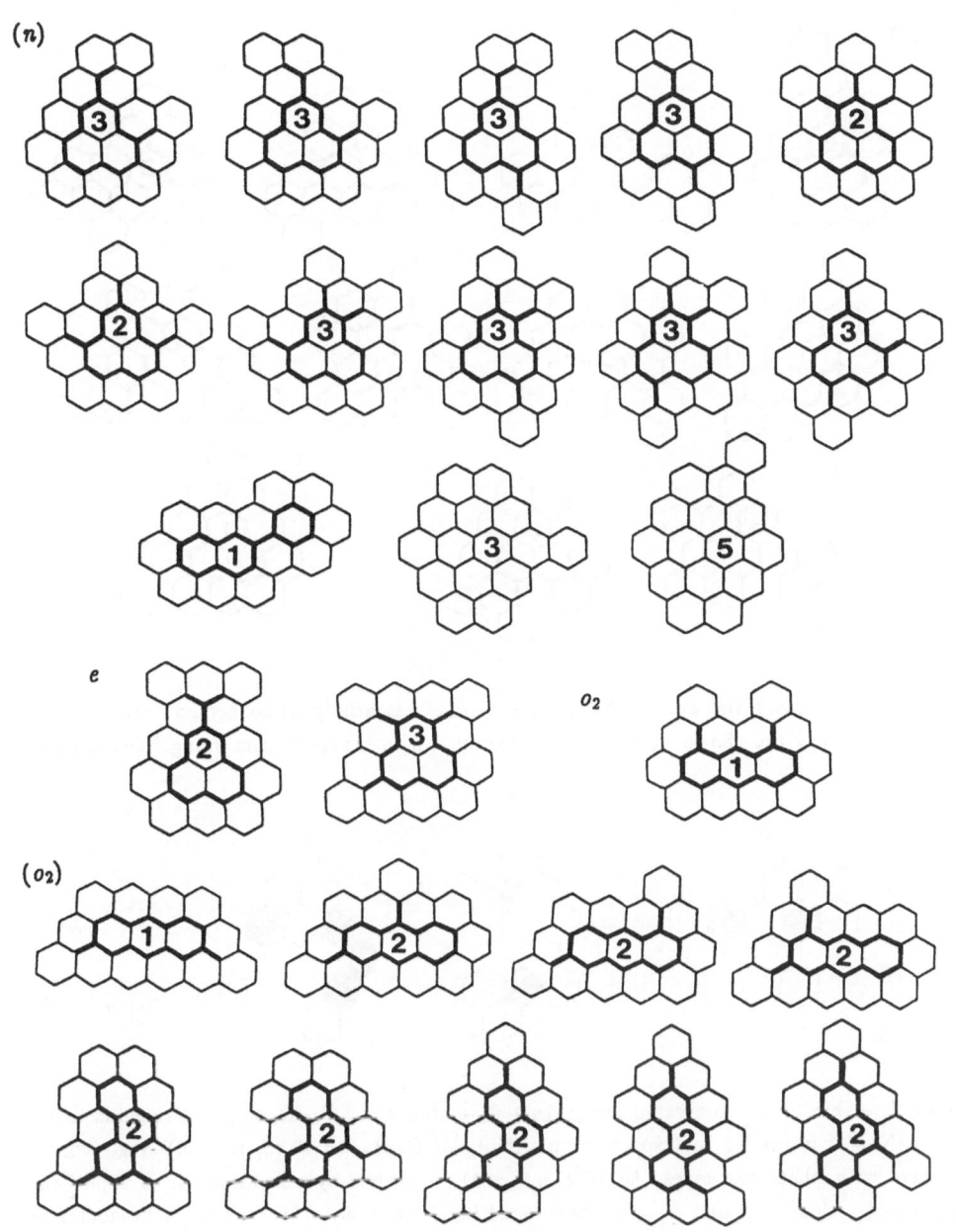

(cont.)

(*o*₂)

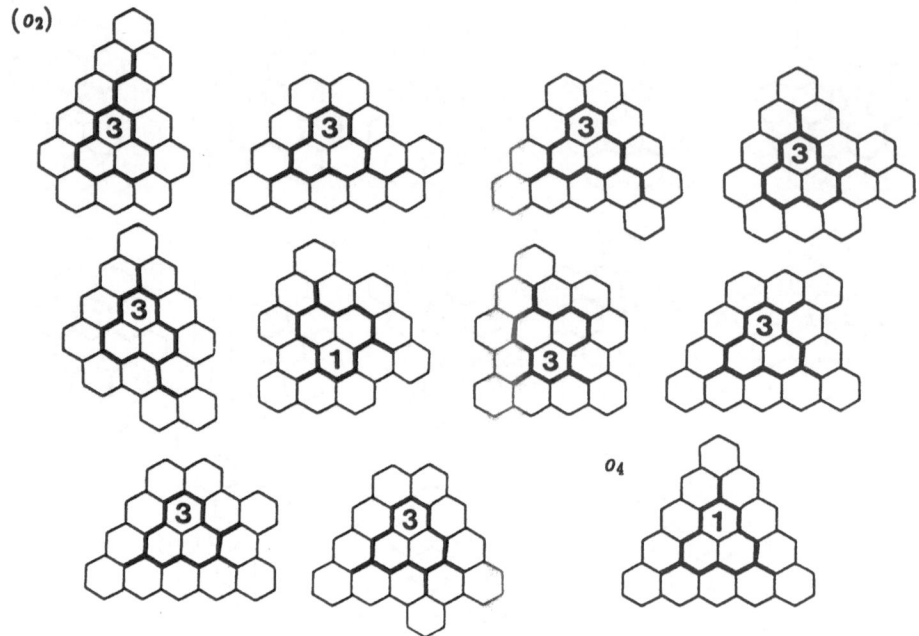

Only in three cases the normal benzenoids become essentially disconnected coronoids when perforated. They are shown in the below diagram, where hexagons containing fixed bonds are black.

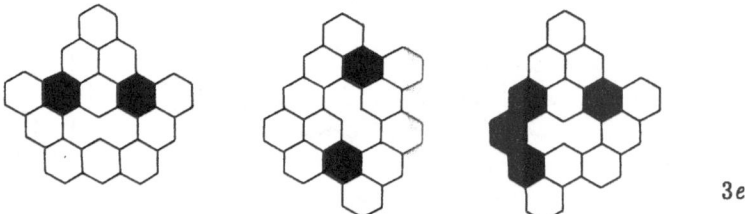

3*e*

In total, we have so far generated 156 ($101n + 8e + 46o_2 + o_4$) $C_{46}H_{20}$ single coronoid isomers.

Now it is needed to inspect the possibility of $C_{46}H_{20}$ ($h = 13$, $n_i = 6$) single coronoid isomers with phenalene holes ($h^o = 3$, $n_i^o = 1$). If such a system exists, it is a perforated benzenoid with $H = 16$ and $N_i = 19$, as is found from eqn. (4.9). This means a benzenoid isomer with the formula $C_{47}H_{17}$. Indeed, one single benzenoid isomer with this formula exists, viz. circumnaphthanthrene, an obvious non–Kekuléan ($\Delta = 1$) system. It is depicted below (left–hand drawing), where the inscribed numeral indicates that two nonisomorphic coronoids can be generated from it by perforating with phenalene holes. The result is (see below) $2\,(n + o_2)$ $C_{46}H_{20}$ single coronoid isomers.

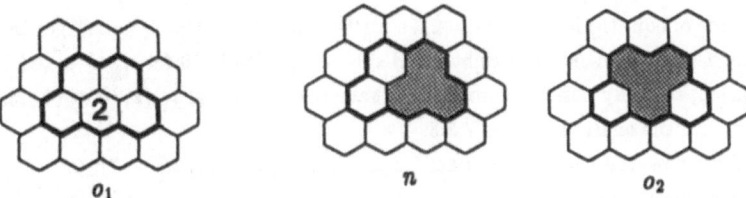

o_1 n o_2

The anthracene and phenanthrene corona holes ($h^o = 3$, $n_i^o = 0$) are the next on the list (Fig. 1). Is one of these holes possible for a $C_{46}H_{20}$ ($h = 13$, $n_i = 6$) single coronoid? According to eqn. (4.9) this would have to be a perforated benzenoid with $H = 16$, $N_i = 20$. But such a benzenoid does not exist; for $H = 16$, $(N_i)_{max} = 19$. It appears also that no other (larger) hole is possible. This will be clear from the general formulations in the following.

Altogether we have generated all the nonisomorphic $C_{46}H_{20}$ single coronoid isomers: 158 ($102n + 8e + 47o_2 + o_4$) in consistency with Table 2. Also the distribution of corona holes is consistent with Table 4: 156 naphthalene holes and 2 phenalene holes.

General Formulations. It is adhered to the notation $B(H,N_i)$ for a benzenoid which is perforated by a hole (h^o, n_i^o) in order to yield a single coronoid $C(h,n_i)$. Furthermore, $B \in \{N; S\}$ and $C \in \{n; s\}$. A useful relation in addition to eqn. (4.9) and in consistency with (4.10) reads

$$(N;\ S) = (n + n_i^o;\ s - 2h^o + n_i^o + 2) \tag{7.4}$$

The relation

$$2h - n_i \geq \lceil (12h + 12h^o - 3)^{1/2} \rceil + \lceil (12h^o - 3)^{1/2} \rceil \tag{7.5}$$

gives an upper limit for the size of the corona hole in terms of h^o. Eqn. (5) emerges directly from a relation which is explained in Volume I, viz. I–(9.6). Here it should be noted that (5) when satisfied with the sign of equality for a certain h^o (while h and n_i are fixed), may still be satisfied for $h^o + 1$. However, when the right–hand side of (5) becomes larger than $2h - n_i$ for a certain h^o, we need not test any higher values; they will certainly not satisfy the relation (5).

After having fixed an h^o value (in addition to the fixed h and n_i values) there is a useful relation which determines a lower limit for n_i^o:

$$2h - n_i \geq 2h^o - n_i^o + 1 + \lceil (12h + 12h^o - 3)^{1/2} \rceil \tag{7.6}$$

This relation (6) emerges directly from I–(9.7), a relation which again is explained in Volume I. The lower limit of n_i^o according to (6) is not always realized. Examples are presented at the end of the next paragraph.

7.3.4 *Examples and Discussion*

Consider the example of Par. 7.3.3: $C_{46}H_{20}$ ($h = 13$, $n_i = 6$); $2h - n_i = 20$. The right–hand side of eqn. (5) is 20 for $h^o = 2$, 20 for $h^o = 3$, 22 for $h^o = 4$, and larger than 22 for $h^o > 4$. Now let $h^o = 3$ be fixed. The right–hand side of eqn. (6) is 20 for $n_i^o = 1$ and 21 for $n_i^o = 0$. In conclusion, the only possible corona holes have (h^o, n_i^o) equal to (2, 0) or (3, 1). The result is consistent with the analysis in Par. 7.3.3.

Consider now the example of Par. 7.3.2: $C_{42}H_{20}$ ($h = 11$, $n_i = 2$); $2h - n_i = 20$. The right–hand side of eqn. (5) is 18 for $h^o = 2$, 19 for $h^o = 3$, 21 for $h^o = 4$, and larger than 21 for $h^o > 4$. When $h^o = 3$ is fixed, then the right–hand side of eqn. (6) is 19 for $n_i^o = 1$ and 20 for $n_i^o = 0$. The possible corona holes have (h^o, n_i^o) equal to (2, 0), (3,1) or (3, 0). The result is consistent with the analysis in Par. 7.3.2.

It was mentioned in Par. 7.3.3 that the application of the fundamental building–up principle to the $C_{46}H_{20}$ isomers would require the inspection of 2 + 37 + 142 systems. This would be practically feasible, but supposedly more tedious than the method of perforating benzenoids, which was applied here (Par. 7.3.3). This does not imply, however, that the latter method always is easier than the application of the fundamental building–up principle. For example, if applied to the $C_{42}H_{20}$ isomers (treated without much trouble in Par. 7.3.2) one would have to inspect: 187 $C_{42}H_{18}$ benzenoid isomers for perforation with the naphthalene hole, 16 $C_{43}H_{17}$ isomers for perforation with the phenalene hole, and finally 1 $C_{42}H_{16}$ isomer (circumpyrene) for perforation with an anthracene or a phenanthrene hole. Then one would discover (as deduced from the analysis of Par. 7.3.2) that all but 14 of the 187 $C_{42}H_{18}$ isomers are too thin to have enough space for a naphthalene hole.

Finally in this paragraph we shall present some examples where eqn. (6) fails to predict exactly the lower limit of n_i^o for given $C(h, n_i)$ isomers and a possible h^o value. This feature occurs only four times among all single coronoids with $h \leq 14$, which are accounted for in Table 4.

(a) $C_{52}H_{26}$ ($h = 13$, $n_i = 0$); for $h^o = 6$, eqn. (6) predicts $n_i^o \geq 2$, but $n_i^o = 2$ is not realized. The reason is that the benzenoid $C_{54}H_{18}$ (circumcoronene, the unique benzenoid with this formula) has not enough space to be perforated by any corona hole with (h^o, n_i^o) = (6, 2) to yield a $C_{52}H_{26}$ single coronoid. However, a degenerate coronoid with such a hole and the prescribed formula ($C_{52}H_{26}$) is possible, as is shown in the below diagram. It has thirteen hexagons and the same formula as a catacondensed ($n_i = 0$) single coronoid with thirteen hexagons.

(b) $C_{51}H_{25}$ ($h = 13$, $n_i = 1$); for $h^o = 6$, eqn. (6) predicts $n_i^o \geq 3$, but $n_i^o = 3$ is not realized. The benzenoid $C_{54}H_{18}$ (circumcoronene) cannot be perforated by a hole with (h^o, n_i^o) = (6, 3); cf. the degenerate coronoid with the formula $C_{51}H_{25}$ below.

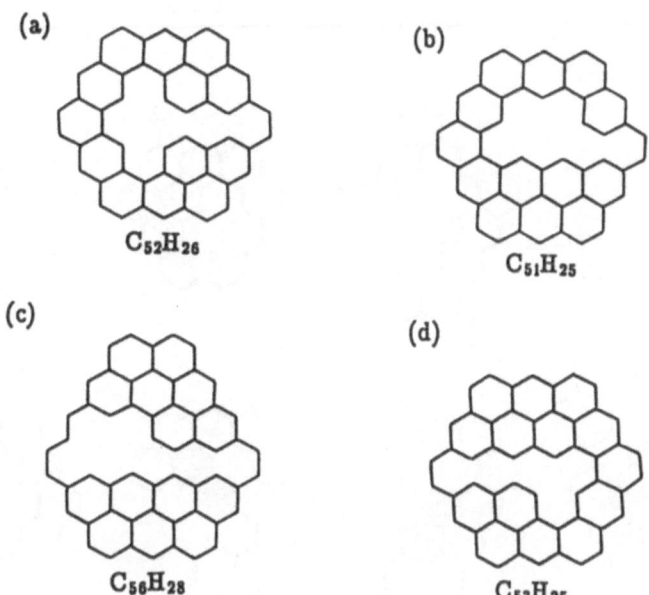

(a) $C_{52}H_{26}$

(b) $C_{51}H_{25}$

(c) $C_{56}H_{28}$

(d) $C_{53}H_{25}$

(c) $C_{56}H_{28}$ ($h = 14$, $n_i = 0$); for $h^o = 7$, eqn. (6) predicts $n_i^o \geq 3$, but $n_i^o = 3$ is not realized. The benzenoid $C_{59}H_{19}$ (circumbenzo[bc]coronene, the unique benzenoid isomer with this formula) cannot be perforated by a hole with (h^o, n_i^o) = (7, 3); cf. the above degenerate coronoid with the formula $C_{56}H_{28}$.

(d) $C_{53}H_{25}$ ($h = 14$, $n_i = 3$); for $h^o = 5$, eqn. (6) predicts $n_i \geq 1$, but $n_i^o = 1$ is not realized. The benzenoid $C_{54}H_{18}$ (circumcoronene) cannot be perforated by a hole with (h^o, n_i^o) = (5, 1); cf. the above degenerate coronoid with the formula $C_{53}H_{25}$.

7.4 Circular Single Coronoids

7.4.1 *Introduction*

A circular single coronoid, O, (Par. 4.6.2) is a circular benzenoid perforated by a naphthalene hole (cf. also Par 5.5.3). Cyvin SJ (1991b) presented a complete mathematical solution in terms of combinatorial formulas for the numbers of O isomers.

Figure 2 includes the forms of some circular single coronoid (O) isomers, viz. those for $16 \leq s \leq 20$. The relevant cardinalities are: $|C_{32}H_{16}|_1 = 1$, $|C_{37}H_{17}|_1 = 1$, $|C_{42}H_{18}|_1 = 2$, $|C_{47}H_{19}|_1 = 4$, $|C_{54}H_{20}|_1 = 2$. In supplement, Fig. 3 shows the forms of the O isomers for $s = 21$ and $s = 22$; $|C_{59}H_{21}|_1 = 8$, $|C_{66}H_{22}|_1 = 6$.

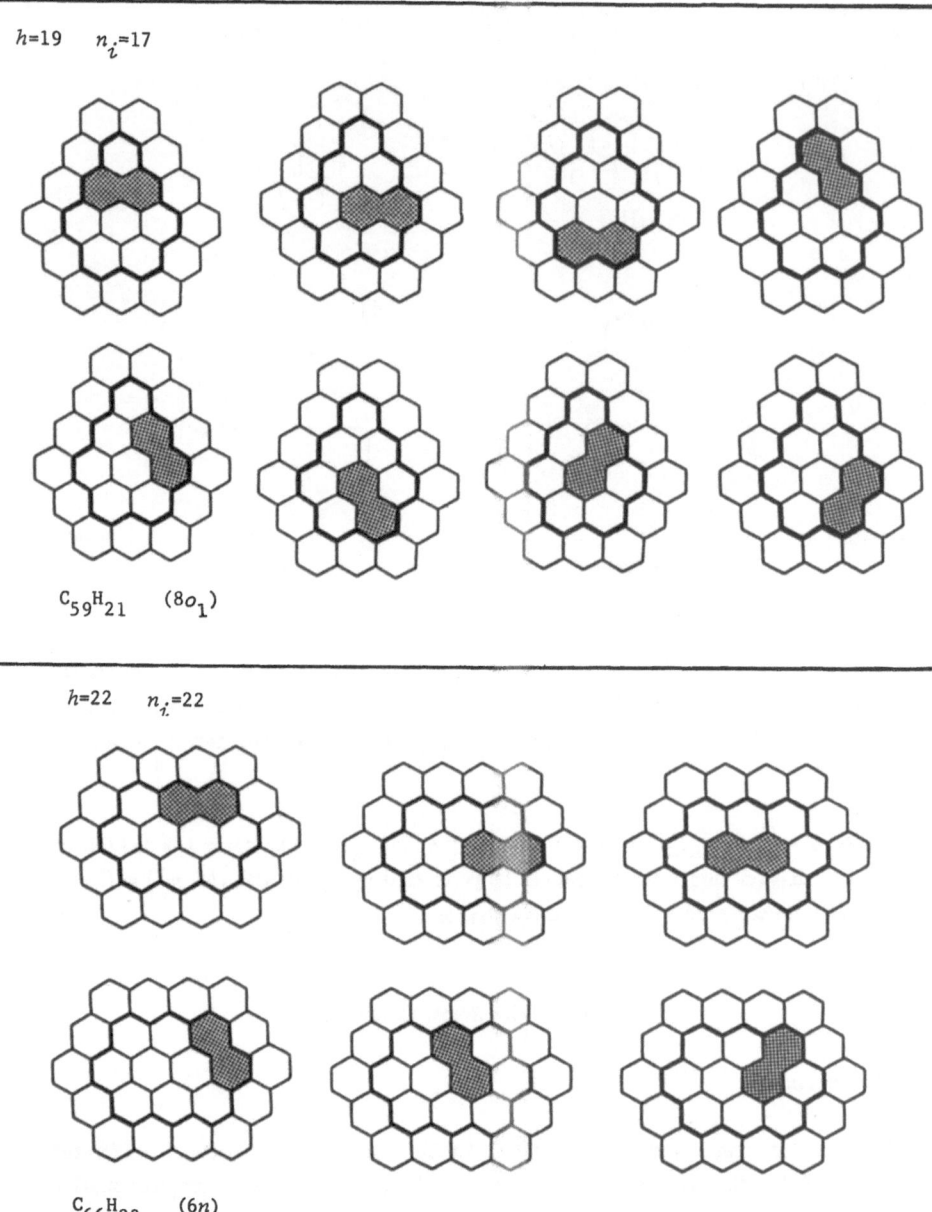

$h=19 \quad n_i=17$

$C_{59}H_{21} \quad (8o_1)$

$h=22 \quad n_i=22$

$C_{66}H_{22} \quad (6n)$

Fig. 7.3. Forms of some circular single coronoid isomers.

The enumeration of O isomers reduces clearly to the counting of distinctly different positions into which a naphthalene hole can be placed within a circular benzenoid. The position of a naphthalene hole is uniquely determined by the central edge of naphthalene. Hence the process ends up with counting of edges.

It is recalled that the circular single coronoid isomers are defined in terms of two parameters by $\{\epsilon, k\}$ (Par. 4.6.4). The formula for $\{\epsilon, k\}$, viz. $(n^O; s^O)$ is given by eqn. (5.67). Then the problem at hand is to determine the cardinalities (of genus 1):

$$|n^O; s^O|_1 \equiv |\{\epsilon, k\}|$$

for $\epsilon = 0, 1, 2, 3, 4, 5$.

7.4.2 Methods

Direct Combinatorics. The simplest method for enumeration of O isomers is a direct counting of appropriate edges under combinatorial considerations. An analytical treatment along these lines (Cyvin SJ 1991b) has led to combinatorial expressions for $|\{\epsilon, k\}|$ as functions of k.

Symmetry Considerations. Symmetry has been exploited systematically many times in different enumeration problems. The references cited in Brunvoll, Cyvin BN and Cyvin (1993c), even when confined to enumerations of polygonal systems (Balaban and Harary 1968; Harary and Read 1970; Balaban 1970; Lunnon 1972), are far from exhaustive. A coronoid (or in general a g–polyhex) may belong to one of eight symmetry groups; cf. Vol. I–3.3.8 and Table 5. Here all

Table 7.5. Symmetry groups (Γ) and the halves of their group orders (γ).

Γ	D_{6h}	C_{6h}	D_{3h}	C_{3h}	D_{2h}	C_{2h}	C_{2v}	C_s
$\gamma(\Gamma)$	12	6	6	3	4	2	2	1

the systems of interest possess a (trivial) plane of symmetry (since they are geometrically planar). Hence the order of any of the eight groups, Γ, is divisible by two, and it is reasonable to define the half of the group order. This quantity, presently denoted by γ (see Table 5), is 2σ for the $D_{\sigma h}$ groups and σ for $C_{\sigma h}$ (note that $C_s \equiv C_{1h}$). Quotients between appropriate γ values are used in the systematic exploitation of symmetry inherent in the method called "stupid sheep counting" (see below), where "stupid" refers to the counting, not to the sheep. This method has recently been applied successfully to the enumeration of polygonal systems representing polycyclic conjugated hydrocarbons with arbitrary ring sizes (Cyvin BN, Brunvoll and Cyvin 1993; Cyvin SJ, Brunvoll and Cyvin 1993; Brunvoll, Cyvin BN and Cyvin 1993a; 1993c; Cyvin SJ, Cyvin and Brunvoll 1993c).

Stupid Sheep Counting. We quote from Cyvin SJ, Cyvin and Brunvoll 1993c: "The method of 'stupid sheep counting' (Brunvoll, Cyvin BN and Cyvin 1993c), in spite of this newly devised name, is not at all new. One of its more recent applications is due to Redelmeier (1981), who says: 'There is a well known way to count cattle in a herd: count the number of legs and divide by four.' Although this (stupid) method is referred to jokingly, it is instructive to pursue the analogy with cattle counting in order to explain an essential part of the present methods of isomer enumeration. Assume that the legs of unknown numbers of sheep and shepherds were counted with the result J altogether (the 'crude total'). Next, assume that the number of shepherds was determined to be M when counted by heads. These two pieces of information are sufficient to give the number of sheep (say U) and the number of shepherds (M) separately, since it is known that each sheep has four legs and a shepherd two. Hence the sum $I = M + U$ is determined, actually as $I = \frac{1}{4}(J + 2M)$."

Generating Functions. Generating functions represent a powerful tool in different enumeration problems (Harary and Prins 1959; Harary and Read 1970; Harary and Schwenk 1973; Harary, Palmer and Read 1975; Read 1978; Cyvin SJ and Brunvoll 1992; Cyvin SJ, Zhang and Brunvoll 1992; Zhang, Guo, Cyvin and Cyvin 1992; Cyvin SJ, Cyvin, Brunvoll, Brendsdal et al. 1993), and have often been employed in connection with symmetry considerations. They have also met with success in conjunction with the method of stupid sheep counting in the enumerations of polygonal systems mentioned above (Cyvin BN, Brunvoll and Cyvin 1993; Brunvoll, Cyvin BN and Cyvin 1993c; Cyvin SJ, Cyvin and Brunvoll 1993c). Some generating functions are encountered under the "new pictures" in Par. 5.5.3 and Par. 6.2.4.

7.4.3 Sample Analysis

Specification of the Problem. In the following, the cardinalities $|\{2, k\}|$ are treated in detail. Hence appropriate edges in ovalene and circumovalenes are to be counted, representing positions of naphthalene holes in the circular coronoids $\{2, k\}$.

Direct Counting. Divide k–circumovalene, which belong to the symmetry D_{2h}, into segments as shown below.

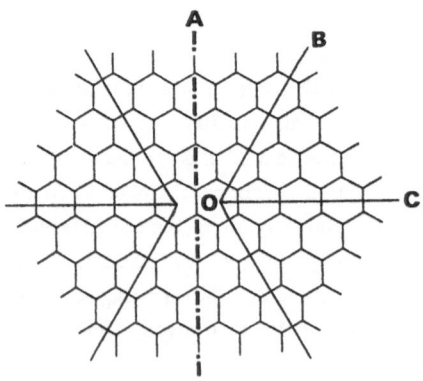

Because of symmetry only the edges in the segment [A, C] should be counted in order to avoid isomorphic systems. This count includes edges on the axis A and those cut by the axis C. In order to accomplish a systematic counting we shall divide the edges into the following five subsets: (a) vertical edges in [A, B> (including those on A); (b) oblique edges in [A, B> (not including those cut by B); (c) edges cut by the axis B; (d) edges parallel with the bisection of angle BOC and in <B, C>; (e) edges not parallel with the bisection of BOC and in <B, C]. The edges cut by C are included in the last subset (e). The below diagram shows the edges (as heavy lines) of the different categories.

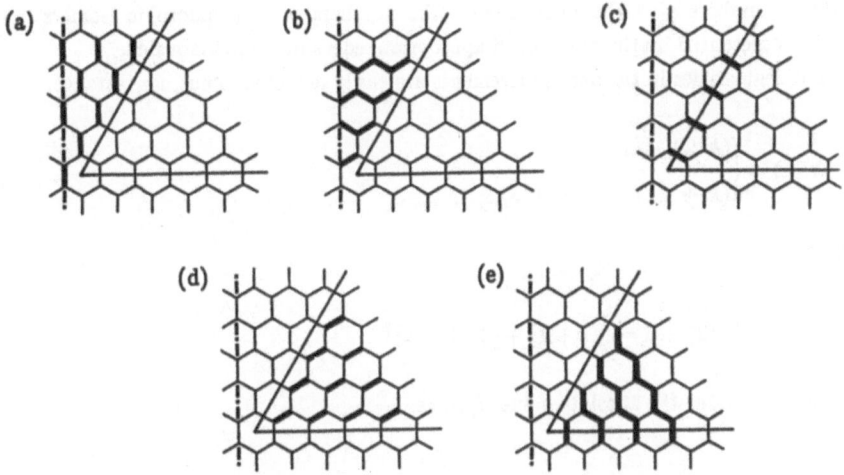

The increments in the number of edges for each k are collected in the below table. In total one obtains the increments in the number of nonisomorphic isomers, say ΔI.

k	$\Delta I(a)$	$\Delta I(b)$	$\Delta I(c)$	$\Delta I(d)$	$\Delta I(e)$	ΔI
0	1	0	0	0	0	1
1	1	1	1	1	1	5
2	2	2	1	2	3	10
3	2	3	1	3	5	14
4	3	4	1	4	7	19

The absolute values, obtained from successive additions of the numbers in the above table, are given below.

k	$I(a)$	$I(b)$	$I(c)$	$I(d)$	$I(e)$	I
0	1	0	0	0	0	1
1	2	1	1	1	1	6
2	4	3	2	3	4	16
3	6	6	3	6	9	30
4	9	10	4	10	16	49

Here the I numbers give the final answer: the numbers of nonisomorphic isomers of $\{2, k\}$ coronoids, represented by the numbers of appropriate edges in k–circumovalene.

The regularities in the above increments are easily detected. Thus, for instance,

$$\Delta I(a) = \begin{cases} (k/2) + 1 \; ; \; k = 0, 2, 4, 6, ... \\ (k + 1)/2 \; ; \; k = 1, 3, 5, 7, ... \end{cases} \tag{7.7}$$

In general:

$$\Delta I(a) = \lfloor k/2 \rfloor + 1 = \tfrac{1}{2}(k + 1) + \tfrac{1}{4}[1 + (-1)^k] \tag{7.8}$$

Herefrom one obtains the absolute value, $I(a)$, as

$$I(a) = \tfrac{1}{2} \sum_{i=0}^{k} (i + 1) + \tfrac{1}{4} \sum_{i=0}^{k} [1 + (-1)^i]$$

$$= \tfrac{1}{4}(k + 1)(k + 2) + \tfrac{1}{4} \sum_{i=0}^{k} [1 + (-1)^i] \tag{7.9}$$

Note that

$$\tfrac{1}{2} \sum_{i=0}^{k} [1 + (-1)^i] = \Delta I(a) \tag{7.10}$$

This yields, in combination with (8),

$$I(\text{a}) = \tfrac{1}{4}(k^2 + 3k + 4 + 2\lfloor k/2 \rfloor) = \tfrac{1}{4}(k+1)(k+3) + \tfrac{1}{8}[1 + (-1)^k] \qquad (7.11)$$

The next steps are simpler. Firstly,

$$\Delta I(\text{b}) = \Delta I(\text{d}) = k \ , \ I(\text{b}) = I(\text{d}) = \tfrac{1}{2}k(k+1) \qquad (7.12)$$

Next we have for the subset (c): $\Delta I(\text{c}) = 0$ for $k = 0$ and $\Delta I(\text{c}) = 1$ for $k > 0$. Then, simply

$$I(\text{c}) = k \qquad (7.13)$$

Finally, for the subset (e): $\Delta I(\text{e}) = 0$ for $k = 0$ and $\Delta I(\text{e}) = 2k - 1$ for $k > 0$. Hence

$$I(\text{e}) = k^2 \qquad (7.14)$$

On adding the appropriate equations and executing some elementary manipulations it was arrived at

$$\Delta I = 4k + 1 + \lfloor k/2 \rfloor = \tfrac{1}{2}(9k + 1) + \tfrac{1}{4}[1 + (-1)^k] \qquad (7.15)$$

and

$$I = |\{2, k\}| = \lfloor (3k + 2)^2/4 \rfloor = \tfrac{1}{4}(3k + 2)^2 - \tfrac{1}{8}[1 - (-1)^k] \qquad (7.16)$$

Now we wish to enumerate the $\{2, k\}$ coronoids of higher symmetries, viz. those belonging to D_{2h}, $C_{2v}(\text{a})$ and $C_{2v}(\text{b})$. It is recalled that the $C_{2v}(\text{a})$ systems have a twofold symmetry axis (C_2) which cuts edges, while in $C_{2v}(\text{b})$ the C_2 axis passes through edges. Let the number of nonisomorphic $\{2, k\}$ isomers belonging to D_{2h}, $C_{2v}(\text{a})$ and $C_{2v}(\text{b})$ be denoted by D, M^a and M^b, respectively. Then the $D_{2h} + C_{2v}(\text{b})$ systems are associated with the edges along the axis A (see above) and are found among those of the category (a). The $C_{2v}(\text{a})$ systems are associated with the edges cut by the axis C and were reckoned to the category (e). In summary one obtains the numbers of the below table.

k	D	M^a	M^b	U
0	1	0	0	0
1	1	1	0	4
2	1	2	1	12
3	1	3	1	25
4	1	4	2	42

The numbers of unsymmetrical (C_s) isomers of $\{2, k\}$, denoted by U (see the above table), were obtained as differences from the total numbers I by virtue of the relation:

$$I = D + M^a + M^b + U \tag{7.17}$$

In general, the following algebraic expressions were deduced.

$$D = 1 \tag{7.18}$$

for every k.

$$M^a = k \tag{7.19}$$

$$M^b = \lfloor k/2 \rfloor = \tfrac{1}{2}k - \tfrac{1}{4}[1 - (-1)^k] \tag{7.20}$$

$$U = \tfrac{1}{4}(9k^2 + 7k - 2\lfloor k/2 \rfloor) = \tfrac{3}{4}k(3k + 2) + \tfrac{1}{8}[1 - (-1)^k] \tag{7.21}$$

The last expression (21) was obtained by means of the expression (16) for the total number of isomers.

Stupid Sheep Counting. It cannot be denied that the above derivation of the numbers of isomers of higher symmetries, eqns. (17) − (19), is much simpler than the derivation of the total number, eqn. (16). The former set of equations, (17) − (19), is a part of the analysis according to the method of stupid sheep counting, but the total number I is obtained without the knowledge of U. Instead, a "crude total" is introduced and denoted by J.

In the present problem of $\{2, k\}$ isomers the crude total (J) is the number of possible positions of a naphthalene corona hole in k–circumovalene, irrespective of symmetry. Denote k–circumovalene by O_k, where $k = 0$ corresponds to ovalene itself: O_0. Then J is the number of internal edges in O_{k-1}, where $k = 0$ corresponds to naphthalene: O_{-1}. Introduce the formulas according to $O_0 \in \{N_0; S_0\}$, where $(N_0; S_0) \equiv C_{32}H_{14}$, and $O_{k-1} \in \{N_{k-1}; S_{k-1}\}$. Then one has from eqn. (5.30)

$$(N_{k-1}; S_{k-1}) = (6(k-1)^2 + N_0 + 2(k-1)S_0; \; 6(k-1) + S_0)$$

$$= (6k^2 + 16k + 10; \; 6k + 8) \tag{7.22}$$

where $N_0 = 32$, $S_0 = 14$ has been inserted. Now we only have to apply the function of (n, s) with $g = 0$ for m_i from Table 3.1 in order to deduce:

$$J = \tfrac{1}{2}(3N_{k-1} - 5S_{k-1}) + 6 = 9k^2 + 9k + 1 \tag{7.23}$$

Example

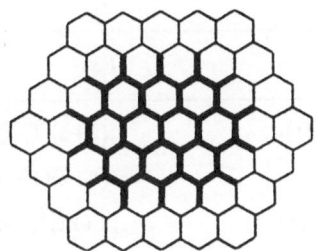

Naphthalene holes in dicircumovalene ($k = 2$) are represented by the internal edges in circumovalene, O_1 ($C_{66}H_{20}$). They are marked by heavy lines in the above diagram, and their number is 55, equal to J for $k = 2$; cf. eqn. (23).

It is evident that J, for a given k, counts the dihedral (D_{2h}) coronoid once, the mirror–symmetrical (C_{2v}) coronoids twice, and the unsymmetrical (C_s) coronoids four times. Therefore:

$$J = D + 2M + 4U \tag{7.24}$$

where

$$M = M^a + M^b = k + \lfloor k/2 \rfloor \tag{7.25}$$

as obtained from eqns. (19) and (20). Furthermore,

$$I = D + M + U \tag{7.26}$$

On eliminating U from (24) and (26) it is attained at

$$I = \tfrac{1}{4}(J + 3D + 2M) \tag{7.27}$$

Now the final result for $I = |\{2, k\}|$ emerges directly in the same form as in the appropriate entry of Table 6 on inserting into (27) the expressions from (18), (23) and (25). The result is equivalent with the expressions of eqn. (16).

Generating Functions. The elementary result

$$(1 - x)^{-1} = 1 + x + x^2 + x^3 + \dots = \sum_{k=0}^{\infty} x^k \tag{7.28}$$

gives already the generating function for the numbers D of eqn. (18). We write:

Table 7.6. Numbers of isomers of circular single coronoids: explicit formulas I; generating functions $I(x)$.

$\{\epsilon, k\}$	I	$I(x)$
$\{0, k\}$	$\frac{1}{4}(3k^2+3k+2\lceil k/2\rceil)$	$x(2+x)(1-x)^{-2}(1-x^2)^{-1}$
$\{1, k\}$	$\frac{1}{2}k(9k+7)$	$x(8+x)(1-x)^{-3}$
$\{2, k\}$	$\frac{1}{4}(9k^2+11k+4+2\lfloor k/2\rfloor)$	$(1+2x)^2(1-x)^{-2}(1-x^2)^{-1}$
$\{3, k\}$	$\frac{1}{2}(k+1)(3k+2)$	$(1+2x)(1-x)^{-3}$
$\{4, k\}$	$\frac{1}{4}(9k^2+17k+8+2\lceil k/2\rceil)$	$(1+2x)(2+x)(1-x)^{-2}(1-x^2)^{-1}$
$\{5, k\}$	$\frac{1}{2}(9k^2+19k+8)$	$(4+6x-x^2)(1-x)^{-3}$

$$D(x) = (1 - x)^{-1} \tag{7.29}$$

The next step is

$$x(1-x)^{-2} = x + 2x^2 + 3x^3 + 4x^4 + \ldots = \sum_{k=0}^{\infty} kx^k \tag{7.30}$$

and we write

$$M^{\mathrm{a}}(x) = x(1-x)^{-2} \tag{7.31}$$

for the generating function which reproduces the numbers M^{a} of eqn. (19). Neither the generating function for M^{b} of eqn. (20) is a big problem. In the first place, one obtains from (30) on replacing x by x^2:

$$x^2(1 - x^2)^{-2} = x^2 + 2x^4 + 3x^6 + 4x^8 + \ldots \tag{7.32}$$

It follows

$$x^3(1 + x^2)^{-2} = x^3 + 2x^5 + 3x^7 + 4x^9 + \ldots \tag{7.33}$$

On adding the two last equations one obtains

$$x^2(1+x)(1-x^2)^{-2} = x^2(1-x)^{-1}(1-x^2)^{-1}$$

$$= x^2 + x^3 + 2x^4 + 2x^5 + 3x^6 + 3x^7 + \dots = \sum_{k=0}^{\infty} \lfloor k/2 \rfloor x^k \tag{7.34}$$

Therefore we write:

$$M^b(x) = x^2(1-x)^{-1}(1-x^2)^{-1} \tag{7.35}$$

The generating function for the numbers M of eqn. (25) is simply the sum of the functions in (31) and (35):

$$M(x) = M^a(x) + M^b(x) = x(1+2x)(1-x)^{-1}(1-x^2)^{-1} \tag{7.36}$$

Now we need the generating function for the J numbers of eqn. (23). Start with

$$(1-x)^{-2} = 1 + 2x + 3x^2 + 4x^3 + \dots = \sum_{k=0}^{\infty} (k+1)x^k \tag{7.37}$$

and

$$(1-x)^{-3} = 1 + 3x + 6x^2 + 10x^3 + \dots = \sum_{k=0}^{\infty} \binom{k+2}{2} x^k$$

$$= \frac{1}{2} \sum_{k=0}^{\infty} (k+1)(k+2)x^k \tag{7.38}$$

When J is rewritten into the form

$$J = 9(k+1)(k+2) - 18(k+1) + 1 \tag{7.39}$$

one finds from eqns. (28), (37) and (38) the desired generating function:

$$J(x) = 18(1-x)^{-3} - 18(1-x)^{-2} + (1-x)^{-1} = (1 + 16x + x^2)(1-x)^{-3} \tag{7.40}$$

Relations in analogy to (24) and (26) are also valid for the pertinent generating functions:

$$J(x) = D(x) + 2M(x) + 4U(x) \tag{7.41}$$

$$I(x) = D(x) + M(x) + U(x) \tag{7.42}$$

Therefore one also has in analogy to (27):

$$I(x) = \tfrac{1}{4}[J(x) + 3D(x) + 2M(x)] \tag{7.43}$$

where $I(x)$ reproduces the numbers I of nonisomorphic $\{2, k\}$ isomers. Hence, as a final result, an explicit expression for $I(x)$ is obtained on inserting the expressions from (29), (36) and (40) into (43). This final result is found in the appropriate entry of Table 6.

7.4.4 Extension to All Circular Single Coronoids

Generalizations. The method of stupid sheep counting, including its adaptation to generating functions, was applied to all the $\{\epsilon, k\}$ isomers. Some generalizations are possible.

Let $(n_0; s_0)$ represent the formulas of the smallest circular single coronoids augmented by two degenerate coronoids as shown in Fig. 4.6. These formulas, which pertain to $k = 0$, are $C_{24}H_{14}$, $C_{27}H_{15}$, $C_{32}H_{16}$, $C_{37}H_{17}$, $C_{42}H_{18}$ and $C_{47}H_{19}$ for $\epsilon = 0, 1, 2, 3, 4$ and 5, respectively. The crude total, J, is obtained as in eqn. (22), where in the general case, $(N_0; S_0)$ symbolizes one of the six following formulas of circular benzenoids: $C_{24}H_{12}$, $C_{27}H_{13}$, $C_{32}H_{14}$, $C_{37}H_{15}$, $C_{42}H_{16}$, $C_{47}H_{17}$. Here $N_0 = n_0$ and $S_0 = s_0 - 2$ should be inserted. This yields

$$(N_{k-1}; S_{k-1}) = (6k^2 + 2k(s_0 - 8) + n_0 - 2s_0 + 10;\ 6k + s_0 - 8) \tag{7.44}$$

The last expression in (22) is the special case of (44) for $n_0 = 32$, $s_0 = 16$ (which pertains to $\epsilon = 2$). Furthermore, on inserting from (44) into the general expression for J in eqn. (23), it is obtained

$$J = 9k^2 + 3k(s_0 - 13) + \tfrac{1}{2}(3n_0 - 11s_0) + 41 \tag{7.45}$$

The last expression of (23) is again the special case for $n_0 = 32$, $s_0 = 16$ of eqn (45). The generating function for the numbers J in the general case of (45) was deduced with the result:

$$J(x) = \frac{3n_0 - 11s_0 + 82 + 2x(14s_0 - 3n_0 - 112) + x^2(3n_0 - 17s_0 + 178)}{2(1-x)^3} \tag{7.46}$$

The special case for $n_0 = 32$, $s_0 = 16$ was encountered in eqn. (40).

A generalization of eqn. (24) reads

$$J = \sum_t \frac{\gamma(\Gamma)}{\gamma(\Gamma_t)} I(\Gamma_t) \tag{7.47}$$

where Γ is the symmetry group of the circular benzenoids, and Γ_t is used to indicate the subgroups of Γ to which the circular coronoids in question belong. The γ values are specified in Table 5. Furthermore, $I(\Gamma_t)$ is the number of the pertinent coronoid isomers belonging to Γ_t. Thus, for instance, with reference to eqn. (24): $I(D_{2h}) = D$, $I(C_{2v}) = M$, $I(C_s) = U$.

Listing of Algebraic Results. Table 6 contains the final results for the total numbers I of nonisomorphic isomers of all circular single coronoids $\{\epsilon, k\}$ as functions of k, and the corresponding generating functions $I(x)$. Some details, especially for the numbers of isomers of specific symmetries, are given in the following.

$\underline{\Gamma = D_{6h}} : \epsilon = 0 , (n_0; s_0) \equiv C_{24}H_{14}$

For $k > 0$:

$$I = M + U = \tfrac{1}{12}(J + 6M) \tag{7.48}$$

where

$$M = M^a + M^b ; \quad M^a = k , \quad M^b = \lceil k/2 \rceil , \quad M(x) = x(2 + x)(1 - x)^{-1}(1 - x^2)^{-1} \tag{7.49}$$

$\underline{\Gamma = D_{3h}} : \epsilon = 3 , (n_0; s_0) \equiv C_{37}H_{17}$

$$I = M + U = \tfrac{1}{6}(J + 3M) \tag{7.50}$$

where

$$M = M^b = k + 1 , \quad M(x) = (1 - x)^{-2} \tag{7.51}$$

$\underline{\Gamma = D_{2h}} : \epsilon = 2 , (n_0; s_0) \equiv C_{32}H_{16}$ and $\epsilon = 4 , (n_0; s_0) \equiv C_{42}H_{18}$

In both cases ($\epsilon = 2, 4$) the relations (17), (24), (26) and (27) from Par. 7.4.3 are valid. This is also the case for the values of D and M^a according to (18) and (19), respectively, and therefore also for the generating functions $D(x)$ and $M^a(x)$ in (29) and (31), respectively. For $\epsilon = 2$, M^b, M, $M^b(x)$ and $M(x)$ are given in (20), (25), (35) and (36), respectively. For $\epsilon = 3$, the expressions for M^b and $M(x)$ as in eqn. (49) are valid.

$\underline{\Gamma = C_{2v}} : e = 1 , (n_0; s_0) \equiv C_{27}H_{15}$ and $\epsilon = 5 , (n_0; s_0) \equiv C_{47}H_{19}$

In both cases, viz. $\epsilon = 1$ for $k > 0$ and $\epsilon = 5$,

$$I = M + U = \tfrac{1}{2}(J + M) \tag{7.52}$$

where the expressions of (51) are applicable.

Numerical Results. The numerical values of I for all circular single coronoids $\{\epsilon, k\}$ with $k \leq 20$ are listed in Table 7.

Table 7.7. Numbers I of $\{\epsilon, k\}$ isomers of circular single coronoids. For the first C_nH_s formulas ($k \leq 6$), see Table 5.2.

k	ϵ 0	1	2	3	4	5
0	–	–	1[a]	1[a]	2[b]	4[b]
1	2[b]	8[b]	6[b]	5[b]	9[c]	18[b]
2	5[b]	25[b]	16[b]	12[b]	20[c]	41[b]
3	10[b]	51[b]	30[b]	22[b]	36[c]	73[b]
4	16[b]	86[b]	49[b]	35[b]	56[c]	114[c]
5	24[b]	130[c]	72[c]	51[c]	81[c]	164[c]
6	33	183[c]	100	70	110	223
7	44	245	132	92	144	291
8	56	316	169	117	182	368
9	70	396	210	145	225	454
10	85	485	256	176	272	549
11	102	583	306	210	324	653
12	120	690	361	247	380	766
13	140	806	420	287	441	888
14	161	931	484	330	506	1019
15	184	1065	552	376	576	1159
16	208	1208	625	425	650	1308
17	234	1360	702	477	729	1466
18	261	1521	784	532	812	1633
19	290	1691	870	590	900	1809
20	320	1870	961	651	992	1994

[a] Cyvin SJ, Brunvoll J (1989). Chem Phys Letters 164: 635
[b] Dias JR (1990). J Chem Inf Comput Sci 30: 251
[c] Cyvin SJ (1991). Coll Sci Papers Fac Sci Kragujevac 12: 95

7.5 Circular Benzenoids Perforated by Phenalene Hole

7.5.1 *Introduction*

Assume that $O(N;S)$ is a circular benzenoid which can be perforated by a phenalene hole to produce a single coronoid $O(n;s)$. Then one has for the formula coefficients: $N = n + 1$, $S = s - 3$. All the isomers of the class O were enumerated. This is not a complete determination of the cardinalities $|C_nH_s|_1$ because there always exist at least naphthalenic single coronoids with the same formula (C_nH_s). However, all the C_nH_s single coronoid isomers with the phenalene hole are accounted for. The method of stupid sheep counting was employed, and generating functions were invoked, as in the case of circular single coronoids, which is treated in details above (Sect. 7.4). The main results are summarized briefly in the following.

7.5.2 *The Systems*

The six characteristic shapes of circular benzenoids are recognized as in the case of circular single coronoids (Fig. 4.6), and are again indicated by the parameter $\epsilon = 0, 1, 2, 3, 4, 5$. The smallest of the systems, which presently are associated with $k = 0$, are displayed in Fig. 4: three coronoids and three degenerate coronoids. All the C_nH_s formulas under consideration are given (in the Harary–Harborth picture) by

$$(n;\ s) = 2\lfloor (1/12)(s^2 + 3)\rfloor - s;\ s) \tag{7.53}$$

where $s = 18, 19, 20, \ldots.$ for the coronoids.

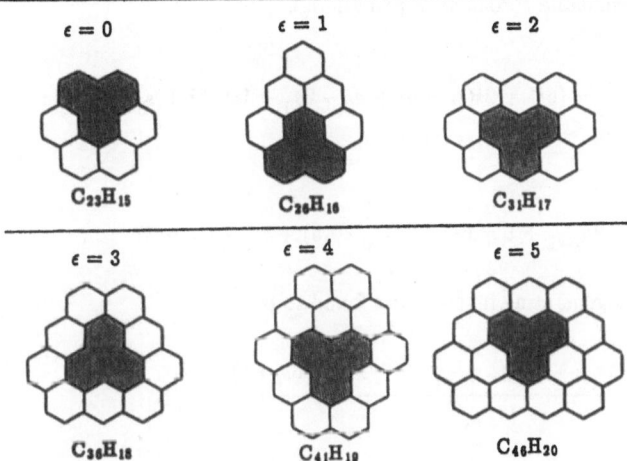

Fig. 7.4. The smallest circular benzenoids perforated by a phenalene hole, augmented by three degenerate coronoids (in the top row).

In Table 8 a number of these formulas are collected, including those for $s = 15, 16, 17$, which pertain to the degenerate coronoids.

7.5.3 *Results of Enumeration*

Crude Total. The symbol $(n_0; s_0)$ is used to identify the six formulas of Fig. 4 (for $k = 0$). When $(N_0; S_0)$ symbolizes the formulas of the six circular benzenoids with $12 \leq S_0 \leq 17$ as in Par. 7.4.4, then: $N_0 = n_0 + 1$, $S_0 = s_0 - 3$. The phenalene hole is completely determined by the central (internal) vertex of phenalene. Hence the crude total (J) for $O(n;s)$ is equal to the number of internal vertices in $O_{k-1}(N_{k-1}; S_{k-1})$, where

Table 7.8. Formulas for circular benzenoids perforated by one phenalene hole each.*

	ϵ					
k	0	1	2	3	4	5
0	$(C_{23}H_{15})$	$(C_{26}H_{16})$	$(C_{31}H_{17})$	$C_{36}H_{18}$	$C_{41}H_{19}$	$C_{46}H_{20}$
1	$C_{53}H_{21}$	$C_{58}H_{22}$	$C_{65}H_{23}$	$C_{72}H_{24}$	$C_{79}H_{25}$	$C_{86}H_{26}$
2	$C_{95}H_{27}$	$C_{102}H_{28}$	$C_{111}H_{29}$	$C_{120}H_{30}$	$C_{129}H_{31}$	$C_{138}H_{32}$
3	$C_{149}H_{33}$	$C_{158}H_{34}$	$C_{169}H_{35}$	$C_{180}H_{36}$	$C_{191}H_{37}$	$C_{202}H_{38}$
4	$C_{215}H_{39}$	$C_{226}H_{40}$	$C_{239}H_{41}$	$C_{252}H_{42}$	$C_{265}H_{43}$	$C_{278}H_{44}$
5	$C_{293}H_{45}$	$C_{306}H_{46}$	$C_{321}H_{47}$	$C_{336}H_{48}$	$C_{351}H_{49}$	$C_{366}H_{50}$
6	$C_{383}H_{51}$	$C_{398}H_{52}$	$C_{415}H_{53}$	$C_{432}H_{54}$	$C_{449}H_{55}$	$C_{466}H_{56}$

* Formulas for degenerate coronoids in parentheses.

$$(N_{k-1}; S_{k-1}) = (6k^2 + 2k(s_0 - 9) + n_0 - 2s_0 + 13; \ 6k + s_0 - 9) \qquad (7.54)$$

Consequently,

$$J = N_{k-1} - 2S_{k-1} + 6 = 6k^2 + 2k(s_0 - 15) + n_0 - 4s_0 + 37 \qquad (7.55)$$

The corresponding generating function was found to be

$$J(x) = \frac{n_0 - 4s_0 + 37 + 2x(5s_0 - n_0 - 49) + x^2(n_0 - 6s_0 + 73)}{(1-x)^3} \qquad (7.56)$$

Listing of Results. A summary of the deduced algebraic expressions for the total numbers of isomers is given in Table 9. Specifications with relevance to symmetry are listed in the following.

Table 7.9. Numbers of isomers of circular benzenoids perforated by one phenalene hole each: explicit formulas I; generating functions $I(x)$.

ϵ	I	$I(x)$
0	$\frac{1}{2}(k^2+k) = \binom{k+1}{2}$	$x(1-x)^{-3}$
1	$k(3k+2)$	$x(5+x)(1-x)^{-3}$
2	$\frac{3}{2}(k^2+k) = 3\binom{k+1}{2}$	$3x(1-x)^{-3}$
3	$(k+1)^2$	$(1+x)(1-x)^{-3}$
4	$\frac{1}{2}(k+1)(3k+2)$	$(1+2x)(1-x)^{-3}$
5	$3k^2+6k+2$	$(2+5x-x^2)(1-x)^{-3}$

$\underline{\Gamma = D_{6h}} : \epsilon = 0, \ (n_0; s_0) \equiv C_{23}H_{15}$

For $k > 0$ the scheme of eqn. (48) is applicable. However, $M = M^b = k$, with the corresponding generating function as in (31).

By passing we wish to point out a feature, which often is encountered in different kinds of enumeration: the total numbers are the same as those of the unsymmetrical systems, but shifted one place on the list; cf., e.g., the results on some polygonal systems representing polycyclic conjugated hydrocarbons with arbitrary ring sizes (Brunvoll, Cyvin BN and Cyvin 1993a). The below diagram and table are supposed to elucidate the feature in question.

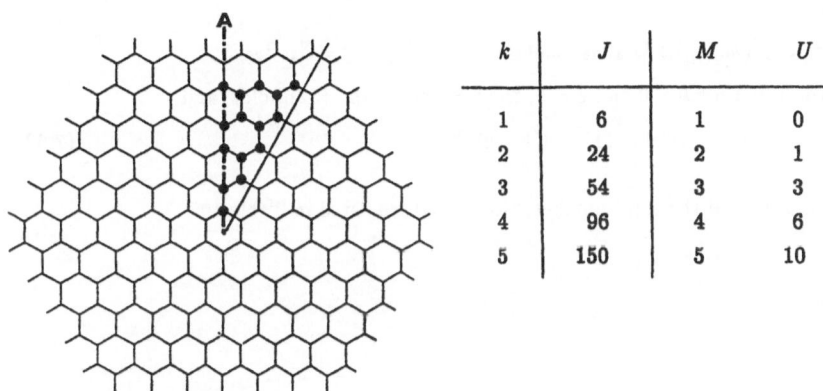

k	J	M	U	I
1	6	1	0	1
2	24	2	1	3
3	54	3	3	6
4	96	4	6	10
5	150	5	10	15

The dots on the A axis are associated with the M mirror–symmetrical (C_{2v}) coronoids, while all the others pertain to the unsymmetrical (C_s) coronoids being U in number. It holds in general that I for a given k is equal to U for $k+1$. Hence (cf. Table 9)

$$U = \tfrac{1}{2}[(k-1)^2 + k - 1] = \binom{k}{2}$$
(7.57)

and the corresponding generating function reads

$$U(x) = xI(x) = x^2(1-x)^3$$
(7.58)

$\Gamma = D_{3h}$: $\epsilon = 3$, $(n_0; s_0) \equiv C_{36}H_{18}$

Introduce $T = I(D_{3h})$ in addition to $M = I(C_{2v})$ and $U = I(C_s)$ as before. Then

$$I = T + M + U = \tfrac{1}{6}(J + 5T + 3M)$$
(7.59)

where

$$T = 1 \ , \ M = M^b = 2k \ , \ M(x) = 2x(1-x)^{-2}$$
(7.60)

$\Gamma = D_{2h}$: $\epsilon = 2$, $(n_0; s_0) \equiv C_{31}H_{12}$ and $\epsilon = 4$, $(n_0; s_0) \equiv C_{41}H_{19}$

In both cases,

$$I = M + U = \tfrac{1}{4}(J + 2M)$$
(7.61)

where, in the case of $\epsilon = 2$ $M = M^b = k$, while in the case of $\epsilon = 4$ $M = M^b = k + 1$. The generating functions are found in eqns. (31) and (51) for $\epsilon = 2$ and 4, respectively.

$\Gamma = C_{2v}$: $\epsilon = 1$, $(n_0; s_0) \equiv C_{26}H_{16}$ and $\epsilon = 5$, $(n_0; s_0) \equiv C_{46}H_{20}$

In both cases, viz. $\epsilon = 1$ for $k > 0$ and $\epsilon = 5$, the pattern of eqn. (52) holds, and

$$M = M^b = 2k + 1 \ , \ M(x) = (1+x)(1-x)^{-2}$$
(7.62)

Numerical values of the total number I for $k \le 20$ are listed in Table 10.

Table 7.10. Numbers I of isomers of circular benzenoids perforated by one phenalene hole each. For the first $C_n H_s$ formulas ($k \leq 6$), see Table 7.8.

k	ϵ 0	1	2	3	4	5
0	–	–	–	1	1	2
1	1	5	3	4	5	11
2	3	16	9	9	12	26
3	6	33	18	16	22	47
4	10	56	30	25	35	74
5	15	85	45	36	51	107
6	21	120	63	49	70	146
7	28	161	84	64	92	191
8	36	208	108	81	117	242
9	45	261	135	100	145	299
10	55	320	165	121	176	362
11	66	385	198	144	210	431
12	78	456	234	169	247	506
13	91	533	273	196	287	587
14	105	616	315	225	330	674
15	120	705	360	256	376	767
16	136	800	408	289	425	866
17	153	901	459	324	477	971
18	171	1008	513	361	532	1082
19	190	1121	570	400	590	1191
20	210	1240	630	441	651	1322

7.6 Circular Benzenoids Perforated by Coronene Hole

7.6.1 *Introduction*

Exactly all the circular benzenoids $O(N;S)$ with $S \geq 18$ can be perforated by a coronene hole to produce single coronoids $O(n;s)$. Then one has $N = n + 6$, $S = s - 6$. A detailed treatment on a complete enumeration of the isomers of the class O has been published (Cyvin BN, Brunvoll, Chen and Cyvin 1993). The smallest system of this class is kekulene. In the cited reference a coronoid with coronene hole(s) was referred to as "coronenic". We quote a nice sentence therefrom: "A coronenic coronoid has a coronene corona hole." We shall not use the term "coronenic" in the following, although it is analogous with our term "naphthalenic" (Definition 3.3). A brief account on the title systems and their enumeration is given in the next two paragraphs.

7.6.2 *The Systems*

The six smallest coronoids of the class under consideration (O) are depicted in Fig. 5. Their formulas $(C_n H_s)$ are given (in the Harary–Harborth picture) by

$$(n;\ s) = (2\lfloor (1/12)(s^2 - 6s)\rfloor - s;\ s) \tag{7.63}$$

where $s = 24, 25, 26, \ldots$; see Cyvin BN, Brunvoll, Chen and Cyvin (1993), where also the expression in the Balaban picture and in the new picture are found. A number of these formulas are collected in Table 11.

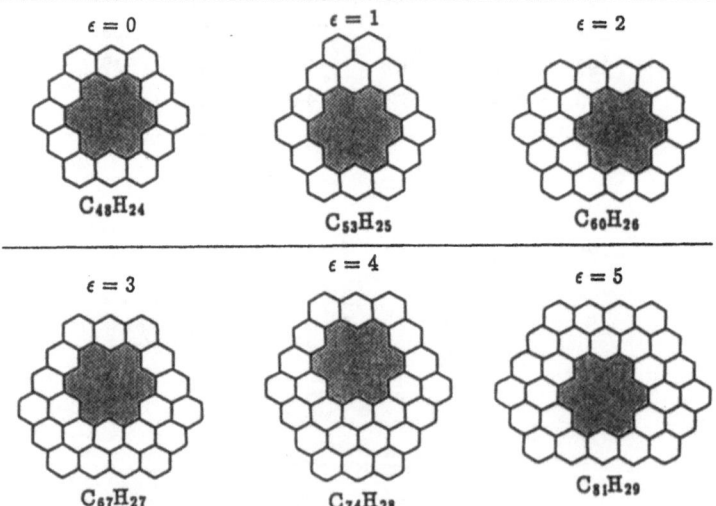

Fig. 7.5. The smallest circular benzenoids perforated by a coronene hole.

Table 7.11. Formulas for circular benzenoids perforated by one coronene hole each.

k	ϵ 0	1	2	3	4	5
0	$C_{46}H_{24}$	$C_{53}H_{25}$	$C_{60}H_{26}$	$C_{67}H_{27}$	$C_{74}H_{28}$	$C_{81}H_{29}$
1	$C_{90}H_{30}$	$C_{97}H_{31}$	$C_{106}H_{32}$	$C_{115}H_{33}$	$C_{124}H_{34}$	$C_{133}H_{35}$
2	$C_{144}H_{36}$	$C_{153}H_{37}$	$C_{164}H_{38}$	$C_{175}H_{39}$	$C_{186}H_{40}$	$C_{197}H_{41}$
3	$C_{210}H_{42}$	$C_{221}H_{43}$	$C_{234}H_{44}$	$C_{247}H_{45}$	$C_{260}H_{46}$	$C_{273}H_{47}$
4	$C_{288}H_{48}$	$C_{301}H_{49}$	$C_{316}H_{50}$	$C_{331}H_{51}$	$C_{346}H_{52}$	$C_{361}H_{53}$
5	$C_{378}H_{54}$	$C_{393}H_{55}$	$C_{410}H_{56}$	$C_{427}H_{57}$	$C_{444}H_{58}$	$C_{461}H_{59}$
6	$C_{480}H_{60}$	$C_{497}H_{61}$	$C_{516}H_{62}$	$C_{535}H_{63}$	$C_{554}H_{64}$	$C_{573}H_{65}$

7.6.3 Results of Enumeration

Crude Total. The symbol $(n_0;\ s_0)$ is now used to identify the six formulas of Fig. 5 (for $k = 0$). Also the symbol $(N_0;\ S_0)$ is redefined inasmuch as it should indicate the six circular benzenoids with $18 \le S_0 \le 23$: $C_{54}H_{18}$, $C_{59}H_{19}$, $C_{66}H_{20}$, $C_{73}H_{21}$, $C_{80}H_{22}$, $C_{87}H_{23}$. Now $N_0 = n_0 + 6$, $S_0 = s_0 - 6$. The coronene hole is completely determined by the central hexagon of coronene. The crude total (J) for $O(n;\ s)$ is therefore equal to the number of hexagons in $O_{k-2}(N_{k-2};S_{k-2})$, where

$$(N_{k-2};\ S_{k-2}) = (6k^2 + 2k(s_0 - 18) + n_0 - 4s_0 + 54;\ 6k + s_0 - 18) \qquad (7.64)$$

Consequently,

$$J = \tfrac{1}{2}(N_{k-2} - S_{k-2}) + 1 = 3k^2 + k(s_0 - 21) + \tfrac{1}{2}(n_0 - 5s_0) + 37 \qquad (7.65)$$

The corresponding generating function reads:

$$J(x) = \frac{n_0 - 5s_0 + 74 + 2x(6s_0 - n_0 - 92) + x^2(n_0 - 7s_0 + 122)}{2(1-x)^3} \qquad (7.66)$$

Listing of Results. A summary of relevant algebraic expressions is found in Table 12. Further specifications are listed in the following.

$\Gamma = D_{6h}$: $\epsilon = 0$, $(n_0;\ s_0) \equiv C_{48}H_{24}$

Introduce $H = I(D_{6h})$ in addition to the above symbols for the numbers of isomers of specific symmetries. Then

Table 7.12. Numbers of isomers of circular benzenoids perforated by one coronene hole each: explicit formulas I; generating functions $I(x)$.

ϵ	I	$I(x)$
0	$\frac{1}{4}(k^2+3k+4+2\lfloor k/2 \rfloor)$	$(1-x)^{-2}(1-x^2)^{-1}$
1	$\frac{1}{2}(k+1)(3k+2)$	$(1+2x)(1-x)^{-3}$
2	$\frac{1}{4}(3k^2+7k+4+2\lceil k/2 \rceil)$	$(1+2x)(1-x)^{-2}(1-x^2)^{-1}$
3	$\frac{1}{2}(k^2+3k+2)=\binom{k+2}{2}$	$(1-x)^{-3}$
4	$\frac{1}{4}(3k^2+9k+8+2\lfloor k/2 \rfloor)$	$(2+x)(1-x)^{-2}(1-x^2)^{-1}$
5	$\frac{3}{2}(k^2+3k+2)=3\binom{k+2}{2}$	$3(1-x)^{-3}$

$$I = H + M + U = \tfrac{1}{12}(J + 11H + 6M) \tag{7.67}$$

where $H = 1$, while eqns. (19), (20), (25), (31), (35) and (36) are valid for the mirror–symmetrical systems.

$\underline{\Gamma = D_{3h}}$: $\epsilon = 3$, $(n_0; s_0) \equiv C_{67}H_{27}$

The scheme of eqn. (50) and the expressions of (51) are valid.

$\underline{\Gamma = D_{2h}}$: $\epsilon = 2$, $(n_0; s_0) \equiv C_{60}H_{24}$ and $\epsilon = 4$, $(n_0; s_0) \equiv C_{74}H_{28}$

In both cases ($\epsilon = 2, 4$), eqn. (61) is applicable. Also in both cases, $M^a = k + 1$, but $M^b = \lceil k/2 \rceil$ for $\epsilon = 2$, while $M^b = 1 + \lfloor k/2 \rfloor$ for $\epsilon = 4$. Therefore

$$M = k + 1 + \lceil k/2 \rceil \quad, \quad M(x) = (1 + 2x)(1 - x)^{-1}(1 - x^2)^{-1} \tag{7.68}$$
for $\epsilon = 2$, and

$$M = k + 2 + \lfloor k/2 \rfloor \quad, \quad M(x) = (2 + x)(1 - x)^{-1}(1 - x^2)^{-1} \tag{7.69}$$
for $\epsilon = 4$.

$\underline{\Gamma = C_{2v}}$: $\epsilon = 1$, $(n_0; s_0) \equiv C_{53}H_{25}$ and $\epsilon = 5$, $(n_0; s_0) \equiv C_{81}H_{29}$

In both cases eqn. (52) holds, where $M = M^b = k + 1$ as in (51).

Numerical values for $k \leq 20$ are presented in Table 13.

Table 7.13. Numbers I of isomers of circular benzenoids perforated by one coronene hole each. For the first C_nH_s formulas ($k \leq 6$), see Table 7.11.

k	ϵ 0	1	2	3	4	5
0	1^a	1^a	1^a	1^a	2^a	3^a
1	2^a	5^a	4^a	3^a	5^a	9^a
2	4^a	12^a	8^a	6^a	10^a	18^a
3	6^a	22^a	14^a	10^a	16^a	30^a
4	9^a	35^a	21^a	15^a	24^a	45^a
5	12^a	51^a	30^a	21^a	33^a	63^a
6	16^a	70^a	40^a	28^a	44^a	84^a
7	20	92	52	36	56	108
8	25	117	65	45	70	135
9	30	145	80	55	85	165
10	36	176	96	66	102	198
11	42	210	114	78	120	234
12	49	247	133	91	140	273
13	56	287	154	105	161	315
14	64	330	176	120	184	360
15	72	376	200	136	208	408
16	81	425	225	153	234	459
17	90	477	252	171	261	513
18	100	532	280	190	290	570
19	110	590	310	210	320	630
20	121	651	341	231	352	693

[a] Cyvin BN, Brunvoll J, Chen RS, Cyvin SJ (1993). Match 29: 131

7.7 Supplementary Enumeration Data

New computations of the numbers of nonisomorphic single coronoids were performed in order to supplement some of the data for $h = 15$. As a result, some new classifications of the 453946 systems (cf. Table 3.2) were achieved.

There are 31177 catacondensed and 422769 pericondensed single coronoids with $h = 15$. This adds one row to Vol.I–Table 5.2. As another coarse classification, it was arrived at 172212 Kekuléan and 281734 non–Kekuléan systems.

The distribution into symmetry groups was determined as $9D_{3h} + 17C_{3h} + 1191C_{2v} + 452729C_s$. Here the last two numbers supplement Vol.I–Table 5.5.

A finer classification of the 172212 Kekuléan single coronoids with $h = 15$ is found in Table 14. The number of concealed non–Kekuléans is $23C_s$ (Cyvin SJ, Brunvoll and Cyvin 1989c).

Notice that the *neo* classification for the single coronoids with $h = 15$ is known: $142388n + 29824e + 281734o$. On the other hand, the corresponding *rio* and *rheo* classifications are not known.

Table 7.14. Classifications of the Kekuléan single coronoids with $h = 15$.

Type*	D_{3h}	C_{3h}	C_{2v}	C_s	Total
prm	2^a	2^a	6^b	58^b	68^c
br	3	5	276	30825	31109
np	1	3	539	110668	111211
e	0	0	54	29770	29824

* Abbreviations: *br* branched catacondensed; *e* essentially disconnected; *np* normal pericondensed; *prm* primitive.

[a] Cyvin SJ, Brunvoll J, Cyvin BN (1988). Acta Chem Scand A42: 434

[b] Brunvoll J, Cyvin BN, Cyvin SJ, Gutman I, Tošić R, Kovačević M (1989). J Mol Struct (Theochem) 184: 165

[c] Balaban AT, Brunvoll J, Cioslowski J, Cyvin BN, Cyvin SJ, Gutman I, He WC, He WJ, Knop JV, Kovačević M, Müller WR, Szymanski K, Tošić R, Trinajstić N (1987). Z Naturforsch 42a: 863

With regard to the classification according to the Δ values, incomplete data for $h = 15$ are listed in Vol.I–Table 7.3. The corresponding complete data are found in Table 15. Here the two largest total numbers (for $\Delta = 0$ and $\Delta = 1$) appear as supplements to I–Table 5.4.

Erratum: In Vol.I–Table 7.3 the number for C_s systems at $h = 15$, $\Delta = 4$ is wrong (cf. Table 14).

Table 7.15. Numbers of single coronoids with different color excess (Δ), classified according to symmetry; complete list for $h = 15$.

h	Δ	D_{3h}	C_{3h}	C_{2v}	C_s	Total
15	0	6	10	875	171344	172235
	1	0	0	157	217896	218053
	2	0	0	121[a]	57384[a]	57505[a]
	3	3[a]	7[a]	33[a]	5952[a]	5995[a]
	4	0	0	5[a]	153	158[a]

[a] Cyvin SJ, Brunvoll J, Cyvin BN (1991). Lecture Notes in Chemistry 54

The data of Tables 1 and 2 for C_nH_s isomers of single coronoids have been extended to $h = 15$; see Tables 16 and 17, respectively.

Finally we give some supplements to the numbers of naphthalenic single coronoids. The 321409 systems of this category with $h = 15$ (Vol.I–Table 8.3; I–Table 8.4) were found to be classified as is shown in Table 18. This table furnishes a supplement to Vol.I–Table 8.5.

Table 7.16. Numbers of single coronoid isomers: complete data for $h = 15$; for $8 \leq h \leq 14$, see Table 7.1.

h	n_i	Formula	Kekuléan	non–Kekuléan	Total
15	0	$C_{60}H_{30}$	31177	0	31177
	1	$C_{59}H_{29}$	0	77935	77935
	2	$C_{58}H_{28}$	79722	30077	109799
	3	$C_{57}H_{27}$	0	100770	100770
	4	$C_{56}H_{26}$	47586	21239	68825
	5	$C_{55}H_{25}$	0	37844	37844
	6	$C_{54}H_{24}$	12174	5637	17811
	7	$C_{53}H_{23}$	0	6996	6996
	8	$C_{52}H_{22}$	1516	719	2235
	9	$C_{51}H_{21}$	0	503	503
	10	$C_{50}H_{20}$	37[a]	14[a]	51[a]

[a] Cyvin SJ, Brunvoll J, Cyvin BN (1991). J Math Chem 8: 63

Table 7.17. Numbers of single coronoid isomers, classified according to *neo* and Δ values, for $h = 15$; for $8 \leq h \leq 14$, and also for abbreviations, see Table 7.2.

h	n_i	Formula	$\Delta = 0$		o (non–Kekuléan)				
			n	e	$\Delta=0$	1	2	3	4
15	0	$C_{60}H_{30}$	31177						
	1	$C_{59}H_{29}$	0			77935			
	2	$C_{58}H_{28}$	64024	15698		0	30077		
	3	$C_{57}H_{27}$	0	0		96791	0	3979	
	4	$C_{56}H_{26}$	35766	11820	18	0	21116	0	105
	5	$C_{55}H_{25}$	0	0	0	36186	0	1658	0
	6	$C_{54}H_{24}$	10013	2161	5	0	5586	0	46
	7	$C_{53}H_{23}$	0	0	0	6662	0	334	0
	8	$C_{52}H_{22}$	1371	145	0	0	712	0	7
	9	$C_{51}H_{21}$	0	0	0	479	0	24	0
	10	$C_{50}H_{20}$	37	0	0	0	14	0	0

Table 7.18. Enumeration and detailed classification of single coronoids with the naphthalene hole and $h = 15$.

h	Type*	Δ	C_{2v}	C_s	Total
15	*br*	0	89	19803	19892
	rp	0	178	53703	53881
	he	0	56	24266	24322
	e	0	19	24791	24810
	o	0	0	23[a]	23[a]
	o	1	12	154736	154748
	o	2	14[b]	39634[b]	39648[b]
	o	3	3[b]	3971[b]	3974[b]
	o	4	2[b]	109[b]	111[b]

* Abbreviations: *br* bridged catacondensed; *e* essentially disconnected; *he* half essentially disconnected; *o* non–Kekuléan; *rp* regular pericondensed.

[a] Cyvin SJ, Brunvoll J, Cyvin BN (1989). J Chem Inf Comput Sci 29: 236

[b] Cyvin SJ, Brunvoll J, Cyvin BN (1991). Lecture Notes in Chemistry 54

Chapter 8

THEOREMS FOR SINGLE CORONOIDS

8.1 Introduction

In this chapter some graph–theoretical proofs are conducted for generalized single coronoids. Some of these proofs are closely connected with the discussion of Sect. 2.2. In particular, the *rheo* conjecture is proved.

8.2 Some Basic Concepts

8.2.1 *Additions*

Benzenoids and coronoids can be generated (or built up) by additions of hexagons to the perimeters (Vol. I–2.1.1; Cyvin BN, Brunvoll and Cyvin 1992b); see also Par. 6.5.1. The *addition modes* are L_1, P_2, L_3, P_4 and L_5, indicating the mode which the added hexagon acquires after the addition (cf. Table 6.4). For coronoids we also need the *corona–condensation*, which is an addition of a hexagon into the mode L_2 or A_2 so that a corona hole is created.

Definitions 8.1: A *normal addition* is an addition (of a hexagon) into the mode L_1, L_3 or L_5. A *regular addition* is either a normal addition or a corona–condensation.

Hence a regular addition is associated with the mode L_1, L_2, A_2, L_3 or L_5. The opposite process of a normal (resp. regular) addition is a *normal* (resp. *regular*) *tearing down*.

8.2.2 *Alternating Cycles*

Let G be a Kekuléan polyhex, and let M represent a Kekulé structure of G. In consistency with the concept of 1–factors (corresponding to Kekulé structures), M is interpreted as a set of edges. They correspond to the double bonds of the Kekulé structure in question, and may therefore appropriately be referred to as M–double bonds. Correspondingly, the edges not belonging to M are sometimes called M–single bonds.

Let the set of edges of G be denoted by E(G).

Definition 8.2: A cycle P of G is an M–*alternating cycle* if the edges of P are alternately in M and E(G) − M.

In other words, P is a conjugated circuit (with alternating single and double bonds) of the Kekulé structure represented by M; see Gutman and Cyvin 1989, and references cited therein, especially Randić 1976a. Sometimes no confusion is caused if we simply speak of an alternating cycle P of G. Then there should exist at least one M so that P is an M–alternating cycle of G.

8.2.3 *Edge Cuts*

Cut segments and (edge) cuts in the context of benzenoids were introduced in the classical paper by Sachs (1984) and taken up quickly by other authors, especially in the likewise classical papers by Zhang, Chen and Guo (1985) and by Kostochka (1985). Since then, several refined definitions of cut segments have been devised and used in graph–theoretical treatments of benzenoids. In particular, varieties of generalized cut (g–cut) segments have been defined (Zhang and Chen 1987; 1989; Zhang and Guo 1988; Sheng 1989; Chen, Cyvin SJ and Cyvin 1990; 1992).

Let G be a single coronoid, and denote its outer and inner perimeter by C' and C", respectively (as in Vol. I–2.1.1).

Definition 8.3: A straight line segment P_1P_2 is called an *elementary cut* (*e–cut*) *segment* of G if –

(1) each of the points P_1 and P_2 is the centre of an edge on C' or C",

(2) P_1P_2 is orthogonal to one of the three edge directions,

(3) P_1P_2 is completely contained in the inside of G except for the two end point P_1 and P_2.

The set of all edges intersected by an e–cut segment P_1P_2 is called an *elementary cut* (*e–cut*) realized by P_1P_2, or simply an e–cut if no confusion is caused.

Definition 8.4: A broken line segment $P_1P_2P_3$ is called a *generalized cut* (*g–cut*) *segment* of G if –

(1) each of the points P_1 and P_3 is the centre of an edge lying on C' or C", and P_2 is the centre of a hexagon of G,

(2) P_1P_2 is orthogonal to one of the three edge directions, while P_1P_2 and P_2P_3 form an angle of $\pi/3$,

(3) $P_1P_2P_3$ is completely contained in G except for the two points P_1 and P_3.

The set of all edges intersected by a g–cut segment $P_1P_2P_3$ is called a *generalized cut* (*g–cut*) realized by $P_1P_2P_3$, or simply a g–cut.

Presently we shall refer to a *special edge cut*, R, as either an e–cut or a g–cut of a single coronoid G. Sometimes also the simpler term "edge cut" is used.

Definition 8.5: Let R be realized by P_1P_2 (resp. $P_1P_2P_3$). Then R is called a *special edge cut of type 1* if both P_1 and P_2 (resp. P_1 and P_3) are simultaneously on C' or C" of G; R is called a *special edge cut of type 2* if one of the points P_1 and P_2 (resp. P_1 and P_3) is on C' and the other is on C".

Examples:

In the above diagram there are five edge cuts: R_a (realized by $P_{1a}P_{2a}$), R_b (realized by $P_{1b}P_{2b}$), R_c (realized by $P_{1c}P_{2c}P_{3c}$), R_d (realized by $P_{1d}P_{2d}$) and R_e (realized by $P_{1e}P_{2e}P_{3e}$). Here R_a, R_b and R_d are e–cuts; R_c and R_e are g–cuts. R_c and R_d are edge cuts of type 1; R_a, R_b and R_e of type 2.

Let R be a special edge cut of G, and denote by G–R the subgraph of G obtained by deleting the edges of R from G. Evidently, if R is of type 2, then G–R is connected. On the other hand, if R is of type 1, then G–R has two components. Let now R_1 and R_2 designate two disjointed special edge cuts of type 2. Then $G-R_1-R_2$, the subgraph of G obtained by deleting all the edges of R_1 and R_2, has two components.

Assume that the vertices of G are colored black and white (cf. Vol. I–3.2.4). Let again R_1 and R_2 designate two special edge cuts of type 2.

Definition 8.6: $\{R_1, R_2\}$ is called a *standard combination* (of edge cuts) if the end vertices of the edges of R_1 and R_2 have the same color when they lie in the same component of $G-R_1-R_2$.

Examples: In the above diagram, $\{R_a, R_b\}$ and $\{R_b, R_e\}$ are standard combinations, whereas $\{R_a, R_e\}$ is not a standard combination.

For a special degenerate single coronoid (Definition 2.2), G, the deletion of an edge belonging to C′ ∩ C″ will not disconnect G, but will open the hole. This is similar to deleting the edges of an e–cut of type 2 for a single coronoid. It is expedient that an edge belonging to C′ ∩ C″ of G also is called an e–cut of type 2.

8.3 A Property of Kekulé Structures

Property 8.1: A Kekuléan single coronoid G has at least one M–alternating cycle for any Kekulé structure M of G.

Proof. From the relations of invariants of single coronoids (Table 4.1) one obtains readily

$$n - 2h - (n_b/2) = 0 \qquad (8.1)$$

Suppose that M contains r edges on the perimeters. Since the total number of edges in M (viz. M–double bonds) is $n/2$, M contains $(n/2) - r$ edges not on the perimeters, viz. internal edges.

Case 1. If $r = n_b/2$, then obviously both the outer and inner perimeter are M–alternating cycles.

Case 2. Suppose now that $r < n_b/2$. We claim that there is at least one hexagon of G being an M–alternating cycle. If not, each hexagon of G would have at most two edges being M–double bonds. Hence

$$2h \geq r + 2(\frac{n}{2} - r) = n - r \qquad (8.2)$$

where it has been taken into account that each internal edge is shared by two hexagons. From (2) together with the initial condition of Case 2 we obtain

$$2h > n - (n_b/2) \qquad (8.3)$$

or

$$n - 2h - (n_b/2) < 0 \qquad (8.4)$$

which contradicts (1). Therefore there is a hexagon of G which is an M–alternating cycle. In other words, a Kekulé structure of G, for which $r < n_b/2$, possesses at least one aromatic sextet (cf. Vol. I–3.4.1).

Examples

1. Consider the below regular (normal) coronoid with $h = 11$, $n = 40$, $n_b = 36$.

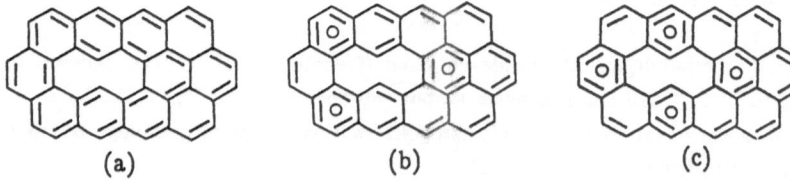

(a)   (b)   (c)

Three Kekulé structures are indicated, each by their $n/2 = 20$ double bonds. In the structures (a) and (b), $r = n_b/2 = 18$. The outer and inner perimeters are alternating cycles in both these structures. This condition (Case 1) does not exclude the possibility for aromatic sextets to be present, as is demonstrated here: in (a) there is no aromatic sextet, but in (b) there are three (marked by small circles). In the structure (c), $r = 16$ and therefore $r < n_b/2$ (Case 2). This condition prescribes with certainty the presence of an aromatic sextet; there are four of them in the structure at hand.

2. Consider the below essentially disconnected coronoid with $h = 11$, $n = 42$, $n_b = 40$.

One Kekulé structure is indicated by its $n/2 = 21$ double bonds. In this case $r = 19$ and therefore $r < n_b/2 = 20$ (Case 2). Hence there must be at least one aromatic sextet present, and there is indeed exactly one (marked with a small circle).

In connection with the above proof and examples it is noted that $2h = n - r$ if and only if $r = n_b/2$ (Case 1), while $2h < n - r$ if and only if $r < n_b/2$ (Case 2).

8.4 Essentially Disconnected Single Coronoids

8.4.1 *Properties and Lemmas*

The following three simple properties of fixed bonds are easily proved.

Property 8.2. An essentially disconnected single coronoid G has a fixed single bond.

Proof. The coronoid G has a fixed bond by definition. If G has a fixed single bond, then there is nothing to prove. If G has a fixed double bond, then the edges incident to this bond must evidently be fixed single bonds. This completes the proof.

In other words, fixed double bonds are always accompanied by fixed single bonds. From this viewpoint we may say that fixed single bonds are the fundamental feature of essentially disconnected systems. The simple reasoning of the above proof is applicable to other essentially disconnected systems than single coronoids as well. It has been applied to benzenoids by Chen, Cyvin SJ and Cyvin (1990).

Property 8.3. A fixed bond cannot be contained in an M–alternating cycle P for a Kekulé structure M of G.

Proof. If E(P) designates the set of edges of the M–alternating cycle P, then M* = (M ∪ E(P)) – (M ∩ E(P)) is another Kekulé structure of G so that P also is an M*–alternating cycle. In simple terms, the single and double bonds of P are just interchanged in M* from M. But this is not possible by the presence of a fixed bond in P. This completes the proof.

The above expression of M* is sometimes called the *symmetric difference* of M and E(P), written M Δ E(P). Property 8.3 applies to a wider class of Kekuléan systems than the single coronoids (Zhang and Zheng 1992). A related property is valid for non–fixed bonds.

Property 8.4: A non–fixed bond of G must be contained in an M–alternating cycle P for a Kekulé structure M of G.

Proof. Let e be a non–fixed bond of G. Then there must be two Kekulé structures M and M' of G such that (without loss of generality) e is an M–double bond and an M'–single bond. It is known that the symmetric difference M Δ M' = (M ∪ M') – (M ∩ M') constitutes several M–alternating cycles, which also are M'–alternating cycles. Evidently one of them contains the edge e. The proof is thus completed.

Properties 8.3 and 8.4 taken together can be expressed as a theorem: for a Kekuléan G, an edge is not fixed if and only if the edge is on an M–alternating cycle for some Kekulé structure M of G. Similarly, for a Kekuléan system G, an edge is a fixed bond if and only if it does not belong to any alternating cycle of G. Similar formulations with proofs are found several places in the literature (He WC and He 1990a; Sheng, Cyvin and Gutman 1991; Hansen and Zheng 1992b; 1993b).

The following lemma (Zhang and Zheng 1992) applies to generalized single coronoids and therefore also to single coronoids G. It deals with fixed single bonds and is therefore relevant to essentially disconnected systems. After some preparative definitions the lemma is formulated, and a proof independent of the one of Zhang and Zheng (1992) is given.

Let e', e and e" be three consecutive edges of a hexagon s in G. Furthermore, consider edges e_1, ... , e_n parallel to e and where e_n is on one of the perimeters C' or C". All e_i's should be contained in the inside of G, i.e. the hexagon s_i which contains the edges e_i and e_{i+1} should belong to G for $i = 1, ..., n-1$. The notation is illustrated in the following diagram.

Lemma 8.1: If e is a fixed single bond in G, and there is a Kekulé structure containing e' and e", then all the edges e_1, ... , e_n are fixed single bonds of G.

Proof. Suppose that e' and e" belong to a Kekulé structure M of G; then e' and e" are referred to as M–double bonds. Since a fixed single bond cannot be contained in an M–alternating cycle (cf. Property 8.3), and e is a fixed single bond, then e_1 cannot belong to M. Consequently, e_1' and e_1" are M–double bonds. If e_2 also was an M–double bond, then e would be contained in an M–alternating cycle consisting of the edges of s and s_1 except e_1 – a contradiction. Hence e_2 does not belong to M, which implies that e_2' and e_2" are M–double bonds. By repeating this argument we show that none of the edges e_1, ... , e_n belong to M, while all the edges e_1', ... , e_{n-1}' and e_1", ... , e_{n-1}" are M–double bonds. Now we want to prove that all the edges e_1, ... , e_n are fixed single bonds of G. Assume that e_1, ... , e_{t-1} (cf. the above diagram) are fixed single bonds, but e_t is not a fixed single bond (if $t = 1$, define $e_{t-1} = e$). Then there is a Kekulé structure M' such that e_t is an M'–double bond. The symmetric difference M Δ M' constitutes several M–alternating cycles, which also are M'–alternating cycles. Suppose that e_t belongs to an M(M')–alternating cycle P*. If both e_{t-1}' and e_{t-1}" (in the case of $t = 1$, define e_{t-1}' = e', e_{t-1}" = e") are contained in P*, then P* is divided by e_{t-1} into two smaller alternating cycles (see the left–hand drawing below). One of these cycles is an M–alternating cycle. Hence e_{t-1} is contained in an M–alternating cycle, contradicting that e_{t-1} is a fixed single bond (by our assumption). If none of e_{t-1}' and e_{t-1}" are contained in P*, then P* can be augmented to form a larger M–alternating cycle which contains the fixed single bond e_{t-1} (see the right–hand drawing in the below diagram) – again a contradiction.

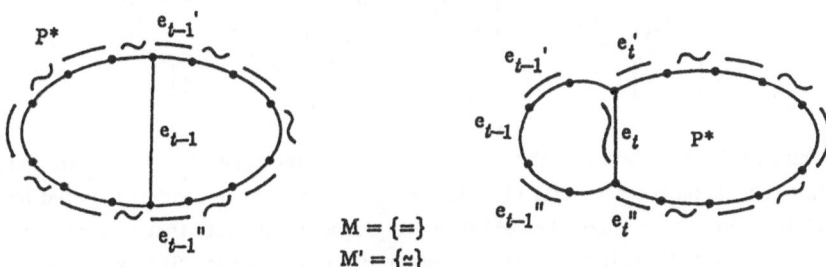

If one and only one of e_{t-1}' and e_{t-1}" is contained in P*, say e_{t-1}", then the edges e_{t-1}, a and b of the hexagon s_{t-1} (see below) and a segment of P* form a cycle with an odd length, which is impossible for G being bipartite.

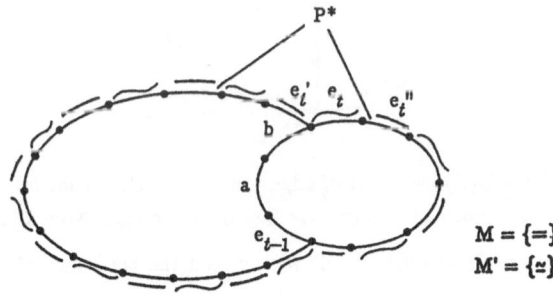

Therefore, e_t cannot be a double bond for any Kekulé structure of G, i.e. e_t is a fixed single bond of G. In analogy, all the edges e_1, ..., e_n are fixed single bonds. Thus the proof is completed.

Also the following lemma (like Lemma 8.1), here formulated for single coronoids G, applies to generalized single coronoids (Zhang and Zheng 1992).

Lemma 8.2: If G has some fixed bonds, then G has at least one fixed single bond on one of its perimeters (C' or C").

Proof. Since G has an M–alternating cycle (Property 8.1), and an edge on an M–alternating cycle cannot be a fixed bond (Property 8.3), we know that G must have some non–fixed bonds. Since G also has some fixed bonds, G must have a fixed single bond (Property 8.2). Therefore we are able to choose a fixed single bond, say e, which is incident to a non–fixed bond, say e'. If e is on C' or C", there is nothing to prove. Assume therefore that e is an internal edge of G. Since e' is a non–fixed bond, there is a Kekulé structure M' so that e' is contained in an M'–alternating cycle P (Property 8.4), and e' is an M'–double bond (see below).

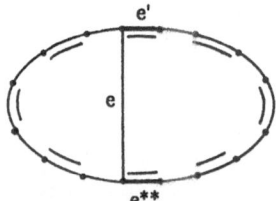

If e" is also an M'–double bond, then by Lemma 8.1 all the edges e_1, ..., e_n are fixed single bonds, where e_n is on the perimeter of G. Hence e_n is a fixed single bond as required by Lemma 8.2. If e" is an M'–single bond, then e** is an M'–double bond. Note that e' is contained in the M'–alternating cycle P. This cycle cannot contain e**. Otherwise the fixed single bond e would also be contained in an M'–alternating cycle, as is demonstrated below, contradicting Property 8.3.

Let M" = M' Δ E(P), where E(P) is the set of edges on P. It is clear that M" is another Kekulé structure of G and such that both e* and e** are M"–double bonds. Now again by Lemma 8.1, all the edges $\tilde{e}_1$, ..., $\tilde{e}_m$ are fixed single bonds with $\tilde{e}_m$ on the perimeter of G. This completes the proof.

8.4.2 *Main Theorem*

In preparation to the following theorem one should recall some definitions and properties of edge cuts (Par. 8.2.3). In the following the term "edge cut" is used as an abbreviation of special edge cut. Let G be a generalized single coronoid, R an edge cut of type 1, and $\{R_1, R_2\}$ a standard combination of edge cuts of type 2. Then G–R, as well as G–R_1–R_2, have two (disconnected) components. Denote such two components by G_1 and G_2. In addition, introduce the numbers of black and white vertices (cf. Vol. I–3.2.4) of G as $n^{(b)}(G)$ and $n^{(w)}(G)$, respectively.

Theorem 8.1: Assume that G has some fixed bonds. Then there is an edge cut R of type 1 or a standard combination $\{R_1, R_2\}$ of edge cuts of type 2 in G, satisfying –

(1) $n^{(b)}(G_1) = n^{(w)}(G_1)$ and $n^{(b)}(G_2) = n^{(w)}(G_2)$,
(2) all the edges of R, R_1 and R_2 are fixed single bonds.

Proof. By Lemma 8.2, G has at least one fixed single bond, say e, on C' or C". If e belongs to C' ∩ C", then G is a degenerate coronoid, and by our convention (Par. 8.2.3), {e} is an e–cut of type 2. Otherwise, e belongs to either C' or C". We distinguish two cases.

Case 1. Edge e' (see the diagram)
is not a fixed double bond, or e' does
not belong to G. Then there is a Kekulé structure of
G, say M, such that e* is an M–double bond. If e** is an M–double bond too, then by Lemma 8.1 all the edges e_1, ... , e_n are fixed single bonds, where e_n is on C' or C". Thus $\{e, e_1, ..., e_n\}$ is an edge cut consisting of fixed single bonds. If e** is an M–single bond, then e" is an M–double bond. We consider the following two subcases.

Subcase 1.1. Edge e" is a fixed double bond. If all the edges e_1", ... , e_n" are fixed double bonds, then $\{e, e_1, ... , e_n\}$ is an edge cut consisting of fixed single bonds. Now suppose that e_1", ... , e_t" are fixed double bonds, but e_{t+1}" is not a fixed double bond. Then there is a Kekulé structure M' of G such that $\tilde{\tilde{e}}$ is an M'–double bond. Edge e_t" is certainly an M'–double bond since it is a fixed double bond. Note that $\tilde{e}$ is a fixed single bond. By Lemma 8.1 all the edges $\tilde{e}_1$, ... , $\tilde{e}_m$ are fixed single bonds. Hence $\{e, e_1, ... , e_t, \tilde{e}, \tilde{e}_1, ... , \tilde{e}_m\}$ is a g–cut consisting of fixed single bonds.

Subcase 1.2. Edge e" is not a fixed double bond. Then there is a Kekulé structure M* ≠ M such that e** is an M*–double bond. It is not difficult to see that the edges in M Δ M* constitute several M–alternating cycles, which also are M*–alternating cycles. Edges e" and e** belong to one of them, say P*. We claim that e* cannot be on P*. Otherwise, an odd–membered cycle P** consisting of a segment of P* and the edge e would be found, contradicting that G is bipartite and therefore has no odd–membered cycle (see below).

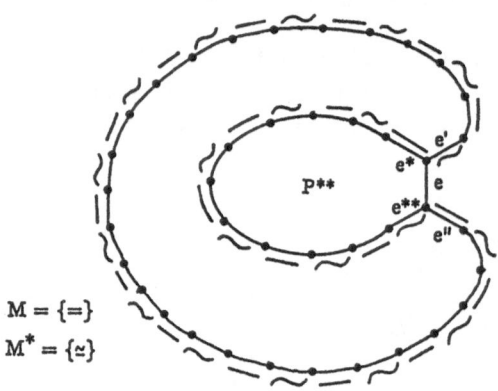

$$M = \{=\}$$
$$M^* = \{\simeq\}$$

Now let $\tilde{M} = M \Delta E(C^*)$. Both e* and e** are $\tilde{M}$–double bonds. Hence by Lemma 8.1 (as in Case 1), $\{e, e_1, ... , e_n\}$ is an edge cut consisting of fixed single bonds.

Case 2. Edge e' is a fixed double bond. This case can be dealt with as Subcase 1.1.

We have proved that a fixed single bond on one of the perimeters (C' or C") determines an edge cut consisting of fixed single bonds including the given fixed single bond.

Case A. If there is a fixed single bond on C' or C" which determines an edge cut R of type 1 consisting of fixed single bonds, then both of the two components G_1 and G_2 of G–R have Kekulé structures. Therefore $n^{(b)}(G_i) = n^{(w)}(G_i)$ for $i = 1, 2$. Hereby Theorem 8.1 is proved for this case.

Case B. Here we assume that all the edge cuts consisting of fixed single bonds which are determined by fixed single bonds on C' or C" are of type 2. We want to prove that there is a standard combination of edge cuts of type 2.

If there is only one fixed single bond e on C' (which we shall prove is impossible), delete all the fixed single bonds in the edge cut of type 2 determined by e, and denote the resulting graph by G'. Then all the edges C' − {e} must be contained in the same subgraph of G'. If G' has no fixed bond, then it is a normal benzenoid by definition; otherwise, delete all the fixed single bonds and the end vertices of all fixed double bonds from G'. The remainder must be the union of some normal benzenoids. If the edges of C' − {e} do not belong to the same normal benzenoid which is a subgraph of G', then some of them must be fixed bonds, contradicting that there is only one fixed single bond on C'. Let the normal benzenoid containing all the edges of C' − {e} be G*. Then all the edges C' − {e} lie on the perimeter of G* and are contained in an M*–alternating cycle, where M* is a Kekulé structure of G*. There is no doubt that M* can be extended to form a Kekulé structure of G. Therefore, we infer that all the edges C' − {e} are contained in an M–alternating cycle for some Kekulé structure M of G. Since e has two end vertices on this M–alternating cycle, it is easy to see that e is also on an M–alternating cycle, contradicting that e is a fixed single bond (see the below diagram). This contradiction implies that there are at least two fixed single bonds on C'.

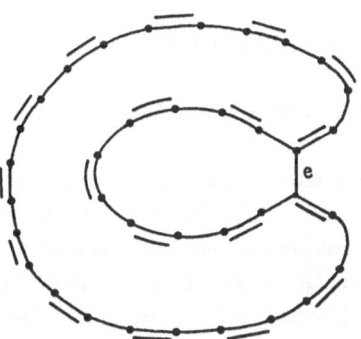

Among the edge cuts determined by fixed single bonds on C', if there is a standard combination $\{R_1, R_2\}$, then both of the two components G_1 and G_2 of $G-R_1-R_2$ have Kekulé structures since all the edges in R_1 and R_2 are fixed single bonds. Hence again $n^{(b)}(G_i) = n^{(w)}(G_i)$ for $i = 1, 2$. We now assume that no two edge cuts of type 2 determined by fixed single bonds on C' form a standard combination. We shall prove that this is impossible for G. Arrange the edges on C' clockwise as $e_1, e_2, \ldots, e_q$ $(q = n_b')$, where e_1 is a fixed single bond. Consider two incident edges e_i and e_{i+1} $(1 \le i \le q$, but for $i = q$, replace $q + 1$ by 1). If the two edges e_i and e_{i+1} are both fixed bonds, then one of them must be a fixed single bond and the other a fixed double bond. Otherwise, if both of these edges were fixed single bonds, then the two edge cuts determined by e_i and e_{i+1} would form a standard combination − a contradiction. Let $e_1, e_2, \ldots, e_f$ be alternately fixed single and fixed double bonds. If $f = q$, then C' is an M–alternating cycle for every Kekulé structure M of G, contradicting that e_1 is a fixed single bond. Hence $f <$ q. Note that e_f is a fixed single bond. Let $e_{f+1}, \ldots, e_{f+t}$ ($f + t \leq q$) be non–fixed bonds. Note that e_{f+t+1} (if $f + t = q$, replace $q + 1$ by 1) is a fixed single bond. By an analogous reasoning as above, $e_{f+1}, \ldots, e_{f+t}$ are contained in a normal benzenoid, say B, defined as the subgraph $G - R_f - R_{f+t+1}$ of G, where R_f and R_{f+t+1} are edge cuts of type 2 determined by e_f and e_{f+t+1}, respectively. The perimeter of B is an M*–alternating cycle for some Kekulé structure M* of B; this follows from a theorem for normal benzenoids (He WC and He 1990a; see also the next section). Since B is a subgraph of G obtained by deleting fixed bonds, M* can certainly be extended to form a Kekulé structure M of G. Since $\{R_f, R_{f+t+1}\}$ is not a standard combination, t must be an odd number. Without loss of generality we may assume that $e_{f+1}, \ldots, e_{f+t}$ are alternately M–double and M–single bonds with the first and the last (viz. e_{f+1} and e_{f+t}) being M–double bonds. Now $e_1, \ldots, e_f, e_{f+1}, \ldots, e_{f+t}, e_{f+t+1}$ are alternately M–single and M–double bonds for some Kekulé structure M of G. If $f + t = q$, then C' is already an M–alternating cycle. If $f + t < q$, we can repeat the above arguments and eventually come to the same conclusion that C' is an M–alternating cycle for some Kekulé structure M of G. Then the fixed single bond e_1 is contained in an M–alternating cycle — a contradiction. This contradiction implies that G must have a standard combination $\{R_1, R_2\}$, where R_1 and R_2 are two edge cuts of type 2 consisting of fixed single bonds and determined by two fixed single bonds on C'.

This completes the proof of Theorem 8.1.

8.4.3 "Essentially Disconnected Coronoids" are Essentially Disconnected

Sheng, Cyvin and Gutman (1991) published a paper with the title: "Essentially Disconnected Benzenoids" are Essentially Disconnected. Therein it is proved that the subgraph obtained from an essentially disconnected benzenoid by deleting all the fixed bonds including the end vertices of the fixed double bonds, is disconnected. The same problem was attacked independently by Hansen and Zheng (1993b), who proved that these disconnected components (also referred to as the effective units; cf. Definitions 2.4) are normal benzenoids. The authors (Hansen and Zheng 1993b) referred to such effective units as "normal components", and they considered more general systems than benzenoids. In the present paragraph a theorem with relevance to these questions is aimed directly at the essentially disconnected single coronoids. The nature of the effective units is accounted for or, in other words, a statement in Par. 2.4.2 is proved rigorously. Presently we shall also use the term "normal component" when the effective unit is a normal benzenoid or a normal generalized single coronoid. Four lemmas are needed in preparation to the theorem about effective units. These lemmas are proved by the same approach as in Property 8.1 (Sect. 8.3).

Consider a benzenoid $B(H, N_i) \equiv B(N; S)$ with the perimeter length N_e, using the same notation as before (cf., e.g., Par. 4.2.1). The relations between different invariants of B are well known and available, e.g., from Table 3.1 on inserting $g = 0$. In the following we shall make use of the relation

$$N - 2H - (N_e/2) = 1 \qquad (8.5)$$

Let A be a set of α nonadjacent vertices on the perimeter of B. Furthermore, B–A designates the subgraph obtained by deleting from B all the vertices of A together with their incident edges. Evidently, in B–A the perimeter of B is broken into several segments. Assume that each of these segments has an even length, i.e. it contains an odd number of vertices.

Lemma 8.3: If B–A has a Kekulé structure M, then B–A has a hexagon which is an M–alternating cycle.

Proof. Suppose that the Kekulé structure M of B–A contains r external edges. Then, under the above conditions, one has certainly

$$r \leq (N_e - 2\alpha)/2 \ , \quad r + \alpha \leq N_e/2 \qquad (8.6)$$

Since the total number of edges in M (viz. M–double bonds) is $(N-\alpha)/2$, M contains $(1/2)(N-\alpha) - r$ internal edges. If none of the hexagons of B–A were an M–alternating cycle, then at most two edges of each hexagon would be M–double bonds. Hence

$$2H \geq r + 2[(1/2)(N-\alpha) - r] = N - (r+\alpha) \qquad (8.7)$$

and by virtue of (6):

$$2H \geq N - (N_e/2) \qquad (8.8)$$

or

$$N - 2H - (N_e/2) \leq 0 \qquad (8.9)$$

which contradicts (5). Therefore, B must have a hexagon being an M–alternating cycle or, in other words, an aromatic sextet.

Consider now a single coronoid G and assume that A is a set of α nonadjacent vertices, all of them on the outer or on the inner perimeter of G. The same restriction is imposed on A as above.

Lemma 8.4: If G–A has a Kekulé structure M, then G–A contains an M–alternating cycle.

Proof. In analogy with (6) one has

$$r \leq (n_b - 2\alpha)/2 = (n_b/2) - \alpha \qquad (8.10)$$

when using the same symbols as in Sect. 8.3.

Case 1. If $r = (n_b/2) - \alpha$, then one of the perimeters (outer or inner) of G is already an M–alternating cycle. More precisely, if the vertices of A are on the inner perimeter of G, then the outer perimeter is an M–alternating cycle, or vice versa.

Case 2. If $r < (n_b/2) - \alpha$, then the relation

$$2h \geq n - (r + \alpha) \tag{8.11}$$

which is obtained in analogy with (7), yields

$$2h > n - (n_b/2) \tag{8.12}$$

by virtue of the condition of the present case. Hence

$$n - 2h - (n_b/2) < 0 \tag{8.13}$$

which contradicts eqn. (1) of Sect. 8.3. This contradiction is caused by the assumption that G has no hexagon being an alternating cycle.

Consequently, G has either a perimeter or a hexagon being an M–alternating cycle. This completes the proof of Lemma 8.4.

Assume now that G is a degenerate generalized single coronoid, and A is a set of nonadjacent vertices, all of them on the outer or on the inner perimeter of G, but not on the intersection of the two perimeters.

Lemma 8.5: If G–A has a Kekulé structure M, then G–A has an M–alternating cycle.

The proof of this lemma can be conducted in full analogy with that of Lemma 8.4. We omit the details.

Let now G be a benzenoid or a generalized single coronoid.

Lemma 8.6: If G has a Kekulé structure M, then G has an M–alternating cycle.

This lemma is obtained from Lemmas 8.3–8.5 by putting A = ∅, i.e. $\alpha = 0$. Property 8.1 (Sect. 8.3) is a special case of Lemma 8.6.

Theorem 8.2: An essentially disconnected single coronoid G has at least two effective units. One of them is a normal benzenoid and an other a normal generalized single coronoid (degenerate or non–degenerate).

Proof. By Theorem 8.1 (Par. 8.4.2), G has an edge cut R of type 1 or a standard combination $\{R_1, R_2\}$ of edge cuts of type 2 satisfying that all the edges of R, R_1 and R_2 are fixed single bonds. Furthermore, G_1 and G_2 have Kekulé structures, where G_1 and G_2 are the two components obtained from G by deleting all the edges of R or of R_1 and R_2, whereby the end vertices should be retained.

If G_i ($i=1,2$) has a pendent edge, delete it together with its end vertices. If there are more pendent edges, repeat this procedure until all of the pendent edges are deleted, and denote the resultant graph by G_i^*. We claim that G_i^* does not equal to the empty graph. In fact, G_i fulfils the condition of one of the Lemmas 8.3–8.6, and by Property 8.3 G_i has non–fixed bonds. But if G_i^* was equal to the empty graph, evidently all the edges of G_i would be fixed bonds — a contradiction. In the following we shall first verify that the conditions of Lemmas 8.3–8.5 are met, taking into account the restriction which was imposed on A.

(1) G has an edge cut R of type 1 satisfying the conditions of Theorem 8.1. If R is an e–cut, then each of G_i has at most one pendent edge. Note that all the pendent edges are fixed double bonds of G. Therefore, if G_i had two pendent edges, then G_i would not have any Kekulé structure — a contradiction. Hence the condition of Lemmas 8.3–8.5 is certainly fulfilled. If R is a g–cut, then G_i has at most two pendent edges. We can check that the condition is fulfilled by means of the below diagram.

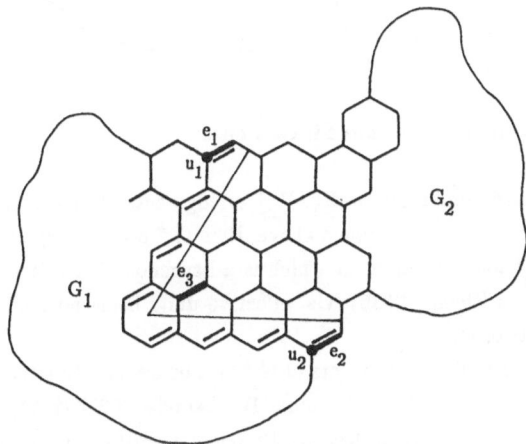

In this case, G_1 has two pendent edges e_1 and e_2 (drawn heavy), while G_2 has one pendent edge e_3 (also heavy). It is easy to see that the segment between u_1 and u_2 (not including u_1 and u_2 themselves) has an odd number of vertices. In the above diagram, $=$ indicates the fixed double bonds of G.

(2) G has a standard combination $\{R_1, R_2\}$ of edge cuts of type 2 satisfying the conditions of Theorem 8.1. In this case, G_i may have at most four pendent edges. Since $\{R_1, R_2\}$ is a standard combination, it is easily ascertained that the end vertex of each pendent edge of G_i has the same color when they do not belong to any hexagon of G_i (see the below diagram). Hence, after deleting the pendent edges together with their end vertices from G_i, the segments of

the perimeter of G_i have odd numbers of vertices. The condition of Lemmas 8.3–8.5 is satisfied again.

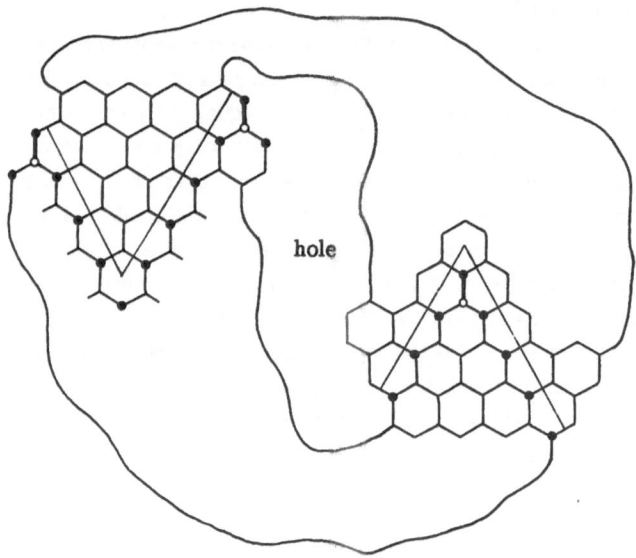

Herefrom the main line of the proof is continued.

Case 1. G has a standard combination $\{R_1, R_2\}$ of edge cuts of type 2, where both R_1 and R_2 consist of fixed single bonds. As mentioned above, then G_i^* does not equal to the empty graph.

Subcase 1.1: G_i^* has a component which is a benzenoid. Then it has at least two normal components (Hansen and Zheng 1993b), viz. effective units being normal benzenoids, which are also normal components of G.

Subcase 1.2: None of the components of G_i^* is a benzenoid. But the 2–connected blocks of a component of G_i^* are evidently benzenoids. By Exercise 3.2.4 (p.47) of Bondy and Murty (1976) there is an end block B^* which has exactly one cut vertex v. Consider the cut edge e = uv being incident to the vertex v. If the number of vertices of B^* is even, then clearly e is a fixed single bond, and by deleting the edge e, it is found similarly as in Subcase 1.1 that B^* has two normal components which also are normal components of G. On the other hand, if the number of vertices of B^* is odd, then clearly e is a fixed double bond, and B^*–u–v has a Kekulé structure. Now B^* is in the same position as G_i^*. Repeating the previous approach for B^*, since B^* is a finite graph and cannot vanish (i.e. become an empty graph), finally a benzenoid must be produced and at least two normal components can be found, which also are normal components of G.

Case 2. G has a special edge cut R of type 1 consisting of fixed single bonds. The proof is similar to that of Case 1, but the components and 2–connected blocks of G_i^* and B^* may contain a generalized single coronoid.

The proof of Theorem 8.2 is completed.

8.5 Normal Single Coronoids

It is known that a Kekuléan benzenoid B is normal if and only if there is a Kekulé structure M of B such that the perimeter of B is an M–alternating cycle. This proposition was first stated by He WC and He (1990a), who claimed it to be valid also for normal coronoids. Zhang, Liu and Chen (1991) pointed out errors in that paper, but confirmed the above proposition for benzenoids as a part of a broad graph–theoretical investigation of such systems (Zhang and Chen 1991). The proposition is not valid for coronoids. Instead, the following theorem has been proved (Zhang and Zheng 1992). It is adhered to the notation C' and C" for the outer and inner perimeter, respectively.

Theorem 8.3: A single coronoid G is normal if and only if there are two Kekulé structures M_1 and M_2 of G such that C' and C" of G are M_1– and M_2–alternating cycles, respectively.

Here M_1 and M_2 may be equal, but need not to be so.

Proof of Sufficiency. Suppose that there are two Kekulé structures M_1 and M_2 of G such that C' and C" are M_1– and M_2–alternating cycles. This implies that none of the edges on C' and C" can be a fixed bond. If G was not normal, i.e. G had some fixed bonds, then by Lemma 8.2 it would be normal and have at least one fixed single bond on C' or C" – a contradiction. Therefore G is normal.

Proof of Necessity. Suppose that G is normal.

First we prove that there is a Kekulé structure M of G such that the outer perimeter (C') of G is an M–alternating cycle. Since G is normal, it has at least two Kekulé structures M' and M"; if G had exactly one Kekulé structure, then all the edges of G would be fixed. It is easy to see that the symmetric difference M' Δ M" constitutes a set of M'(M")–alternating cycles. Let P' be an M'–alternating cycle. If P' is just C', then there is nothing to prove. Now suppose that P' is smaller than C'. Then there is an edge e of G which is not on P' and has one end vertex on P'. Since G is normal, e is not a fixed bond. Hence there is another Kekulé structure M_1 such that e is an M_1–double bond. Then e must be contained in an M'(M_1)–alternating cycle P_1. Note that P' $\neq$ P_1 since e is on P_1, but is not on P'. Therefore, we can find a path on P_1 with the original and terminal vertices on P' (see below). In the case (a) of the below diagram, an M'–alternating cycle P* consisting of a segment of P' and the path mentioned above is found. Evidently, P* is larger than P'. On the other hand, in the case as shown in (b), let M* = M' Δ E(P'). As before, we find an M*–alternating cycle P* which is larger than P'. If P* is just C', there is nothing to prove. If there is still an edge of P* not on C', repeat the above reasoning, and one eventually comes to the conclusion that the outer perimeter (C') of G is an M–alternating cycle for some Kekulé structure M of G since G is finite.

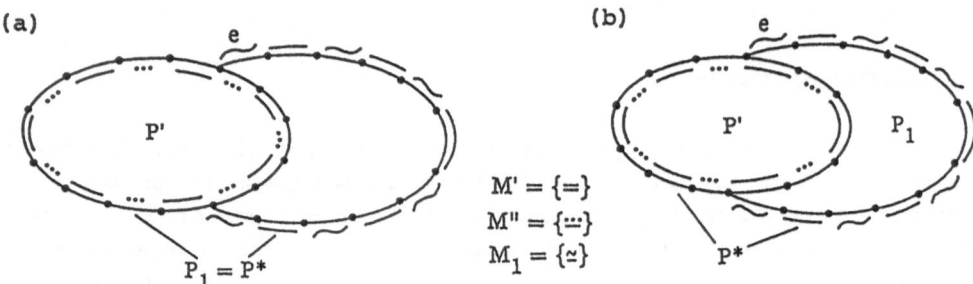

(a)

e

P'

$P_1 = P^*$

(b)

e

P'

P_1

$M' = \{=\}$
$M'' = \{\cdots\}$
$M_1 = \{\simeq\}$

P^*

In the following we prove that the inner perimeter (C'') of G is an $\bar{M}$–alternating cycle for some Kekulé structure $\bar{M}$ of G. Let e* be an edge not on C' (the outer perimeter) and with one end vertex on C'. We know that C' is an M–alternating cycle for some Kekulé structure M of G. Evidently, e* is an M–single bond. Then there is a Kekulé structure M_2 such that e* is an M_2–double bond since e* is not a fixed bond. Therefore, e* is contained in an $M(M_2)$–alternating cycle P_2. The cycle P_2 is not equal to C'. Hence P_2 has a segment with the original and terminal vertices on C' (see below).

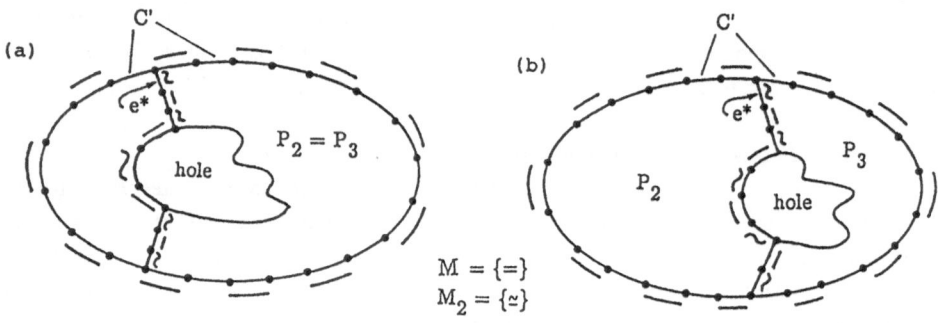

(a)

C'

e*

$P_2 = P_3$

hole

(b)

C'

e*

P_3

P_2

hole

$M = \{=\}$
$M_2 = \{\simeq\}$

In the case (a) of the above diagram, let $M_3 = M$; in the case (b), let $M_3 = M \; \Delta \; E(C')$. It is not difficult to see that C'' is inside an M_3–alternating cycle P_3 which is smaller than C'. Note that P_3 and C'' may have some edges in common. If P_3 is not equal to C'', i.e. if there is at least one edge of P_3 not on C'', we can repeat the above reasoning and find an M_4–alternating cycle, P_4, which is smaller than P_3 and contains C'' as its interior; P_4 may have some edges in common with C''. Since G is finite, we will eventually reach the conclusion that the inner perimeter (C'') of G is an $\bar{M}$–alternating cycle for some Kekulé structure $\bar{M}$ of G.

Theorem 8.3 ': A generalized single coronoid G is normal if and only if there are two Kekulé structures M_1 and M_2 of G such that C' and C" of G are M_1- and M_2-alternating cycles, respectively.

This is a generalization of Theorem 8.3, and also the proof of Theorem 8.3' can be conducted similarly to that of Theorem 8.3. For details, see Zhang and Zheng (1992).

8.6 Regular Single Coronoids

8.6.1 *Definition*

A definition of regular single coronoids is illustrated in Sect. 2.2. Here we give a precise formulation.

Definition 8.7: A single coronoid is regular if it can be subjected to a regular tearing down, hexagon by hexagon, down to benzene.

Under this process, the removing of an L_2- or A_2-mode hexagon, in order to open the corona hole, should be executed only once.

It follows that any regular single coronoid can be built up by regular additions starting from benzene, whereby a corona–condensation should be executed only once.

It is evident that the tearing down or building up of regular single coronoids as described above, passes exclusively through normal benzenoids and regular coronoids in every step.

8.6.2 *A Criterion for a Single Coronoid to be Regular*

Theorem 8.4: A single coronoid G is regular if and only if there is a Kekulé structure M of G such that both C' and C" of G are M–alternating cycles.

Proof of Necessity. Suppose that G is a regular single coronoid. By definition, G can be obtained from benzene by regular additions in three steps (cf. Definition 8.7):

Step 1. A benzenoid G_1 is obtained by normal additions to benzene. Each time only one hexagon is added (into a mode L_1, L_3 or L_5).

Step 2. A coronoid G_2 is formed by a corona–condensation. One hexagon is added (into the mode L_2 or A_2) so that a corona hole is created.

Step 3. The final coronoid G is obtained by normal additions to G_2.

It is not difficult to see that if a benzenoid B has a Kekulé structure M such that the perimeter of B is an M–alternating cycle, then a benzenoid B' obtained by exactly one normal addition to B will have a Kekulé structure M' such that the perimeter of B' is an M'–alternating cycle. This is consistent with known properties of normal benzenoids (cf. Sect. 2.2 and Sect. 8.3), in particular the fact that a normal addition to a normal benzenoid creates another normal benzenoid. Hence all the benzenoids generated in Step 1 are normal, and in particular G_1 is normal.

Let M_1 be a Kekulé structure of G_1 such that the perimeter of G_1 is an M_1–alternating cycle. In the below diagram the edges on the perimeter of G_1 which belong to M_1 are indicated as double bonds.

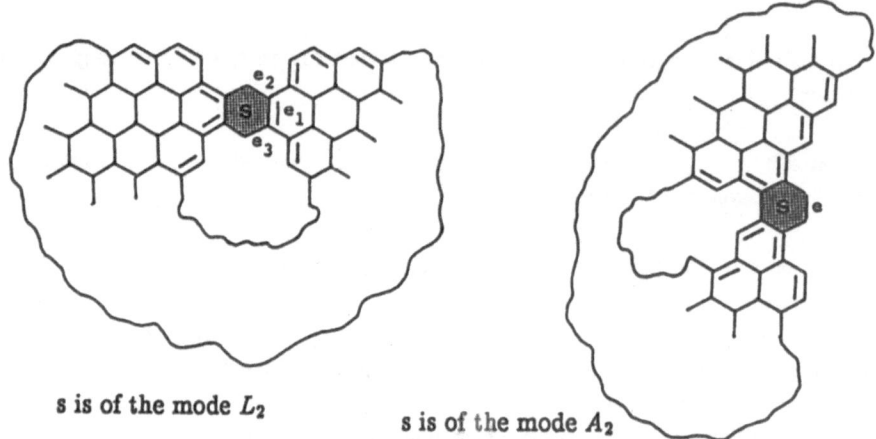

s is of the mode L_2

s is of the mode A_2

Suppose that the hexagon s is added in Step 2. If s is of mode L_2, let $M_2 = (M_1 - \{e_1\}) \cup \{e_2, e_3\}$; if s is of mode A_2, let $M_2 = M_1 \cup \{e\}$ (see the above diagram). Evidently, in both cases M_2 is a Kekulé structure of G_2, and both C' and C'' are M_2–alternating cycles.

In Step 3 it suffices to check (similarly to Step 1) the normal additions to G_2. If the added hexagon s* acquires the mode L_1, let $M = M_2 \cup \{e_1, e_2\}$; if s* acquires the mode L_3, let $M = M_2 \cup \{e\}$; if s* acquires the mode L_5, let $M = M_2$ (see below).

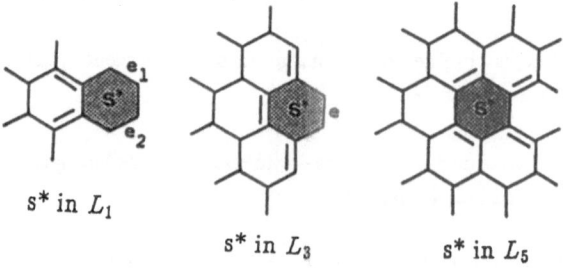

s* in L_1

s* in L_3

s* in L_5

In each case C' and C" of G are M–alternating cycles.

This completes the proof of necessity.

Proof of Sufficiency. Suppose that a single coronoid G has a Kekulé structure M such that both C' and C" are M–alternating cycles. We want to prove that G can be subjected to a regular tearing down, hexagon by hexagon, down to benzene; in other words that G is regular (cf. Definition 8.7). We distinguish three cases.

Case 1. G has a hexagon of mode L_2 or A_2. Note that if a cycle P is an M–alternating cycle for some Kekulé structure M, and E(P) designates the set of edges of P, then $M^* = M \Delta E(P)$ is another Kekulé structure, and P is also an M*–alternating cycle. Hence if G has a Kekulé structure such that both C' and C" are M–alternating cycles, then G has four Kekulé structures with the same property. Without loss of generality we may assume that M is one of the Kekulé structures as indicated in the below diagram.

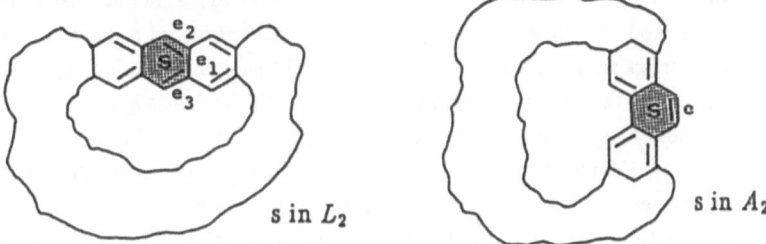

s in L_2 ... s in A_2

Delete hexagon s from G, and denote the resulting benzenoid by G'. If s is of the mode L_2, set M' $= (M - \{e_2, e_3\}) \cup \{e_1\}$; if s is of the mode A_2, set M' $= M - \{e\}$ (see the above diagram). It is evident that M' is a Kekulé structure of G', and the perimeter of G' is an M'–alternating cycle. Hence G' is a normal benzenoid (cf. the opening of Sect. 8.5) and can be subjected to a normal tearing down. Then altogether G' can be subjected to a regular tearing down.

Case 2. G has a hexagon of mode L_1. Since G has a Kekulé structure such that both C' and C" are M–alternating cycles, we may assume without loss of generality that M_1 is one of the Kekulé structures as indicated below.

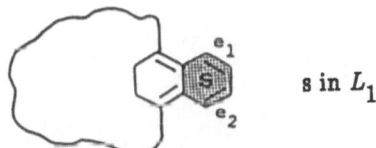

s in L_1

After a normal tearing down of hexagons s from G the resulting coronoid has a Kekulé structure M' such that both C' and C" are M–alternating cycles where M' $= M - \{e_1, e_2\}$.

Case 3. G has no hexagon of mode L_1 or L_2 or A_2. For each top row of G we can label a series of hexagons $s_1, s_2, ... s_t$ on C' of G as shown below.

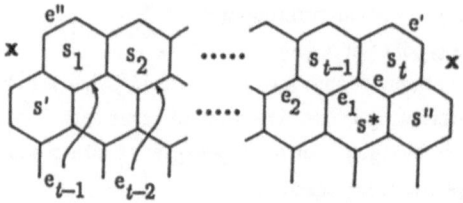

The hexagons s_i and s_{i+1} $(i = 1, 2, ..., t-1)$ have one edge in common, while there is no hexagon belonging to G at the left–hand side of s_1 and at the right–hand side of s_t (marked by crosses).

Subcase 3.1: $t \geq 2$. If neither s' nor s'' belongs to G, then G − C' has no Kekulé structure, contradicting that C' should be an M–alternating cycle. Here G − C' indicates the subgraph obtained by deleting from G all the vertices of C' together with their incident edges. Then, without loss of generality, we may assume that s'' belongs to G. By our assumption that G has no hexagon of the mode A_2, the edge e cannot lie on C'' of G, which implies that s* belongs to G. Hence s_t is of the mode L_3. Now we may assume, again without loss of generality, that e' belongs to M. When e belongs to M, let G' = G − {e'} denote the subgraph obtained by deleting from G the edge e' together with the two end vertices of e' and their incident edges. In other words, G' is the coronoid obtained by deleting s_t from G, which represents a normal tearing down. Evidently, M' = M − {e'} is a Kekulé structure of G', and both the outer and inner perimeters of G' are M'–alternating cycles. If e does not belong to M, then all the edges e_1, e_2, ..., e_{t-1} belong to M. Set M* = M △ E(C'), where E(C') is the set of edges on C' of G. Then M* is another Kekulé structure of G such that both C' and C'' are M*–alternating cycles. Now both e'' and e_{t-1} belong to M*. Similarly to the above argument, G − {e''}, viz. the coronoid resulting from a normal tearing down, has a Kekulé structure M* − {e''} such that both the outer and inner perimeter of G − {e''} are (M* − {e''}) − alternating cycles.

Subcase 3.2: For each top row of G we have $t = 1$. Consider the hexagon s_1 representing a top row at the extreme right. When going clockwise from s_1 along C' we must come to a hexagon s_2 where the perimeter turns left since G is finite. Suppose we have the situation as shown below, where $s_p^* = s_2$.

Then it is possible to reduce the case to Subcase 3.1 by rotating the system so that the hexagons $s_1^* \ldots s_p^*$ ($p \geq 2$) come on the top. Assume now that a top row of this kind (for $p \geq 2$) cannot be achieved by rotating the system. Then we have one of the situations (A) or (B) as shown below.

(A) There is no pair of adjacent vertices of valence two on the perimeter between u and v.

(B) There is at least one pair of adjacent vertices of valence two on the perimeter between u and v.

Since G has a Kekulé structure M such that both C' and C'' are M–alternating cycles, we may assume without loss of generality that M is one of the Kekulé structures with double bonds as indicated in the above diagram. Note that, since s_1' and s_2' are hexagons of G, the row (resp. column) of hexagons from s_2' to s' (resp. from s_1' to s') must belong to G. Otherwise C' or C'' could not be an M–alternating cycle. In case (A) we interchange the double and single bonds of C'; then after a normal tearing down of hexagon s from G, the resulting coronoid has a Kekulé structure M' such that both its outer and inner perimeter are M'–alternating cycles, where M' = M Δ C'. In case (B), if e_1 is an M–double bond, we rotate the double bonds as in case (A) and tear down hexagon s from G. If e_1 is an M–single bond, then e_2, ..., e_l must all be M–double bonds. According to the same reasoning as in Subcase 2.1, after a tearing down of s* from G, where s* has the mode L_3, the resulting coronoid has a Kekulé structure M' such that both C' and C'' are M'–alternating cycles.

In both subcases (Subcase 3.1 and Subcase 3.2) it was demonstrated that a coronoid G' is generated from G by a normal tearing down so that both the outer and inner perimeters of G' are M'–alternating cycles for some Kekulé structure M' of G'. Repeat the argument for G'. Eventually one will arrive at a coronoid with a hexagon of mode L_2 or A_2, whereby the case is reduced to Case 1. Thus the proof of sufficiency is completed.

In summary, Theorem 8.4 is now proved.

An immediate corollary follows from Theorem 8.4. Let G–C'–C" denote the graph obtained by deleting all the vertices on C' and C" of G together with all their incident edges.

Corollary 8.1: A single coronoid G is regular if and only if G–C'–C" has a Kekulé structure.

Remark: It was demonstrated in Sect. 2.2 that the order of deleting hexagons in a regular tearing down is not arbitrary. In particular, the deleting of a hexagon of mode L_3 may cause the resulting coronoid not to be regular. In the proof of sufficiency for Theorem 8.4 we have actually indicated a rule for deleting an L_3 hexagon in a regular tearing down: if G–C'–C" has a Kekulé structure such that e (as indicated on the below diagram) is a double bond, then the deleting of the hexagon s (which is in mode L_3) yields another regular coronoid.

8.7 Half Essentially Disconnected Single Coronoids

8.7.1 *Introductory Remark*

In Sect. 2.2 the half essentially disconnected (HED) single coronoids are defined (provisionally) as the normal single coronoid which are not regular. A direct definition of the HED single coronoids, compatible with the provisional definition of Sect. 2.2, is furnished through the theorem of the next paragraph.

8.7.2 *Main Theorem*

The main theorem for HED single coronoids is formulated in the following.

Theorem 8.5: A normal single coronoid G is not regular if and only if –
(1) the set K of Kekulé structures of G can be divided into two disjointed subsets K_1 and K_2,
(2) K_i ($i = 1, 2$) contains some fixed single bonds which form an edge cut R_i of type 2,
(3) $\{R_1, R_2\}$ is a standard combination.

Here the term "fixed single bond" is used in relation to the subset K_i of Kekulé structures, not the whole set.

Proof of Sufficiency. Suppose that the conditions (1) – (3) are fulfilled for a normal single coronoid G. Then we shall prove by contradiction that G cannot be regular. Assume that the set K of G can be divided into K_1 and K_2 such that K_i ($i = 1, 2$) contains some fixed single bonds $e_{i1}, \ldots, e_{it_i}$ forming an edge cut of type 2, and $\{R_1, R_2\}$ is a standard combination. Suppose that G is regular. Then, by Theorem 8.4, there is a Kekulé structure M of G such that both perimeters C' and C" are M–alternating cycles. Let E(C') and E(C") denote the sets of edges of C' and C", respectively. Without loss of generality we may assume that e_{11} and e_{1t_1} are M–double bonds (see below).

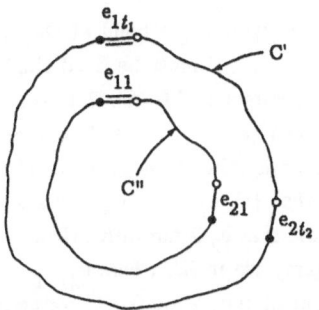

Then e_{21} and e_{2t_2} are M–single bonds. Let $M_1 = M \triangle E(C')$ and $M_2 = M \triangle E(C")$. Thus both M_1 and M_2 are Kekulé structures of G. In M_1, e_{11} and e_{2t_2} are double bonds. Hence M_1 belongs to neither K_1 nor K_2 since e_{11} is a single bond in K_1, and e_{2t_2} is a single bond in K_2. This contradicts that $K = K_1 \cup K_2$. The proof of sufficiency is completed.

Proof of Necessity. Suppose that G is a normal single coronoid which is not regular. Then we shall prove that the conditions (1) – (3) are fulfilled. Delete from G – C' one of the pendent edges (if any) together with its two end vertices repeatedly until no pendent edge is found. By the same approach as in the proof of Theorem 8.2, after deleting all the fixed double bonds together with their end vertices, the effective units of G–C' can be produced, or the remaining subgraph is empty. Let K' be the set of fixed double bonds of G–C'. We distinguish two cases.

Case 1: G–C'–K' is empty, or in G–C' there is a fixed double bond e incident with a vertex of C''. We infer that in the process of breaking the hole of G–C', e is a pendent edge at a certain stage, and $\tilde{e}$ must be a fixed single bond. Furthermore, for each Kekulé structure M of G–C' either e_1 or e_2 (see the above diagram) must be a fixed double bond. Suppose that e_2 belongs to K'. Since $\tilde{e}$ (see above) becomes a fixed single bond of G–C', while e and e_2 are fixed double bonds of G–C', all the edges e_1', ..., e_m' must be fixed single bonds by virtue of Lemma 8.1. Here e_m' is on the outer or inner perimeter of G–C'. Note that G–C' and G have the same inner perimeter, C''. If e_m' is on C'', then $\{e', e_1', ..., e_m'\}$ is a g–cut of type 1 of G, consisting of fixed single bonds of G, contradicting that G is normal. Therefore e_m' must be on the outer perimeter of G–C'. This means that $\{e', e_1', ..., e_m'\}$ is a g–cut of type 2, consisting of fixed single bonds of G–C'. Now suppose that e_2 is not in K'; then e_1 must be in K'. If all the edges e_3, ..., e_{2r-1} (see the above diagram) are in K', where e_{2r-1} is on the outer perimeter of G–C', then $\{e', e_2, ..., e_{2r}\}$ is an e–cut of type 2 for G–C', consisting of fixed single bonds of G–C'. Now assume that $e_1, ..., e_{2t-1}$ are in K', but e_{2t+1} is not in K'. Then e_{2t+2} must be in K'. A similar reasoning as above shows that all the edges $e_1^*, ..., e_p^*$ (see above) are fixed single bonds of G–C'. Thus $\{e', e_2, ..., e_{2t}, e_1^*, ..., e_p^*\}$ is a g–cut of type 2 consisting of fixed single bonds of G–C'. Similarly, e'' also determines (like e') an edge cut of type 2 of G–C', consisting of fixed single bonds of G–C'.

For the sake of convenience we denote by $R_{e'} = \{e', ..., e_u'\}$ and $R_{e''} = \{e'', ..., e_v''\}$ the two edge cuts of G–C' determined by e' and e'', respectively, where e_u' and e_v'' are on the outer perimeter of G–C'. Let e_u (resp. e_v) be the edge on C' of G which is parallel to e_u' (resp. e_v'') and is in the same hexagon with e_u' (resp. e_v''). Let Let $\tilde{R}_{e'} = R_{e'} \cup \{e_u\}$ and $\tilde{R}_{e''} = R_{e''} \cup \{e_v\}$. Then it is not difficult to see that $\{\tilde{R}_{e'}, \tilde{R}_{e''}\}$ is a standard combination for G. Suppose that M* is a Kekulé structure of G–C' and M_* is a Kekulé structure of C'; then $M = M^* \cup M_*$ is a Kekulé structure of G. Moreover, C' is an M–alternating cycle. Since $\{\tilde{R}_{e'}, \tilde{R}_{e''}\}$ is a standard combination, one of the edges e_u and e_v is an M–double bond, while the other is an M–single bond. Bear in mind that all the edges of $R_{e'}$ and $R_{e''}$ are fixed single bonds of G–C' and hence M*–single bonds; therefore they are also M–single bonds. This means that there is only one M–double bond which is contained in $\tilde{R}_{e'} \cup \tilde{R}_{e''}$. Since $\{\tilde{R}_{e'}, \tilde{R}_{e''}\}$ is a standard combination, the above statement is true for all Kekulé structures of G. Thus for an arbitrary Kekulé structure of G, if it has a double bond in $\tilde{R}_{e'}$ (resp. $\tilde{R}_{e''}$), then all the edges in $\tilde{R}_{e''}$ (resp. $\tilde{R}_{e'}$) are single bonds in that structure.

Now denote by K the set of all Kekulé structures of G, by K_1 the set of Kekulé structures of G with no double bond in $\tilde{R}_{e'}$, and by K_2 the set of Kekulé structures of G with no double bond in $\tilde{R}_{e''}$. Evidently, K_1 and K_2 are disjointed subsets of K, and $K = K_1 \cup K_2$. Moreover, K_1 has at least the edges of $\tilde{R}_{e'}$ as its fixed single bonds, and K_2 has at least the edges of $\tilde{R}_{e''}$ as its fixed single bonds. The edge cuts $\tilde{R}_{e'}$ and $\tilde{R}_{e''}$ are of type 2 and form a standard combination.

Case 2: In G–C' there is no fixed double bond incident to any vertex of C". Thus C" must be contained in an effective unit G* which is a generalized coronoid and contains C" as its inner perimeter. By Theorem 8.3 or 8.3', G*–C" has a Kekulé structure M_1'. Let the other effective units of G–C' be G_2, ... , G_r , which have Kekulé structure M_2', ... , M_r', respectively. Clearly ($\bigcup_{i=1}^{r} M_i'$) ∪ K' is a Kekulé structure of G–C'–C" by Corollary 8.1. Then G would be regular – a contradiction. This completes the proof of necessity.

Hereby the proof of Theorem 8.5 is accomplished.

8.7.3 *New Definition*

Now we are able to give a definition of HED single coronoids as it was foreshadowed in the introductory remark (Par. 8.7.1).

Definition 8.8: A normal single coronoid G is half essentially disconnected (HED) if and only if –
  (1) the set of Kekulé structures of G can be divided into two disjointed subsets K_1 and K_2,
  (2) K_i ($i = 1, 2$) contains some fixed single bonds which form an edge cut R_i of type 2,
  (3) $\{R_1, R_2\}$ is a standard combination.

These conditions are identical with those of Theorem 8.5. In particular, the usage of the term "fixed single bond" as referring to the Kekulé structures of a subset (K_i) only, should be noted.

Notice also that the conditions (2) and (3) are indispensable. A regular single coronoid may also have two subsets of Kekulé structures, which together constitute the complete set of Kekulé structures, and where each subset possesses fixed single bonds.

Example (1)

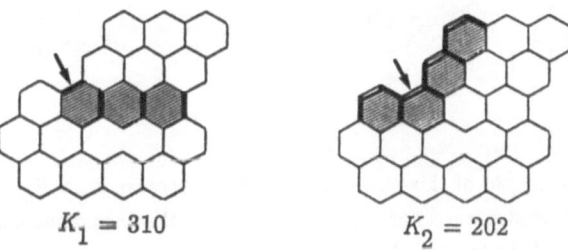

$$K_1 = 310 \qquad K_2 = 202$$

Two subsets of Kekulé structures, K_1 and K_2 with numbers K_1 and K_2, respectively, were produced by assuming one edge (marked by an arrow in the above depictions) to be single and then double. Consequently, in accord with the method of fragmentation (Hall 1973; Randić 1976b; Cyvin SJ and Gutman 1988) it is guaranteed that the two subsets together constitute the complete set of Kekulé structures ($K = K_1 + K_2 = 512$). The fixed single and double bonds for each subset are indicated as heavy and double lines, respectively, within hatched hexagons. The fixed single bonds do not form any special edge cut, and certainly not one of type 2. The coronoid of this example is regular.

Perhaps an even more instructive example is treated in the following.

Example (2)

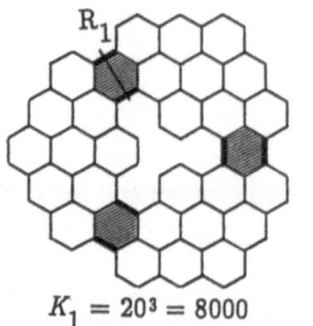

$$K_1 = 20^3 = 8000$$

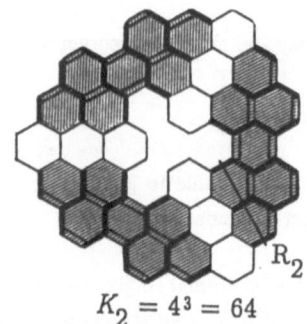

$$K_2 = 4^3 = 64$$

Here again two subsets of Kekulé structures (K_1 and K_2) are indicated. Fixed single bonds within each subset are found so that they form edge cuts of type 2, say R_1 and R_2 associated with K_1 and K_2, respectively, e.g. as indicated above. Moreover, $\{R_1, R_2\}$ is a standard combination:

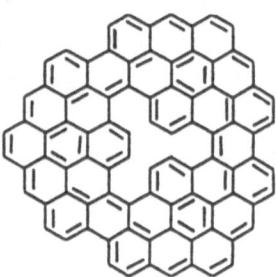

Nevertheless, the coronoid of this example is not HED because $K_1 \cup K_2$ does not constitute the total of 82688 Kekulé structures for this system; in fact, the formidable amount of 74624 Kekulé structures do not belong to $K_1 \cup K_2$. One of them is shown in the right–hand drawing above. This Kekulé structure was selected among those where both the outer and inner perimeter are alternating cycles, confirming that the system is a regular coronoid.

8.7.4 *Example*

After the two examples treated above it will be instructive to treat a coronoid which actually is HED.

Example (3)

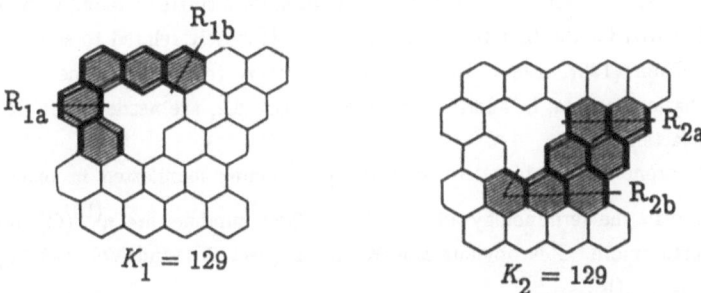

The two subsets (K_1 and K_2) of Kekulé structures are indicated. Two edge cuts containing fixed single bonds are selected within each of subsets K_1 and K_2, viz. (R_{1a}, R_{2a}) and (R_{1b}, R_{2b}), respectively. Here the total number of Kekulé structures is indeed $K_1 + K_2 = 258$, and there exist standard combinations as for instance (when using the edge cuts selected above): {R_{1a}, R_{2a}}, {R_{1b}, R_{2a}}, {R_{1a}, R_{2b}}, {R_{1b}, R_{2b}}. Two of these standard combinations are illustrated below.

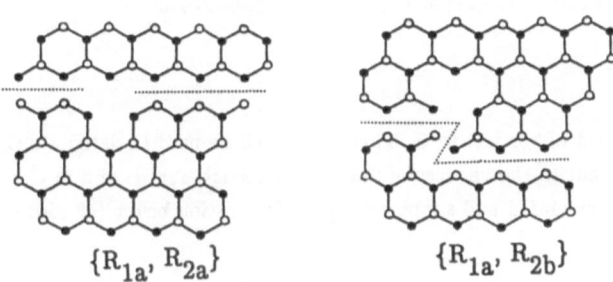

{R_{1a}, R_{2a}}   {R_{1a}, R_{2b}}

8.8 Kekuléan and Non–Kekuléan Single Coronoids

8.8.1 *Introduction*

Necessary and sufficient conditions for a benzenoid to be Kekuléan, and the recognition of Kekuléan/non–Kekuléan benzenoids, are topics of many investigations (Sachs 1984; Zhang, Chen and Guo 1985; Kostochka 1985; He WC and He 1985; Hosoya 1986b; Gutman and Cyvin 1986; Sheng 1987; 1989; Cyvin SJ and Gutman 1987; 1991; He WJ and He 1987; Zhang and Chen 1987; 1989; Zhang and Guo 1988; Sheng, Cyvin and Gutman 1989; Guo and Zhang 1989; 1990; Chen, Cyvin SJ and Cyvin 1990; 1992; Hansen and Zheng 1991. Reviews: Gutman and Cyvin 1988; Hall and Dias 1989; Zhang, Guo and Chen 1990; He WC and He 1990c; Sheng 1990). Zhang and Lin (1992) supplied an algorithm for recognizing Kekulé structures in benzenoids and coronoids. In the present section, a necessary and sufficient condition for a single coronoid to be Kekuléan is formulated and proved for the first time. The treatment is closely related to some of the works by Zhang and Chen (1987; 1989). In preparation to this treatment, some definitions and terminology of basic concepts, in addition to those of Sect. 8.2, are needed. But first we recall some important facts about G.

A single coronoid G is bipartite with the partitioning manifested in black and white vertices according to the terminology adopted here. Their numbers are $n^{(b)}(G)$ and $n^{(w)}(G)$, respectively. Furthermore, G is obvious non–Kekuléan (Sect. 2.1 and Vol. I–3.3.3) when, by definition, $n^{(w)}(G) \neq n^{(b)}(G)$ or

$$\Delta(G) = |n^{(b)}(G) - n^{(w)}(G)| > 0 \tag{8.14}$$

These systems are easily recognized. In the following we shall assume

$$n^{(w)}(G) = n^{(b)}(G) \ , \ \Delta(G) = 0 \tag{8.15}$$

This property tells that G is Kekuléan or concealed non–Kekuléan.

8.8.2 *Old Theorem*

Let S be a set of vertices in G. Then N(S) shall be used to designate the set of vertices in G which are adjacent to the vertices of S, called the *neighbour set* of S in G. A useful theorem of Hall P (1935) is formulated and supported by an illustration below (cf. also Chen, Cyvin SJ and Cyvin 1990).

Theorem (Hall P 1935): Let G be a bipartite graph with bipartition (X, Y). Then G has a matching that saturates every vertex in X if and only if $|N(S)| \geq |S|$ for all $S \subseteq X$.

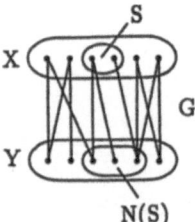

8.8.3 *Additional Basic Concepts and Properties*

Let again G be used to identify a single coronoid, and denote its sets of black and white vertices by B(G) and W(G), respectively. Accordingly, the bipartition of G is (W(G), B(G)).

When $\langle Z \rangle$ is used to designate the graph induced by a set of vertices Z, we may write G = $\langle W(G), B(G) \rangle$. If X is a set of white (resp. black) vertices, then N(X), the neighbour set of X, is clearly a set of black (resp. white) vertices.

Consider the edge cuts R of type 1 or (R_1, R_2) of type 2. Furthermore, $\{R_1, R_2\}$ should be a standard combination. Introduce as previously G_1 and G_2 as the two components of G–R or G–R_1–R_2. The set of black and white vertices in G_1 are $B(G_1)$ and $W(G_1)$, respectively, and correspondingly for G_2, viz. $B(G_2)$ and $W(G_2)$.

Now introduce for G_1 and G_2 either (a) or (b) as follows.

(a) $\quad X_1 = W(G_1)$, $\quad N(X_1) = B(G_1)$, $\quad X_2 = B(G_2)$, $\quad N(X_2) = W(G_2)$;

(b) $\quad X_1 = B(G_1)$, $\quad N(X_1) = W(G_1)$, $\quad X_2 = W(G_2)$, $\quad N(X_2) = B(G_2)$;

In both cases,
$$G_1 = \langle X_1 \cup N(X_1) \rangle , \quad G_2 = \langle X_2 \cup N(X_2) \rangle .$$
But,

(a) $\quad X_1 \cup N(X_2) = W(G)$, $\quad X_2 \cup N(X_1) = B(G)$;

(b) $\quad X_1 \cup N(X_2) = B(G)$, $\quad X_2 \cup N(X_1) = W(G)$.

The alternatives (a) and (b) are not arbitrary; one of them is determined when the coloring is established and the edge cuts R or (R_1, R_2) are given.

Examples

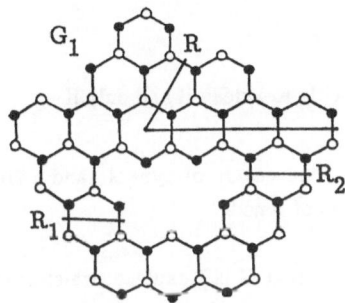

1. Consider the edge cut R of type 1 in the above system, and let G_1 be the larger component (which includes the top hexagon). Then X_1 must be the set of black vertices in G_1, which makes the neighbour set $N(X_1)$ contain the white vertices in G_1. This is the alternative

(b). Here we cannot choose X_1 as the set of white vertices in G_1 because its neighbour set trespasses the border of G_1 into G_2 (the smaller component).

2. Consider the standard combination $\{R_1, R_2\}$ of edge cuts of type 2, and let G_1 be the upper component. Then X_1 must be the set of white vertices in G_1 so that $N(X_1)$ becomes the set of black vertices in G_1. This is the alternative (a).

Now introduce for $i = 1, 2$:

$$D(G_i) = |N(X_i)| - |X_i| \tag{8.16}$$

where $|X_i|$ and $|N(X_i)|$ denote the numbers of vertices in X_i and $N(X_i)$, respectively. When G obeys eqn. (15) one finds

$$D(G_2) - D(G_1) = |X_1| + |N(X_2)| - (|X_2| + |N(X_1)|) \tag{8.17}$$

which is either (a) $n^{(w)}(G) - n^{(b)}(G)$ or (b) $n^{(b)}(G) - n^{(w)}(G)$, in both cases zero. Therefore:

$$n^{(w)}(G) = n^{(b)}(G) \ , \ D(G_1) = D(G_2) \tag{8.18}$$

Since G_1 and G_2 are associated with edge cuts R or (R_1, R_2), it is reasonable to introduce the notation

$$D(R) = D(G_1) = D(G_2) \ , \ D(R_1, R_2) = D(G_1) = D(G_2) \tag{8.19}$$

when eqn. (15) is valid.

8.8.4 *Main Theorem*

Theorem 8.6: A single coronoid G is Kekuléan if and only if –

(1) $n^{(w)}(G) = n^{(b)}(G)$;

(2) $D(R) \geq 0$ for every edge cut R of type 1, and $D(R_1, R_2) \geq 0$ for every standard combination $\{R_1, R_2\}$ of edge cuts of type 2.

Proof of Necessity. Suppose that G is Kekuléan, which is equivalent to saying that G has a perfect matching (in order to approach the terminology in Hall's theorem as quoted above). Since G is bipartite, condition (1) holds. Let $G_1 = \langle X_1 \cup N(X_1) \rangle$ be one of the components of $G-R$, where R is an edge cut of type 1. The bipartition of G is $(W(G), B(G))$, and $X_1 \subset W(G)$ or $X_1 \subset B(G)$ in accord with the alternatives (a) or (b) in Par. 8.8.3. By Hall's theorem, $|N(X_1)| \geq X_1$; hence $D(R) = D(G_1) = |N(X_1)| - |X_1| \geq 0$. The same is true for $D(R_1, R_2)$, according to the same reasoning, when $\{R_1, R_2\}$ is a standard combination of edge cuts of type 2.

Proof of Sufficiency. It suffices to prove that, if G has no Kekulé structure and condition (1) holds, then there exists an edge cut R of type 1 in G satisfying $D(R) < 0$, or a standard combination $\{R_1, R_2\}$ of edge cuts of type 2 satisfying $D(R_1, R_2) < 0$.

Since G has no Kekulé structure (perfect matching), by Hall's theorem there exists a subset $S \subset B(G)$ or $S \subset W(G)$ such that $|S| > |N(S)|$. Let $G' = \langle S \cup N(S) \rangle$, and introduce $\bar{G}' = G - G'$, viz. the subgraph of G obtained by deleting all the vertices of G' together with their incident edges. We claim that there is an $S \subset B(G)$ or $S \subset W(G)$ satisfying $|S| > |N(S)|$ such that both G' and $\bar{G}'$ are connected. Suppose that G' was not connected and had t components $G_1', \ldots, G_t'$ $(t \geq 2)$. Then $G_i' = \langle S_i \cup N(S_i) \rangle$ $(i = 1, \ldots, t)$, while $S = S_1 \cup S_2 \cup \ldots \cup S_t$, and $S_i \cap S_j = \emptyset$ for $i \neq j$; furthermore $N(S) = N(S_1) \cup N(S_2) \cup \ldots \cup N(S_t)$, and $N(S_i) \cap N(S_j) = \emptyset$ for $i \neq j$. Since $|S| > |N(S)|$, there is at least one i $(1 \leq i \leq t)$ such that $|S_i| > |N(S_i)|$. We replace S by S_i and ensure that $\langle S_i \cup N(S_i) \rangle$ is connected. Now suppose that G' is connected. If also $\bar{G}'$ is connected, there is nothing to prove. If $\bar{G}'$ is not connected, $\bar{G}'$ has $h \geq 2$ components $\bar{G}_1', \ldots, \bar{G}_h'$. Use the notation $\bar{G}' = \langle T \cup N(T) \rangle$, $\bar{G}_j' = \langle T_j \cup N(T_j) \rangle$, $j = 1, \ldots, h$; where $T = T_1 \cup \ldots \cup T_h$, $N(T) = N(T_1) \cup \ldots \cup N(T_h)$, $T_i \cap T_j = \emptyset$ and $N(T_i) \cap N(T_j) = \emptyset$ for $i \neq j$. By our assumption, condition (1) of Theorem 8.6 holds, hence $|S| + |N(T)| = |N(S)| + T$. Since $|S| > |N(S)|$, we have $|T| > |N(T)|$ and $\sum_j |T_j| > \sum_j |N(T_j)|$. Therefore there is at least one j satisfying $1 \leq j \leq h$ such that $|T_j| > |N(T_j)|$. Now both $\langle T_j \cup N(T_j) \rangle$ and $G - \langle T_j \cup N(T_j) \rangle$ are connected. Take T_j as S, which completes this part of the proof. In conclusion, we have shown that there is an $S \subset B(G)$ or $S \subset W(G)$ with $|S| > |N(S)|$ such that both $G' = \langle S \cup N(S) \rangle$ and $\bar{G}'$ are connected.

Suppose that J is a Jordan curve (closed, non–self–intersecting curve) in the plane separating G' from $\bar{G}'$. Denote by $(G', \bar{G}')$ the set of edges of G which each has one end vertex in G' and the other in $\bar{G}'$. Then J must intersect all the edges of $(G', \bar{G}')$. We may without loss of generality assume that $(G', \bar{G}')$ contains some edges on the outer or inner perimeter of G. Let J intersect each edge of $(G', \bar{G}')$ at the midpoint of the edge and turn at the centre of a hexagon of G. If $(G', \bar{G}')$ should not contain any edge on a perimeter of G, we could add as many vertices to S as possible such that the added vertices would have the same color as the vertices of S, and the enlarged subset, say S^*, would still satisfy that $\langle S^* \cup N(S^*) \rangle$ and $G - \langle S^* \cup N(S^*) \rangle$ are connected. Now we find that the Jordan curve J can only be a triangle as shown in the below diagram.

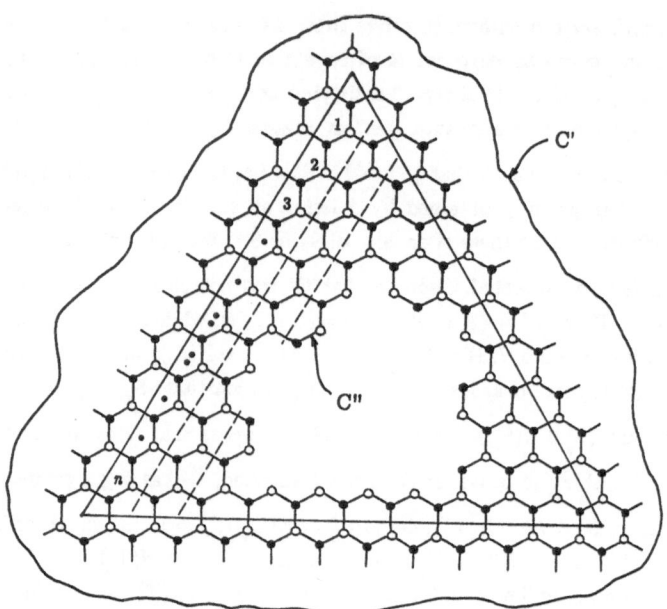

Note that at each turning point of J the angle must primarily be 60° or 300°, but by the maximality of S*, all the angles should be the same, viz. 60°. Starting from S*, delete some vertices (cf. the vertices 1, 2, 3, ... , n found within a trapezoid in the above diagram and being of the same color, here white) to get a new S** such that the corresponding (new) Jordan curve intersects some edges of the inner perimeter of G; then S** maintains the properties of S*, i.e. $|S^{**}| > |N(S^{**})|$, while both $\langle S^{**} \cup N(S^{**})\rangle$ and $G - \langle S^{**} \cup N(S^{**})\rangle$ are connected. Introduce the logical symbols:

$A = \{G' \mid G' = \langle S \cup N(S)\rangle$ such that $S \subset B(G)$ or $S \subset W(G)$, $|S| > |N(S)|$, both G' and $\bar{G}' = G - G'$ are connected, $(G', \bar{G}')$ has some edges on the outer or inner perimeter of G$\}$,

$\alpha = \min\{D(G') = |N(S)| - |S| \mid G' \in A\}$,

$A_{\alpha} = \{G' \mid G \in A, \; D(G') = \alpha\}$,

$\beta = \max\{n(G') \mid G' \in A_{\alpha}\}$,

$A_{\beta} = \{G' \mid G' \in A_{\alpha}, \; n(G') = \beta\}$.

Let $G' \in A_{\beta}$. Assume that the edges $(G', \bar{G}')$ are arranged in the order $e_1, e_2, ... , e_n$ such that the corresponding Jordan curve intersects them consecutively. Only e_1 and e_n are simultaneously on the outer perimeter of G while e_{r-1} and e_r $(2 < r < n)$ are simultaneously on the inner perimeter of G. We distinguish two cases.

Case 1: $(G', \bar{G}')$ has exactly e_1 and e_n simultaneously on the outer or inner perimeter of G.

Subcase 1.1: the edges e_1 and e_n are on the outer perimeter of G. If $e_1, \dots, e_n$ are on the outer perimeter of G. If $e_1, \dots, e_n$ are mutually parallel, then $R = \{e_1, \dots, e_n\}$ is an e–cut of type 1 satisfying $D(R) = D(G') = |N(S)| - |S| < 0$ as required. Now suppose that $e_1, \dots, e_{m-1}$ and e_m $(m < n)$ are mutually parallel, but e_{m+1} is not parallel to them. Bear in mind that the end vertices of e_i $(i = 1, \dots, n)$ have the same color when they belong to the same component G' or $\bar{G}'$. Hence $e_{m+1} = e_1^*$ or $e_{m+1} = e_2^*$ (see Fig. 1). Without loss of generality we may assume that $e_{m+1} = e_1^*$. If $e_{m+1}, e_{m+2}, \dots, e_{n-1}$ and e_n are mutually parallel, then it is not difficult to see that $R = \{e_1, \dots, e_m, \dots, e_n\}$ is a g–cut of type 1 satisfying $D(R) < 0$. Now suppose that $e_{m+1}, \dots, e_{m+t-1}$ and e_{m+t} are parallel, but e_{m+t+1} is not parallel to them. We need to consider two possibilities.

Subcase 1.1.1: $e_{m+t+1} = e_3^*$ (see Fig. 1); Consider the edges $b_1 w_1$ and $b_1^* w_1$; if $b_1 w_1 \in$ G', then $b_1^* w_1^* \in \bar{G}'$. Introduce $G'' = G' \cup \{b_1^*, w_1^*\}$. If $\bar{G}'' = G - G''$ is connected, then $G'' \in$ A. Furthermore, $G'' \in A_\alpha$ since $D(G'') = D(G') = \alpha$. But $n(G'') = n(G') + 2$, contradicting $G' \in A_\beta$. Therefore $\bar{G}''$ is disconnected. This implies that w_1^* is on one of the perimeters of G. Since e_1 and e_n are on the outer perimeter of G by assumption, the inner perimeter of G entirely belongs to G' or $\bar{G}'$. If w_1^* was on the inner perimeter of G, $\bar{G}'' = G - G'' = G' - \{b_1^*, w_1^*\}$ would not be disconnected. Hence w_1^* is on the outer perimeter of G. Let the component of $\bar{G}''$ connected to $e_1, \dots, e_{m-1}$ be $\bar{G}_1'' = \langle T_1 \cup N(T_1) \rangle$. We claim that $|T_1| > |N(T_1)|$. Otherwise let $S^* = S \cup N(T_1) \cup \{b_1^*\}$ (see Fig. 1). Then $N(S^*) = N(S) \cup T_1 \cup \{w_1^*\}$. Introduce $G^* = \langle S^* \cup N(S^*) \rangle$. It is not difficult to see that $G^* \in A_\alpha$. But $n(G^*) > n(G') = \beta$, a contradiction. Now set $R = \{e_1, \dots, e_{m-1}, e^*\}$ (see Fig. 1). Here R is a g–cut and satisfies $D(R) = D(\bar{G}_1'') = |N(T_1)| - |T_1| < 0$. Consider again the edge $b_1 w_1$. If $b_1 w_1 \in \bar{G}'$, an analogous reasoning as above shows that $\{b_1, w_1\}$ is a vertex cut of $\bar{G}'$, and b_1 is on the outer perimeter of G. We distinguish two subcases of Subcase 1.1.1.

Subcase 1.1.1.1: e_5^* is on one of the perimeters of G (see again Fig. 1).

First we suppose that e_5^* is on the outer perimeter of G. Then $R = \{e_1, \dots, e_m, e_5^*\}$ is an e–cut. Let the two components of $G - R$ be G^* and $\bar{G}^*$, where G^* is entirely contained in $\bar{G}'$. Introduce $G^* = \langle S^* \cup N(S^*) \rangle$. If $|S^*| > |N(S^*)|$, then $D(R) = D(G^*) = |N(S^*)| - |S^*| < 0$. Hence R fulfils the required conditions for an edge–cut of type 1. If $|S^*| < |N(S^*)|$, set $S^{**} = S \cup N(S^*)$. Then $N(S^{**}) = N(S) \cup S^* \cup \{b_t\}$. Introduce $G^{**} = \langle S^{**} \cup N(S^{**}) \rangle$. It is evident that $G^{**} \in A_\alpha$. But $n(G^{**}) > n(G')$, contradicting $G' \in A_\beta$. Therefore $|S^*| = |N(S^*)|$. If all edges $b_i w_i$ $(i = 1, \dots, t)$ are on the outer perimeter of G, set $S^{***} = S \cup N(S^*) \cup \{w_1, \dots, w_t\}$. Then $N(S^{***}) = N(S) \cup S^* \cup \{b_1, \dots, b_t\}$. Introduce $G^{***} = \langle S^{***} \cup N(S^{***}) \rangle$. It is easy to see that

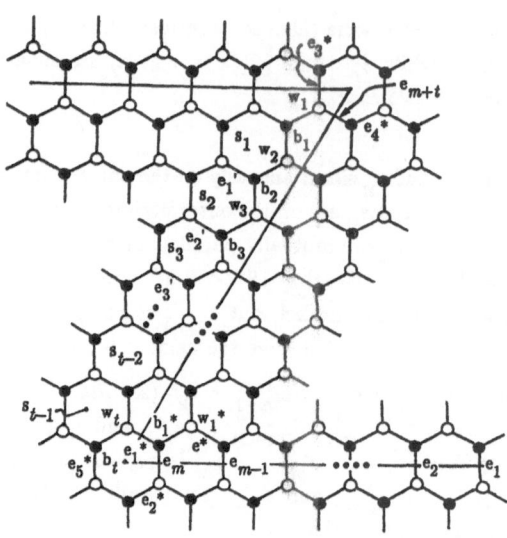

Fig. 8.1. Diagram to Case 1 and Case 2 in the text. Notice the subdivisions of Case 1: Subcase 1.1.1 (including Subcases 1.1.1.1 and 1.1.1.2) and Subcase 1.1.2.

$G^{***} \in A_\alpha$. But $n(G^{***}) > n(G')$, again a contradiction, which implies that there exist $d_f \geq 0$ and $i_f (f = 1, ..., p)$ satisfying $i_1 > 1$, $i_2 > i_1 + d_1 + 1$, ..., $i_p > i_{p-1} + d_{p-1} + 1$, $t > i_p + d_p$, such that one has for the appropriate hexagons (cf. Fig. 1): $s_{i_f} \in G$, $s_{i_f + 1} \in G$, ..., $s_{i_f + d_f} \in G$ ($f = 1, ..., p$) and $s_j \notin G$ for $j \neq i_f$, $j \neq i_f + 1$, $j \neq i_f + d_f (f = 1, ..., p)$. Consider $R_f = \{e_{i_f - 1}', e_{i_f}',$ ...., $e_{i_f + d_f}'\}$ ($f = 1, ..., p$). Note that at most one pair $\{R_{f_1}, R_{f_2}\}$ ($1 \leq f_1 < f_2 \leq p$) is a standard combination of type 2, while all the other R_f's are edge cuts of type 1 (cf. the diagram at the top of the next page). Let the component of $G - R_f$ or $G - R_{f_1} - R_{f_2}$ contained in $\bar{G}'$ be denoted by G_f^* or $G_{f_1 f_2}^*$. Introduce $G_f^* = \langle S_f \cup N(S_f) \rangle$ and $G_{f_1 f_2} = \langle S_{f_1 f_2} \cup N(S_{f_1 f_2}) \rangle$. If $|S_f| \leq |N(S_f)|$ for all $f \neq f_1, ..., f \neq f_p$ ($f = 1, ..., p$) and $|S_{f_1 f_2}| \leq |N(S_{f_1 f_2})|$, set

$$S'' = S \cup N(S^*) \cup \{\bigcup_f N(S_f)\} \cup \{w_1, ..., w_t\} \cup N(S_f).$$

Then

$$N(S'') = N(S) \cup S^* \cup \{\bigcup_f N(S_f)\} \cup \{b_1, ..., b_t\} \cup S_{f_1 f_2}.$$

Introduce $G'' = \langle S'' \cup N(S'') \rangle$. It is not difficult to see that $G'' \in A_\alpha$ and $n(G'') > n(G')$, again a contradiction. This contradiction indicates that there is a k ($1 \leq k \leq p$) such that $|S_k| > |N(S_k)|$, or there is a standard combination $\{R_r, R_s\}$ ($1 \leq r < s \leq p$) such that $|S_{rs}| > |N(S_{rs})|$. Consequently, a special edge–cut of type 1 or a standard combination of edge–cuts of type 2 satisfying condition (2) of Theorem 8.6 is found.

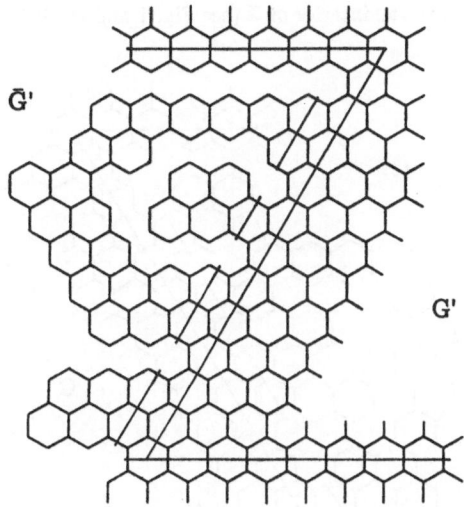

Now we turn to the case when e_5^* is on the inner perimeter of G (see Fig. 1 and the below diagram).

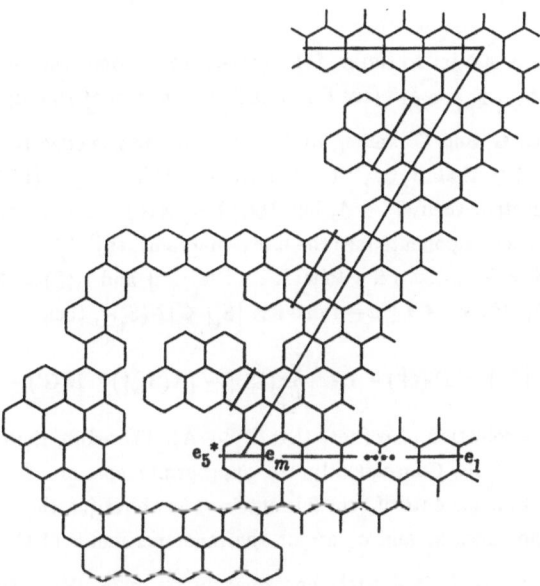

By an analogous reasoning as above we can find a special edge–cut R of type 1 satisfying $D(R) < 0$ or a standard combination $\{R_1, R_2\}$ of edge–cuts of type 2 satisfying $D(R_1, R_2) < 0$, where $R_1 = \{e_1, \ldots, e_m, e_5^*\}$.

Subcase 1.1.1.2: $e_5{}^*$ is in the interior of G (see Fig. 1 and the below diagram).

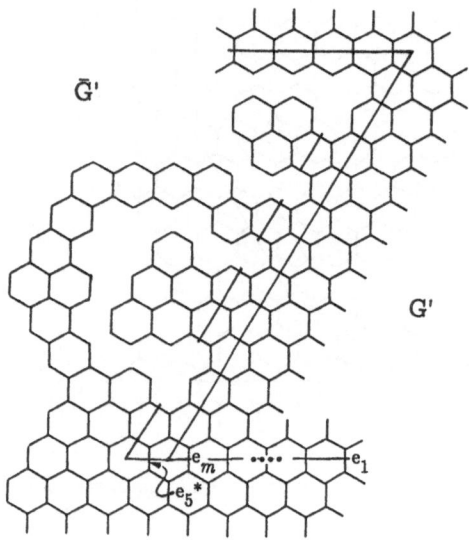

The arguments are quite similar to those of the above case; we omit the details.

Subcase 1.1.2: $e_{m+t+1} = e_4{}^*$ (see Fig. 1 and the diagram at the top of the next page). As mentioned above, both G' and $\bar{G}'$ belong to A_α. Considering the edge $b_1{}^*w_1{}^*$, we may assume without loss of generality that $b_1{}^*w_1{}^* \in G'$. Introduce $G'' = G' - \{b_1{}^*, \ldots , b_t{}^*, w_1{}^*, \ldots, w_{t-1}{}^*\}$. If G'' is connected, then $G'' \in A$, but $D(G'') = D(G') - 1 < \alpha$ contradicts the selection of α. Hence G'' has $d \geq 2$ components (see the below diagram), say G_i'' $(i = 1, \ldots , d)$. Set $G_i'' = \langle S_i \cup N(S_i)\rangle$. Then $S = S_1 \cup \ldots \cup S_d \cup \{w_1{}^*, \ldots , w_{t-1}{}^*\}$ and $N(S) = N(S_1) \cup N(S_2) \cup \ldots \cup N(S_d) \cup \{b_1{}^*, \ldots ,b_t{}^*\}$. If for $1 \leq i \leq d-1$ one has $|S_i| \leq |N(S_i)|$, then

$$D(G_d'') = |N(S_d)| - |S_d| = D(G') - 1 + \sum_{i=1}^{d-1} (|S_i| - |N(S_i)|) \leq D(G') - 1 < \alpha,$$

again contradicting the selection of α (note that $G_d'' \in A$). Therefore, there is a q $(1 \leq q \leq d-1)$ such that $|S_q| > |N(S_q)|$. Consequently, an appropriate special edge–cut of type 1 or a standard combination of edge–cuts of type 2 is obtained as above (cf. the below diagram).

Subcase 1.2: the edges e_1 and e_n are on the inner perimeter of G. We note that in this case the outer perimeter of G is entirely contained in G' or $\bar{G}'$. By a similar reasoning as in Subcase 1.1 we can find an appropriate special edge–cut of type 1. Note that in this case, in contrast to Subcase 1.1, we cannot find an appropriate standard combination of edge–cuts of type 2.

Case 2: $(G', \bar{G}')$ has e_1 and e_n on the outer perimeter of G, while e_{r-1} and e_r are on the inner perimeter of G. We want to show that $R_1 = \{e_1, \ldots, e_{r-1}\}$ and $R_2 = \{e_r, \ldots, e_n\}$ both are special edge–cuts of type 2, while $\{R_1, R_2\}$ is a standard combination of them. If R_1 is not a special edge–cut of type 2, then there are m and t satisfying $m + t < r - 1$ such that $e_1, \ldots, e_m$ are mutually parallel while e_{m+1} is not parallel to these edges; furthermore $e_{m+1}, \ldots, e_{m+t}$ are parallel, while e_{m+t+1} is not parallel to them (cf. Fig. 1). Note that $G - R_1$ is connected if R_1 is an edge–cut of type 2. It is evident that $G - R_1 - \{b_1, w_1\}$ also is connected, and the same is true for $G - R_1 - \{b_1{}^*, w_1{}^*\}$. Hence, in a similar way as in Case 1 we will find a contradiction. Therefore, R_1 is a special edge–cut of type 2. By the same reasoning the same is true for R_2. Finally, by the definition of $(G', \bar{G}')$, $\{R_1, R_2\}$ must be a standard combination.

This completes the proof of Theorem 8.6. It implies that, if its condition (1) is fulfilled, but not (2), then G is a concealed non–Kekuléan single coronoid.

BIBLIOGRAPHY

Agranat I, Hess BA Jr, Schaad LJ (1980) Aromaticity of Non–Alternant Annuleno–annulenes and of Corannulenes. Pure & Appl Chem 52: 1399

Aihara J (1976) On the Number of Aromatic Sextets in a Benzenoid Hydrocarbon. Bull Chem Soc Japan 49: 1429

Aihara J (1992) Is Superaromaticity a Fact or an Artifact? The Kekulene Problem. J Am Chem Soc 114: 865

Aihara J (1993) General Graph Theory of Superaromaticity. Bull Chem Soc Japan 66: 57

Aihara J, Hosoya H (1988) Spherical Aromaticity of Buckminsterfullerene. Bull Chem Soc Japan 61: 2657

Alder RW, Sessions RB (1985) Force Field Calculations on Molecular Belts built from Cyclohexane –1,4–diene Rings. J Chem Soc Perkin Trans II: 1849

Allinger NL, Li FB, Yan LQ, Tai JC (1990) Molecular Mechanics (MM3) Calculations on Conjugated Hydrocarbons. J Comput Chem 11: 868

Angus RO Jr, Johnson RP (1988) Columnar Homoconjugation. J Org Chem 53: 314

Ashton PR, Isaacs NS, Kohnke FH, Slawin AMZ, Spencer CM, Stoddart JF, Williams DJ (1988) Towards the Making of [12]Collarene. Angew Chem Int Ed Engl 27: 966

Babić D (1993) Isospectral Benzenoid Graphs with an Odd Number of Vertices. J Math Chem 12: 137

Babić D, Graovac A (1986) Enumeration of Kekulé Structures in One–dimensional Polymers. Croat Chem Acta 59: 731

Babić D, Gutman I (1992) On Isospectral Graphs. J Math Chem 9: 261

Bakowies D, Thiel W (1991) Theoretical Infrared Spectra of Large Carbon Clusters. Chem Phys 151: 309

Balaban AT (1969) Chemical Graphs – VII – Proposed Nomenclature of Branched *Cata*–Condensed Benzenoid Hydrocarbons. Tetrahedron 25: 2949

Balaban AT (1970) Chemical graphs – XI – (Aromaticity – IX) Isomerism and Topology of Non–Branched *Cata*–Condensed Polycyclic Conjugated Non–Benzenoid Hydrocarbons. Rev Roumaine Chim 15: 1251

Balaban AT (1971) Chemical Graphs – XII – Configurations of Annulenes. Tetrahedron 27: 6115

Balaban AT (1976) Enumeration of Cyclic Graphs [in] Chemical Applications of Graph Theory (Balaban AT, Edit). Academic Press, London: 63

Balaban AT (1980) Is Aromaticity Outmoded? Pure & Appl Chem 52: 1409

Balaban AT (1982) Challenging Problems Involving Benzenoid Polycyclics and Related Systems. Pure & Appl Chem 54: 1075

Balaban AT (1988) Chemical Graphs – Part 49 – Open Problems in the Area of Condensed Polycyclic Benzenoids – Topological Stereoisomers of Coronoids and Congeners. Rev Roumaine Chim 33: 699

Balaban AT (1991) Enumeration of Isomers [in] Chemical Graph Theory – Introduction and Fundamentals (Bonchev D, Rouvray DH, Edit). Abacus Press/Gordon and Breach, New York; Mathematical Chemistry – Volume 1: 177

Balaban AT, Biermann D, Schmidt W (1985) Dualist Graph Approach for Correlating Diels–Alder Reactivities of Polycyclic Aromatic Hydrocarbons. Nouv J Chim 9: 443

Balaban AT, Brunvoll J, Cioslowski J, Cyvin BN, Cyvin SJ, Gutman I, He WC, He WJ, Knop JV, Kovačević M, Müller WR, Szymanski K, Tošić R, Trinajstić N (1987) Enumeration of Benzenoid and Coronoid Hydrocarbons. Z Naturforsch 42a: 863

Balaban AT, Brunvoll J, Cyvin SJ (1991) Chemical Graphs – Part 54 – Enumeration of Unbranched Catacondensed Polyhexes with Equidistant Linearly Condensed Segments. Rev Roumaine Chim 36: 145

Balaban AT, Harary F (1968) Chemical Graphs – V – Enumeration and Proposed Nomenclature of Benzenoid *Cata*–Condensed Polycyclic Aromatic Hydrocarbons. Tetrahedron 24: 2505

Balasubramanian K (1987) Computational Graph Theory [in] Graph Theory and Topology in Chemistry (King RB, Rouvray DH, Edit). Elsevier, Amsterdam; Studies in Physical and Theoretical Chemistry 51: 514

Barth WE, Lawton RG (1966) Dibenzo[*ghi,mno*]fluoranthene. J Am Chem Soc 88: 380

Barth WE, Lawton RG (1971) The Synthesis of Corannulene. J Am Chem Soc 93: 1730

Baumgartner P, Paioni R, Jenny W (1971) Höhere, kondensierte Ringsysteme – 6. Mitteilung – NBS–Dehydrierungen an 9,10–Dihydrophenanthrensystemen – Synthese von [2³](2,7)Phenanthrenophan–trien. Helv Chim Acta 54: 266

Bell TW, Jousselin H (1991) Expanded Heterohelicenes – Molecular Coils That Form Chiral Complexes. J Am Chem Soc 113: 6283

Bergan JL, Cyvin BN, Cyvin SJ (1987) The Fibonacci Numbers, and Kekulé Structures of Some Corona–Condensed Benzenoids (Corannulenes). Acta Chim Hung 124: 299

Bergan JL, Cyvin SJ, Cyvin BN (1986) Number of Kekulé Structures of Single–Chain Corona–Condensed Benzenoids (Cycloarenes). Chem Phys Letters 125: 218

Bonchev D, Balaban AT (1981) Topological Centric Coding and Nomenclature of Poly-cyclic Hydrocarbons – 1 – Condensed Benzenoid Systems (Polyhexes, Fusenes). J Chem Inf Comput Sci 21: 223

Bondy JA, Murty USR (1976) Graph Theory with Applications. Macmillan, London

Borchardt A, Fuchicello A, Kilway KV, Baldridge KK, Siegel JS (1992) Synthesis and Dynamics of the Corannulene Nucleus. J Am Chem Soc 114: 1921

Brendsdal E, Cyvin SJ (1989) Kekulé Structures of Footballene. J Mol Struct (Theochem) 188: 55

Brown RL (1983) Counting of Resonance Structures for Large Benzenoid Hydrocarbons. J Comput Chem 4: 556

Brunvoll J, Cyvin BN, Cyvin SJ (1987) Enumeration and Classification of Coronoid Hydrocarbons. J Chem Inf Comput Sci 27: 14

Brunvoll J, Cyvin BN, Cyvin SJ (1990) Enumeration and Classification of Double Coronoid Hydrocarbons — Appendix — Triple Coronoids. Croat Chem Acta 63: 585

Brunvoll J, Cyvin BN, Cyvin SJ (1992a) Pauling Bond Orders in Coronoid Hydrocarbons — A General Solution. Struct Chem 3: 103

Brunvoll J, Cyvin BN, Cyvin SJ (1992b) Benzenoid Chemical Isomers and Their Enumeration [in] Advances in the Theory of Benzenoid Hydrocarbons II (Gutman I, Edit). Springer—Verlag, Berlin; Topics in Current Chemistry 162: 181

Brunvoll J, Cyvin BN, Cyvin SJ (1993a) Isomers of Polycyclic Conjugated Hydrocarbons With Arbitrary Ring Sizes — Generation and Enumeration. Computers & Chemistry 17: 291

Brunvoll J, Cyvin BN, Cyvin SJ (1993b) More About Extremal Animals. J Math Chem 12: 109

Brunvoll J, Cyvin BN, Cyvin SJ (1993c) Enumeration of Chemical Isomers of Polycyclic Conjugated Hydrocarbons with Different Ring Sizes. Z Naturforsch 48a: 1017

Brunvoll J, Cyvin BN, Cyvin SJ, Gutman I (1988) Essentially Disconnected Benzenoids — Enumeration and Classification of Benzenoid Hydrocarbons — IX. Match 23: 209

Brunvoll J, Cyvin BN, Cyvin SJ, Gutman I, Tošić R, Kovačević M (1989) Enumeration and Classification of Coronoid Hydrocarbons — Part V — Primitive Coronoids. J Mol Struct (Theochem) 184: 165

Brunvoll J, Cyvin BN, Cyvin SJ, Knop JV, Müller WR, Szymanski K, Trinajstić N (1990) Enumeration and Classification of Coronoid Hydrocarbons — Note to a Note. J Mol Struct (Theochem) 207: 131

Brunvoll J, Cyvin SJ (1990) What do We Know about the Numbers of Benzenoid Isomers? Z Naturforsch 45a: 69

Brunvoll J, Cyvin SJ, Cyvin BN (1991) On the Enumeration and Classification of Benzenoid and Coronoid Hydrocarbons. J Mol Struct (Theochem) 235: 147

Burri K, Jenny W (1967) Stereoselektive Umwandlung von [2.2.2.2.2.2]Metacyclophan in ein all–trans–Hexaen. Helv Chim Acta 50: 2542

Cayley A (1875) Ueber die analytische Figuren, welche in der Mathematik Bäume genannt werden und ihre Anwendung auf die Theorie chemischer Verbindungen. Ber Dtsch Chem Ges 8: 1056

Chen RS, Cyvin SJ (1989) Enumeration of Kekulé Structures — Perforated Rectangles. J Mol Struct (Theochem) 200: 251

Chen RS, Cyvin SJ, Cyvin BN (1990) Recognition of Essentially Disconnected Benzenoids. Match 25: 71

Chen RS, Cyvin SJ, Cyvin BN (1992) Recognition of Essentially Disconnected Benzenoids. Acta Math Appl Sinica 8: 377

Chen RS, Cyvin SJ, Cyvin BN, Brunvoll J, Klein DJ (1990) Methods of Enumerating Kekulé Structures, Exemplified by Applications to Rectangle—Shaped Benzenoids [in] Advances in the Theory of Benzenoid Hydrocarbons (Gutman I, Cyvin SJ, Edit). Springer—Verlag, Berlin; Topics in Current Chemistry 153: 227

Chen RS, Guo XF (1993) k–Coverable Coronoid Systems. J Math Chem 12: 147

Cioslowski J (1991) A Conjecture on Benzenoid Graphs. J Math Chem 6: 111

Cioslowski J, Mixon ST, Edwards WD (1991) Weak Bonds in the Topological Theory of Atoms in Molecules. J Am Chem Soc 113: 1083

Cioslowski J, O'Connor PB, Fleischmann ED (1991) Is Superbenzene Superaromatic? J Am Chem Soc 113: 1086

Clar E (1964) Polycyclic Hydrocarbons, Vols. 1, 2. Academic Press, London/Springer–Verlag, Berlin

Clar E (1972) The Aromatic Sextet. Wiley, London

Cvetković D, Gutman I, Trinajstić N (1974) Graph Theory and Molecular Orbitals – VII – The Role of Resonance Structures. J Chem Phys 61: 2700

Cyvin BN, Brunvoll J, Chen RS, Cyvin SJ (1993) Coronenic Coronoids – A Course in Chemical Enumeration [in] Topological Aspects of Benzenoid Hydrocarbons and Related Structures (Cyvin SJ, Edit). Max–Planck–Institut für Strahlenchemie, Mülheim a d Ruhr; Match 29: 131

Cyvin BN, Brunvoll J, Cyvin SJ (1992a) Notes on Fully Benzenoid Hydrocarbons and Their Constant–Isomer Series. J Chem Inf Comput Sci 32: 72

Cyvin BN, Brunvoll J, Cyvin SJ (1992b) Enumeration of Benzenoid Systems and Other Polyhexes [in] Advances in the Theory of Benzenoid Hydrocarbons II (Gutman I, Edit). Springer–Verlag, Berlin; Topics in Current Chemistry 162: 65

Cyvin BN, Brunvoll J, Cyvin SJ (1993) Numbers of Isomers of Some Classes of Pentacyclic Conjugated Hydrocarbons. Bull Chem Soc Japan 66: 2187

Cyvin BN, Brunvoll J, Cyvin SJ, Gutman I (1986) Distribution of K, the Number of Kekulé Structures in Benzenoid Hydrocarbons – Part III – Kekulé Structure Statistics. Match 21: 301

Cyvin BN, Cyvin SJ, Brunvoll J (1988) Number of *Kekulé* Structures for Circumkekulene and its Homologs. Monatsh Chem 119: 563

Cyvin BN, Cyvin SJ, Brunvoll J (1993) A New Picture for Constant–Isomer Series of Benzenoids and Related Systems. J Chem Inf Comput Sci 33: 745

Cyvin BN, Guo XF, Cyvin SJ, Zhang FJ (1992) Enumeration of Helicenes. Chem Phys Letters 188: 537

Cyvin SJ (1983) Symmetry of Kekulé Structures. J Mol Struct 100: 75

Cyvin SJ (1988a) Resolution of Hosoya's Mystery? Bull Chem Soc Japan 61: 4445

Cyvin SJ (1988b) The Number of Kekulé Structures for Primitive Coronoids (Cycloarenes). Chem Phys Letters 147: 384

Cyvin SJ (1989) Enumeration of Kekulé Structures for Some Coronoid Hydrocarbons – "Waffles". Monatsh Chem 120: 243

Cyvin SJ (1990) The Lucas Numbers and Kekulé Structures of Some Coronoid and Coronoid–Like–Hydrocarbons. Acta Chim Hung 127: 849

Cyvin SJ (1991a) Note on the Series of Fully Benzenoid Hydrocarbons With a Constant Number of Isomers. Chem Phys Letters 181: 431

Cyvin SJ (1991b) Enumeration and Classification of Coronoid Hydrocarbons – Part XIII – Circular Coronoids. Coll Sci Papers Fac Sci Kragujevac 12: 95

Cyvin SJ (1991c) Enumeration and Classification of Benzenoid Hydrocarbons – 17 – Further Developments for Constant–Isomer Series. J Chem Inf Comput Sci 31: 340

Cyvin SJ (1992a) Generalization of Extremal Hexagonal Animals (Polyhexes). J Math Chem 9: 389

Cyvin SJ (1992b) Graph–Theoretical Studies on Fluoranthenoids and Fluorenoids – Part 1. J Mol Struct (Theochem) 262: 219

Cyvin SJ (1992c) Constant–Isomer Series of Benzenoid Hydrocarbons. Theor Chim Acta 81: 269

Cyvin SJ, Bergan JL, Cyvin BN (1987) Benzenoids and Coronoids with Hexagonal Symmetry ("Snowflakes"). Acta Chim Hung 124: 691

Cyvin SJ, Brendsdal E, Brunvoll J, Skaret M (1991) Corannulene as a Member of Circulenes – Its Topological Properties and Molecular Vibrations. J Mol Struct 247: 119

Cyvin SJ, Brunvoll J (1989) Chemical Isomers of Coronoid Hydrocarbons. Chem Phys Letters 164: 635

Cyvin SJ, Brunvoll J (1990) The Number of Catacondensed Benzenoids with Thirteen Hexagons, and Enumerations of Some Chemical Benzenoid and Coronoid Isomers. Chem Phys Letters 170: 364

Cyvin SJ, Brunvoll J (1991) Series of Benzenoid Hydrocarbons with a Constant Number of Isomers. Chem Phys Letters 176: 413

Cyvin SJ, Brunvoll J (1992) Generating Functions for the Harary–Read Numbers Classified According to Symmetry. J Math Chem 9: 33

Cyvin SJ, Brunvoll J, Cyvin BN (1988) Molecular Vibrations of Cyclo[d.e.e.d.e.e.d.e.e]–nonakisbenzene and the Topology of Primitive Coronoids with Trigonal Symmetry. Acta Chem Scand A42: 434

Cyvin SJ, Brunvoll J, Cyvin BN (1989a) Topological Aspects of Benzenoids and Coronoids, Including "Snowflakes" and "Laceflowers" [in] Symmetry 2 Unifying Human Understanding (Hargittai I, Edit). Pergamon Press, Oxford; Computers Math Applic 17: 355

Cyvin SJ, Brunvoll J, Cyvin BN (1989b) Distribution of K, the Number of Kekulé Structures, in Benzenoid Hydrocarbons – Normal Benzenoids with K to 110. J Chem Inf Comput Sci 29: 79

Cyvin SJ, Brunvoll J, Cyvin BN (1989c) Search for Concealed Non–Kekuléan Benzenoids and Coronoids. J Chem Inf Comput Sci 29: 236

Cyvin SJ, Brunvoll J, Cyvin BN (1989d) Kekulé Structure Counts of Special Coronoid Hydrocarbons – Hollow Hexagons [in] MATH/CHEM/COMP 1988 (Graovac A, Edit), Elsevier, Amsterdam; Studies in Physical and Theoretical Chemistry 63: 127

Cyvin SJ, Brunvoll J, Cyvin BN (1990a) Enumeration and Classification of Coronoid Hydrocarbons – 10 – Double Coronoids. J Chem Inf Comput Sci 30: 210

Cyvin SJ, Brunvoll J, Cyvin BN (1990b) The Hunt for Concealed Non–Kekuléan Polyhexes. J Math Chem 4: 47

Cyvin SJ, Brunvoll J, Cyvin BN (1990c) Kekulé Structure Counts in Coronoid Hydrocarbons – A General Solution. Struct Chem 1: 429

Cyvin SJ, Brunvoll J, Cyvin BN (1991a) A Periodic Table for All–Benzenoid Hydro-carbons and Enumeration of Some Polyhex Isomers. J Math Chem 8: 63

Cyvin SJ, Brunvoll J, Cyvin BN (1991b) On the Topological Properties of Benzenoid Systems – Some Recent Advances. J Serb Chem Soc 56: 369

Cyvin SJ, Brunvoll J, Cyvin BN (1991c) Theory of Coronoid Hydrocarbons. Springer–Verlag, Berlin; Lecture Notes in Chemistry 54

Cyvin SJ, Brunvoll J, Cyvin BN (1991d) Forms of Benzenoid Chemical Isomers. Match 26: 27

Cyvin SJ, Brunvoll J, Cyvin BN (1991e) Enumeration and Classification of Coronoid Hydrocarbons – Part 12 – Triple Coronoids. Rev Roumaine Chim 36: 313

Cyvin SJ, Brunvoll J, Cyvin BN (1993) Isomers of Polycyclic Conjugated Hydrocarbons with Different Ring Sizes. Chem Phys Letters 205: 343

Cyvin SJ, Brunvoll J, Cyvin BN, Bergan JL, Brendsdal E (1991) Enumeration and Classification of Coronoid Hydrocarbons – Hollow Hexagons. Struct Chem 2: 555

Cyvin SJ, Brunvoll J, Cyvin BN, Brendsdal E (1988) Condensed Aromatics – XXIII – Cyclo[d.e.d.e.d.e.d.e.d.e]Dodekakisbenzene (kekulene). Spectrochim Acta 44A: 975

Cyvin SJ, Brunvoll J, Cyvin BN, Tošić RD (1990) Enumeration of Benzenoids According to Their Perimeter Lengths. Review of Research Fac Sci Univ Novi Sad 20: 5

Cyvin SJ, Brunvoll J, Cyvin BN, Tošić R, Kovačević M (1989) Waffles. J Mol Struct (Theochem) 200: 261

Cyvin SJ, Brunvoll J, Gutman I (1990) Annulenes and Primitive Coronoids – Some Topological Properties and Enumerations. Rev Roumaine Chim 35: 985

Cyvin SJ, Cyvin BN (1993) Theory of Helicenic Hydrocarbons – Part 2 – Chemical Formulas. Struct Chem 4: 303

Cyvin SJ, Cyvin BN, Brunvoll J (1987) Half Essentially Disconnected Coronoid Hydrocarbons. Chem Phys Letters 140: 124

Cyvin SJ, Cyvin BN, Brunvoll J (1989a) Kekulé Structure Counts and Multiple Coronoid Hydrocarbons. Chem Phys Letters 156: 595

Cyvin SJ, Cyvin BN, Brunvoll J (1989b) Benzenoids with Hexagonal Symmetry, Including the Rare Animals "All–Flakes". J Mol Struct 198: 31

Cyvin SJ, Cyvin BN, Brunvoll J (1991) Kekulé Structure Counts for Two Classes of All–Coronoid Hydrocarbons. Acta Chim Hung 128: 419

Cyvin SJ, Cyvin BN, Brunvoll J (1993a) General Formulations for Some Polycyclic Hydrocarbons – Constant–Isomer Series and Formula Index. Chem Phys Letters 201: 273

Cyvin SJ, Cyvin BN, Brunvoll J (1993b) Graph–Theoretical Studies of Fluoranthenoids and Fluorenoids – Part 2. J Mol Struct (Theochem) 281: 229

Cyvin SJ, Cyvin BN, Brunvoll J (1993c) Polycyclic Conjugated Hydrocarbons with Arbitrary Ring Sizes. J Mol Struct 300: 9

Cyvin SJ, Cyvin BN, Brunvoll J (1993d) Contributions to the Theory of Benzenoid Isomers – Some Properties of Extremal Benzenoids. Monatsh Chem 124: 477

Cyvin SJ, Cyvin BN, Brunvoll J (1993e) Enumeration of Benzenoid Chemical Isomers with a Study of Constant–Isomer Series [in] Computer Chemistry (Ugi I, Edit). Springer–Verlag, Berlin; Topics in Current Chemistry 166: 65

Cyvin SJ, Cyvin BN, Brunvoll J, Bergan JL (1987) Coronoid Hydrocarbons with Hexagonal Symmetry. Coll Sci Papers Fac Sci Kragujevac 8: 137

Cyvin SJ, Cyvin BN, Brunvoll J, Brendsdal E, Zhang FJ, Guo XF, Tošić R (1993) Theory of Polypentagons. J Chem Inf Comput Sci 33: 466

Cyvin SJ, Cyvin BN, Brunvoll J, Chen RS (1989) The Number of Kekulé Structures for Rectangle–Shaped Benzenoids, Part VIII – Some Perforated Rectangles. Monatsh Chem 120: 833

Cyvin SJ, Cyvin BN, Brunvoll J, Gutman I (1991) Kekulé Structure Counts for Some Classes of All–Benzenoid and All–Coronoid Hydrocarbons. Monatsh Chem 122: 771

Cyvin SJ, Cyvin BN, Brunvoll J, Gutman I, John P (1993) Towards a Theoretical Basis of the Constant–Isomer Series of Benzenoid Hydrocarbons. J Serb Chem Soc 58: 1

Cyvin SJ, Cyvin BN, Brunvoll J, Hosoya H, Zhang FJ, Klein DJ, Chen RS, Polansky OE (1991) Kekulé Structure Counts – General Formulations for Primitive Coronoid Hydrocarbons. Monatsh Chem 122: 435

Cyvin SJ, Cyvin BN, Brunvoll J, Zhang FJ, Guo XF (1992) Number of Polyhex Isomers of Particular Chemical Interest. J Mol Struct (Theochem) 258: 331

Cyvin SJ, Cyvin BN, Brunvoll J, Zhang FJ, Guo XF, Tošić R (1993) Graph–Theoretical Studies on Fluoranthenoids and Fluorenoids – Enumeration of Some Catacondensed Systems. J Mol Struct (Theochem) 285: 179

Cyvin SJ, Gutman I (1986) Number of Kekulé Structures as a Function of the Number of Hexagons in Benzenoid Hydrocarbons. Z Naturforsch 41a: 1079

Cyvin SJ, Gutman I (1987) Topological Properties of Benzenoid Hydrocarbons – Part XLIV – Obvious and Concealed Non–Kekuléan Benzenoids. J Mol Struct (Theochem) 150: 157

Cyvin SJ, Gutman I (1988) Kekulé Structures in Benzenoid Hydrocarbons. Springer–Verlag, Berlin; Lecture Notes in Chemistry 46

Cyvin SJ, Zhang FJ, Brunvoll J (1992) Enumeration of Perifusenes with One Internal Vertex – A Complete Mathematical Solution. J Math Chem 11: 283

Cyvin SJ, Zhang FJ, Cyvin BN, Guo XF (1993) Theory of Helicenic Hydrocarbons – Part 1 – Invariants and Symmetry. Struct Chem 4: 149

Cyvin SJ, Zhang FJ, Cyvin BN, Guo XF, Brunvoll J (1992) Enumeration and Classification of Benzenoid Systems – 32 – Normal Perifusenes with Two Internal Vertices. J Chem Inf Comput Sci 32: 532

Davies RE, Freyd PJ (1989) $C_{167}H_{336}$ Is the Smallest Alkane with More Realizable Isomers than the Observed Universe Has "Particles". J Chem Ed 66: 278

Davis CC, Cross K, Ebel M (1971) Computer Calculation of Alkane Isomers. J Chem Educ 48: 675

Derflinger G, Sofer H (1968) Die HMO–Koeffizienten der linearen Polyacene in geschlossener Form. Monatsh Chem 99: 1866

Dewar MJS, Longuet–Higgins HC (1952) The Correspondence Between the Resonance and Molecular Orbital Theories. Proc Roy Soc A214: 482

Dias JR (1982a) A Periodic Table for Polycyclic Aromatic Hydrocarbons – Isomer Enumeration of Fused Polycyclic Aromatic Hydrocarbons – 1. J Chem Inf Comput Sci 22: 15

Dias JR (1982b) Periodic Table for Polycyclic Aromatic Hydrocarbons – 2 – Polycyclic Aromatic Hydrocarbons Containing Tetragonal, Pentagonal, Heptagonal, and Octagonal Rings. J Chem Inf Comput Sci 22: 139

Dias JR (1983) A Periodic Table for Polycyclic Aromatic Compounds – Part 3 – Enumeration of all the Polycyclic Conjugated Isomers of Pyrene Having Ring Sizes Ranging from 3 to 9. Match 14: 83

Dias JR (1984a) Isomer Enumeration of Nonradical Strictly Peri–Condensed Polycyclic Aromatic Hydrocarbons. Can J Chem 62: 2914

Dias JR (1984b) A Periodic Table for Polycyclic Aromatic Hydrocarbons – 4 – Isomer Enumeration of Polycyclic Conjugated Hydrocarbons – 2. J Chem Inf Comput Sci 24: 124

Dias JR (1985a) A Periodic Table for Polycyclic Aromatic Hydrocarbons – PartV – 1–Factorable, 2–Factorable, and Dewar Graph Structures Associated with Benzenoid Hydrocarbons. J Macromol Sci–Chem A22: 335

Dias JR (1985b) A Periodic Table for Polycyclic Aromatic Hydrocarbons – Part VII – 1–Factors and 2–Factors of Benzenoid Hydrocarbons. Nouv J Chim 9: 125

Dias JR (1986) A Periodic Table for Polycyclic Aromatic Hydrocarbons – Part IX – Isomer Enumeration and Properties of Radical Strictly Peri–Condensed Polycyclic Aromatic Hydrocarbons. J Mol Struct (Theochem) 137: 9

Dias JR (1987) Handbook of Polycyclic Hydrocarbons – Part A – Benzenoid Hydrocarbons. Elsevier, Amsterdam; Physical Sciences Data 30A

Dias JR (1988) Handbook of Polycyclic Hydrocarbons – Part B – Polycyclic Isomers and Heteroatom Analogs of Benzenoid Hydrocarbons. Elsevier, Amsterdam; Physical Sciences Data 30B

Dias JR (1989a) A Theoretical Study of C_{60} Benzenoids. J Mol Struct (Theochem) 185: 57; (1990) erratum. Ibid 207: 141

Dias JR (1989b) A Formula Periodic Table for Benzenoid Hydrocarbons and the Aufbau and Excised Internal Structure Concepts in Benzenoid Enumerations. Z Naturforsch 44a: 765

Dias JR (1990a) Benzenoid Series Having a Constant Number of Isomers. J Chem Inf Comput Sci 30: 61

Dias JR (1990b) Benzenoid Series Having a Constant Number of Isomers – 2 – Topological Characteristics of Strictly Peri–Condensed Constant–Isomer Benzenoid Series. J Chem Inf Comput Sci 30: 251

Dias JR (1990c) A Formula Periodic Table for Benzenoid Hydrocarbons and the Aufbau and Excised Internal Structure Concepts in Benzenoid Enumerations. J Math Chem 4: 17

Dias JR (1990d) The Current Status of Isomer Enumeration of Useful Benzenoids and New Topological Paradigm. J Phys Org Chem 3: 765

Dias JR (1990e) Constant–Isomer Benzenoid Series and Their Topological Charac–teristics. Theor Chim Acta 77: 143

Dias JR (1990f) A Periodic Table for Benzenoid Hydrocarbons [in] Advances in the Theory of Benzenoid Hydrocarbons (Gutman I, Cyvin SJ, Edit). Springer–Verlag, Berlin; Topics in Current Chemistry 153: 123

Dias JR (1990g) Topological Characteristics of Strictly Pericondensed Constant–Isomer Benzenoid Series. Z Naturforsch 45a: 1335

Dias JR (1991a) Enumeration of Benzenoid Series Having a Constant Number of Isomers. Chem Phys Letters 176: 559

Dias JR (1991b) Series of Fluorenoid/Fluoranthenoid Hydrocarbons Having a Constant Number of Isomers. Chem Phys Letters 185: 10

Dias JR (1991c) Current Status of Isomer Enumeration of Practical Benzenoids. J Mol Struct (Theochem) 230: 155; (1993) erratum. Ibid 279: 325

Dias JR (1991d) Strictly Pericondensed Benzenoid Isomers. Match 26: 87

Dias JR (1991e) Constant–Isomer Benzenoid Series and Their Polyradical Subsets. Theor Chim Acta 81: 125

Dias JR (1992a) Studies in Deciphering the Information Content of Chemical Formulas – A Comprehensive Study of Fluorenes and Fluoranthenes. J Chem Inf Compt Sci 32: 2

Dias JR (1992b) Benzenoid Isomer Enumeration and a New Topological Paradigm. Struct Chem 3: 389

Dias JR (1993) Notes on Constant–Isomer Series. J Chem Inf Comput Sci 33: 117

Diederich F, Staab HA (1978) Benzenoid *versus* Annulenoid Aromaticity – Synthesis and Properties of Kekulene. Angew Chem Inf Ed Engl 17: 372

Diederich F, Whetten RL (1992) Beyond C_{60} – The Higher Fullerenes. Acc Chem Rev 25: 119

Diercks R, Vollhardt KPC (1986) Novel Synthesis of the Angular [3]Phenylene (Terphenylene) by Cobalt–Catalyzed Cyclization of Bis(2–ethynylphenyl)ethyne – a Molecule with an Internal Cyclohexatriene Ring. Angew Chem Int Ed Engl 25: 266

Ege G, Vogler H (1974) Zur Konjugation in makrocyclischen Bindungssystemen – IV – Berechnung des nichtlokalen Anteils der magnetischen Suszeptibilität und der chemischen Verschiebung bei entarteten Eigenwerten – Chemische Verschiebung bei Circumpolyacenen. Theor Chim Acta 35: 189

El–Basil S (1988) Binomial–Combinatorial Properties of Clar Structures. Discrete Appl Math 19: 145

El–Basil S (1992) Combinatorial Properties of Some Macrocyclic Polyhex Systems. Chem Phys Letters 200: 179

El–Basil S, Hussean AE (1993) Clar Sextet Theory of Regular Corannulenes [in] Topological Aspects of Benzenoid Hydrocarbons and Related Structures (Cyvin SJ, Edit). Max–Planck–Institut für Strahlenchemie, Mülheim a d Ruhr; Match 29: 167

Elk SB (1985) Effect of Taxonomy Class and Spanning Set on Identifying and Counting Rings in a Compound. J Chem Inf Comput Sci 25: 11

Elk SB (1993) A Canonical Ordering and Nomenclature for the Kekule Structures of Polybenzenes [in] Topological Aspects of Benzenoid Hydrocarbons and Related Structures (Cyvin SJ, Edit). Max–Planck–Institut für Strahlenchemie, Mülheim a d Ruhr; Match 29: 19

Elser V, Brendsdal E, Cyvin SJ, Brunvoll J, Cyvin BN, Klein DJ (1990) Buckminster-fullerene, Part D – Kekulé Structures [in] Quasicrystals, Networks, and Molecules of Fivefold Symmetry (Hargittai I, Edit). VCH, New York

Farrell EJ (1979) An Introduction to Matching Polynomials. J Combin Theory B27: 75

Feng JK, Li J, Wang ZZ, Zerner MC (1990) Quantum–chemical Investigation of Buckminsterfullerene and Related Carbon Clusters (I) – The Electronic Structure and UV Spectra of Buckminsterfullerene, and Other C_{60} Cages. Int J Quant Chem 37: 599

Fisher ME (1961) Statistical Mechanics of Dimers on a Plane Lattice. Phys Rev (Ser 2) 124: 1664

Flavinskii F (1876) Bemerkung zur der Abhandlung des Herrn Hugo Schiff – "Zur Statistik chemischer Verbindungen". Ber Dtsch Chem Ges 9: 267

Fukushima K (1992) Calculation of Heats of Formation of Intermediate Radicals Expected in the Formation Process of Buckminsterfullerene. J Mol Struct (Theochem) 258: 159

Funhoff DJH, Staab HA (1986) Cyclo[*d.e.d.e.e.d.e.d.e.e*]decakisbenzene, a New Cyclo-arene. Angew Chem Int Ed Engl 25: 742

Gleicher GJ (1967) Calculations on the Corannulene System. Tetrahedron 23: 4257

Godsil CD, Gutman I (1981) Some Remarks on the Matching Polynomial and Its Zeros. Croat Chem Acta 54: 53

Goodson AL (1991) Nomenclature of Chemical Compounds [in] Chemical Graph Theory – Introduction and Fundamentals (Bonchev D, Rouvray DH, Edit). Abacus Press/Gordon and Breach, New York; Mathematical Chemistry – Volume 1: 97

Graovac A, Babić D, Strunje M (1986) Enumeration of Kekulé Structures in Polymers. Chem Phys Letters 123: 433

Graovac A, Gutman I, Trinajstić N (1977) Topological Approach to the Chemistry of Conjugated Molecules. Springer–Verlag, Berlin; Lecture Notes in Chemistry 4

Guo XF, Zhang FJ (1989) A Construction Method for Concealed Non–Kekuléan Benzenoid Systems with h = 12, 13. Match 24: 85

Guo XF, Zhang FJ (1990) Recognizing Kekuléan Benzenoid Systems by $C–P–V$ Path Elimination. J Math Chem 5: 157

Guo XF, Zhang FJ (1992) Mathematical Properties and Structures of Sets of Sextet Patterns of Generalized Polyhexes. J Math Chem 9: 279

Guo XF, Zhang FJ (1993) An Efficient Algorithm for Generating all Kekulé Patterns of a Generalized Benzenoid System. J Math Chem 12: 163

Gutman I (1974) Some Topological Properties of Benzenoid Systems. Croat Chem Acta 46: 209

Gutman I (1979) The Matching Polynomial. Match 6: 75

Gutman I (1982) Topological Properties of Benzenoid Molecules. Bull Soc Chim Beograd 47: 453

Gutman I (1985) Topological Properties of Benzenoid Systems – XLI – Carbon–Carbon Bond Types and Connectivity Indices of Benzenoid Hydrocarbons. J Serb Chem Soc 50: 451

Gutman I (1992) Topological Properties of Benzenoid Systems [in] Advances in the Theory of Benzenoid Hydrocarbons II (Gutman I, Edit). Springer–Verlag, Berlin; Topics in Current Chemistry 162: 1

Gutman I, Cyvin SJ (1986) Recognizing Kekuléan Benzenoid Molecules. J Mol Struct (Theochem) 138: 325

Gutman I, Cyvin SJ (1988) Kekuléan and Non–Kekuléan Benzenoid Hydrocarbons. J Serb Chem Soc 53: 391

Gutman I, Cyvin SJ (1989) Introduction to the Theory of Benzenoid Hydrocarbons. Springer–Verlag, Berlin

Gutman I, Cyvin SJ (1993) Number of Kekulé Structures in Antikekulene and Its Homologs. J Mol Struct (Theochem) 288: 85

Gutman I, Dias JR (1990) The Excised Internal Structure of Hexagonal Systems [in] Contemporary Methods in Graph Theory (Bodendiek R, Edit). BI–Wissenschaftsverlag, Mannheim: 249

Gutman I, El–Basil S (1984) Topological Properties of Benzenoid Systems – XXIV – Computing the Sextet Polynomial. Z Naturforsch 39a: 276

Gutman I, Lee SL (1993) Cyclic Conjugation in Circulenes. Bull Inst Chem Academia Sinica 40: 47

Gutman I, Marković S, Grbović V (1991) The Smallest Pair of Isospectral Benzenoid Systems. J Serb Chem Soc 56: 553

Gutman I, Trinajstić N, Wilcox CF Jr (1975) Graph Theory and Molecular Orbitals – X – The Number of Kekulé Structures and the Thermodynamic Stability of Conjugated Systems. Tetrahedron 31: 143

Hall GG (1973) A Graphical Model of a Class of Molecules. Int J Math Educ Sci Technol 4: 233

Hall GG (1988) Molecules with Holes. Theor Chim Acta 73: 425

Hall GG, Dias JR (1989) The Recognition of Polyhex Hydrocarbon Radicals. J Math Chem 3: 233

Hall P (1935) On Representatives and Subsets. J London Math Soc 10: 26

Hansen P, Zheng ML (1991) A Revised Peeling Algorithm for Determining if a Hexagonal System is Kekuléan. J Mol Struct (Theochem) 235: 293

Hansen P, Zheng ML (1992a) Upper Bounds for the Clar Number of a Benzenoid Hydrocarbon. J Chem Soc Faraday Trans 88: 1621

Hansen P, Zheng ML (1992b) A Linear Algorithm for Fixed Bonds in Hexagonal Systems. J Mol Struct (Theochem) 257: 75

Hansen P, Zheng ML (1993a) Sharp Bounds on the Order, Size, and Stability Number of Graphs. Networks 23: 99

Hansen P, Zheng ML (1993b) Normal Components of Benzenoid Systems. Theor Chim Acta 85: 335

Hanson JC, Nordman CE (1976) The Crystal and Molecular Structure of Corannulene, $C_{20}H_{10}$. Acta Cryst B32: 1147

Harary F, Harborth H (1976) Extremal Animals. J Combin Inf & System Sci 1: 1

Harary F, Palmer EM, Read RC (1975) On the Cell—Growth Problem for Arbitrary Polygons. Discrete Math 11: 371

Harary F, Prins G (1959) The Number of Homeomorphically Irreducible Trees, and Other Species. Acta Math 101: 141

Harary F, Read RC (1970) The Enumeration of Tree—Like Polyhexes. Proc Edinburgh Math Soc (Ser II) 17: 1

Harary F, Schwenk AJ (1973) The Number of Caterpillars. Discrete Math 6: 359

He WC, He WJ (1985) A Novel Nomenclature of Polycyclic Aromatic Hydrocarbons Without Using Graph Centre. Theor Chim Acta 68: 301

He WC, He WJ (1990a) Some Topological Properties of Normal Benzenoids and Coronoids. Match 25: 225

He WC, He WJ (1990b) Some Topological Properties and Generation of Normal Benzenoids. Match 25: 237

He WC, He WJ (1990c) Peak—Valley Path Method on Benzenoid and Coronoid Systems [in] Advances in the Theory of Benzenoid Hydrocarbons (Gutman I, Cyvin SJ, Edit). Springer—Verlag, Berlin; Topics in Current Chemistry 153: 195

He WJ, He WC (1986) One—to—One Correspondence Between Kekulé and Sextet Patterns. Theor Chim Acta 70: 43

He WJ, He WC (1987) On Kekulé Structure and P—V Path Method [in] Graph Theory and Topology in Chemistry (King RB, Rouvray DH, Edit). Elsevier, Amsterdam; Studies in Physical and Theoretical Chemistry 51: 476

He WJ, He WC, Wang QX, Brunvoll J, Cyvin SJ (1988) Supplements to Enumeration of Benzenoid and Coronoid Hydrocarbons. Z Naturforsch 43a: 693

Heilbronner E (1954) Molecular Orbitals in homologen Reihen mehrkerniger aromatischer Kohlenwasserstoffe — I — Die Eigenwerte von LCAO—MO's in homologen Reihen. Helv Chim Acta 37: 921

Hellwinkel D (1970) Das Corannulen—Konzept. Chemiker—Zeitung 94: 715

Henze HR, Blair CM (1931) The Number of Isomeric Hydrocarbons of the Methane Series. J Am Chem Soc 53: 3077

Herndon WC (1973) Enumeration of Resonance Structures. Tetrahedron 29: 3

Herndon WC (1988) New Dimensions in Polynuclear Aromatic Compounds [in] Polynuclear Aromatic Compounds (Ebert LB, Edit), American Chemical Society, Washington DC; Advances in Chemistry Series 217: 1

Herndon WC (1990) On Enumeration and Classification of Condensed Polycyclic Benzenoid Aromatic Hydrocarbons. J Am Chem Soc 112: 4546

Herrmann F (1880) Ueber das Problem, die Anzahl der isomeren Paraffine der Formel C_nH_{2n+2} zu bestimmen. Ber Dtsch Chem Ges 13: 792

Herrmann F (1897) Ueber das Problem, die Anzahl der isomeren Paraffine von der Formel C_nH_{2n+2} zu bestimmen. Ber Dtsch Ges 30: 2423

Herrmann F (1898) Entgegnung. Ber Dtsch Chem Ges 31: 91

Hosoya H (1986a) Matching and Symmetry of Graphs [in] Symmetry Unifying Human Understanding (Hargittai I, Edit.) Pergamon Press, New York; Computers Math Applic 12B: 271

Hosoya H (1986b) How to Design Non–Kekulé Polyhex Graphs? Croat Chem Acta 59: 583

Hosoya H, Aida M, Kumagai R, Watanabe K (1987) Analysis of the π–Electronic Structure of Infinitely Large Networks – I – Some Remarks on the Characteristic Polynomial and Density of States of Large Polycyclic Aromatic Hydrocarbons. J Comput Chem 8: 358

Janata J, Gendell J, Ling CY, Barth W, Backes L, Mark HB Jr, Lawton RG (1967) Concerning the Anion and Cation Radicals of Corannulene. J Am Chem Soc 89: 3056

Jenny W, Burri K (1967) [2.2.2.2.2.2]Metacyclophan–hexaen. Chimia 21: 534

Jenny W, Paioni R (1969) [2.2.2](2,7)Phenanthrenophan–trien. Chimia 23: 41

Jenny W, Peter R (1965) Coronaphene, eine neue Gruppe cyclischer Kohlenwasserstoffe. Angew Chem 77: 1027

Joesten MD, Johnston DO, Netterville JT, Wood JL (1991). World of Chemistry. Saunders College Publishing, Philadelphia

John P (1990) Über ein einfaches Wachstum hexagonaler Systeme und das Verhalten der Paulingschen Bindungsordnung. Wiss Z Friedrich Schiller Universität Jena, Naturwiss Reihe 39: 192

John P (1992a) Über Anzahlen von Linearfaktoren in hexagonalen Bändern. Match 28: 181

John P (1992b) Two Codes for Hexagonal Systems. Match 28: 209

John P (1993) Linearfaktoren in defekten hexagonalen Systemen. Bayreuther Math Schriften 43: 35

Jovanovic AD (1988) Combinatorial Characterization of Hexagonal Systems. Discrete Appl Math 19: 259

Kao J (1987) A Molecular Orbital Based Molecular Mechanics Approach To Study Conjugated Hydrocarbons. J Am Chem Soc 109: 3817

Kao J, Allinger NL (1977) Conformational Analysis – 122 – Heats of Formation of Conjugated Hydrocarbons by the Force Field Method. J Am Chem Soc 99: 975

Kasteleyn PW (1961) The Statistics of Dimers on a Lattice – I – The Number of Dimer Arrangements on a Quadratic Lattice. Physica 27: 1209

Kasteleyn PW (1963) Dimer Statistics and Phase Transitions. J Math Phys 4: 287

Kataoka M, Nakajima T (1986) Geometrical Structures and Spectra of Corannulene and Icosahedral C_{60}. Tetrahedron 23: 6437

Katritzky AR, Marson CM (1983) Synthesis of a Dodecahydro–18,21–dioxoniakekulene. J Am Chem Soc 105: 3279

Kirby EC (1990) Why can so few Benzenoids be Completely Drawn with Clar's Resonant Sextets? An Analysis using 'Branching Graphs' and a 'Coiled–hexagon Code'. J Chem Soc Faraday Trans 86: 447

Kivelson S, Chapman OL (1983) Polyacene and a New Class of Quasi–One–Dimensional Conductors. Phys Rev B 28: 7236

Klarner DA (1965) Some Results Concerning Polyominoes. Fibonacci Quarterly 3: 9

Klein DJ, Cravey MJ, Hite GE (1991) Fractal Benzenoids. Polycyclic Aromatic Compounds 2: 163

Klein DJ, Schmalz TG, Hite GE, Seitz WA (1986) Resonance in C_{60}, Buckminster–fullerene. J Am Chem Soc 108: 1301

Klein DJ, Seitz WA, Schmalz TG (1986) Icosahedral Symmetry Carbon Cage Molecules. Nature 323: 703

Klein DJ, Seitz WA, Schmalz TG (1990) Conjugated–Circuit Computations for Conjugated Hydrocarbons [in] Computational Chemical Graph Theory (Rouvray DH, Edit). Nova, New York

Klein DJ, Trinajstić N (1990) Pascal Recurrence Algorithm for Kekulé–Structure Counts of Benzenoid and Coronoid Hydrocarbons. J Mol Struct (Theochem) 206: 135

Klein DJ, Živković TP, Balaban AT (1993) The Fractal Family of Coro[n]enes. [in] Topological Aspects of Benzenoid Hydrocarbons and Related Structures (Cyvin SJ, Edit). Max–Planck–Institut für Strahlenchemie, Mülheim a d Ruhr; Match 29: 107

Knop JV, Müller WR, Jeričević Ž, Trinajstić N (1981) Computer Enumeration and Generation of Trees and Rooted Trees. J Chem Inf Comput Sci 21: 91

Knop JV, Müller WR, Szymanski K, Nikolić S, Trinajstić N (1990) Computer–Oriented Molecular Codes [in] Computational Chemical Graph Theory (Rouvray DH, Edit). Nova, New York

Knop JV, Müller WR, Szymanski K, Trinajstić N (1985) Computer Generation of Certain Classes of Molecules. SKTH/Kemija u industriji (Association of Chemists and Technologists of Croatia), Zagreb

Knop JV, Müller WR, Szymanski K, Trinajstić N (1986) A Note on the Number of Circulenes. Match 20: 197

Knop JV, Müller WR, Szymanski K, Trinajstić N (1990a) Use of Small Computers for Large Calculations – Enumeration of Polyhex Hydrocarbons. J Chem Inf Comput Sci 30: 159

Knop JV, Müller WR, Szymanski K, Trinajstić N (1990b) A Note on the Classification and Enumeration of Coronoid Hydrocarbons. J Mol Struct (Theochem) 205: 361

Knop JV, Müller WR, Szymanski K, Trinajstić N (1990c) Enumeration of Planar Polyhex Hydrocarbons. Reports on Molecular Theory 1: 95

Knop JV, Szymanski K, Jeričević Ž, Trinajstić N (1983) Computer Enumeration and Generation of Benzenoid Hydrocarbons and Identification of Bay Regions. J Comput Chem 4: 23

Knop JV, Szymanski K, Jeričević Ž, Trinajstić N (1984) On the Total Number of Polyhexes. Match 16: 119

Kornilov MYu, Zamkovii VI (1981) Обчислення на ЕОМ кількості структуних ізомерів алканів до $C_{100}H_{201}$ і алканолів до $C_{100}H_{201}OH$. Вісник Київ ун-та, Хімія 22С: 38

Kostochka AV (1985) Критерий существования совершенных паросочетаний в шестиугольных системах. Proc 30 Intern Wiss Koll TH Ilmenau, Vortragsreihe F: 49

Krätschmer W, Lamb LD, Fostiropoulos K, Huffman DR (1990) Solid C_{60} – A New Form of Carbon. Nature 347: 1990

Krieger C, Diederich F, Schweitzer D, Staab HA (1979) Molecular Structure and Spectroscopic Properties of Kekulene. Angew Chem Int Ed Engl 18: 699

Kroto HW, Heath JR, O'Brien SC, Curl RF, Smalley RE (1985) C_{60} – Buckminster-fullerene. Nature 318: 162

Lahti PM (1988) Localization of Aromaticity in Fused–Ring Cycloarene Systems – Prediction by an Effective Molecular Mechanics Model. J Org Chem 53: 4590

Lederberg J, Sutherland GL, Buchanan BG, Feigenbaum EA, Robertson AV, Duffield AM, Djerassi C (1969) Application of Artificial Intelligence for Chemical Inference – I – The Number of Possible Organic Compounds – Acyclic Structures Containing C, H, O, and N. J Am Chem Soc 91: 2973

Losanitsch SM (1897a) Die Isomerie–Arten bei den Homologen der Paraffin–Reihe. Ber Dtsch Chem Ges 30: 1917

Losanitsch SM (1897b) Bemerkung zu der Hermannschen Mittheilung – Die Anzahl der isomeren Paraffine. Ber Dtsch Chem Ges 30: 3059

Lunnon WF (1972) Counting Hexagonal and Triangular Polyominoes [in] Graph Theory and Computing (Read RC, Edit). Academic Press, New York : 87

Mašulović D, Tošić R, Cyvin BN, Cyvin SJ (1993) Supplement to the Düsseldorf–Zagreb Numbers for Polyhexes [in] Topological Aspects of Benzenoid Hydrocarbons and Related Structures (Cyvin SJ, Edit). Max–Planck–Institut für Strahlenchemie, Mülheim a d Ruhr; Match 29: 165

Mitchell RH, Carruthers RJ, Mazuch L, Dingle TW (1982) Toward the Understanding of Benzannelated Annulenes – Synthesis and Properties of [a]–Ring Monobenzannelated Dihydropyrenes. J Am Chem Soc 104: 2544

Mitchell RH, Williams RV, Dingle TW (1982) Toward the Understanding of Benz–annelated Annulenes – Synthesis and Properties of [a,h]– and [a,i]–Ring Dibenzannelated Dihydropyrenes. J Am Chem Soc 104: 2560

Mitchell RH, Williams RV, Mahadevan R, Lai YH, Dingle TW (1982) Toward the Understanding of Benzannelated Annulenes – A Simple Correlation of the Diatropicity of Several Benzannelated Dihydropyrenes in Terms of Bond Order Deviations with Predictions for Other Benzannulenes. J Am Chem Soc 104: 2571

Mitchell RH, Yan JSH, Dingle TW (1982) Toward the Understanding of Benzannelated Annulenes — Synthesis and Properties of an [e]–Ring Monobenzannelated Dihydropyrene. J Am Chem Soc 104: 2551

Morikawa T, Balaban AT (1992) Topological Formulas and Upper/Lower Bounds in Chemical Polygonal Systems, Particularly in Benzenoid Polyhexes. Match 28: 235

Müller WR, Szymanski K, Knop JV, Nikolić S, Trinajstić N (1989) On Counting Polyhex Hydrocarbons. Croat Chem Acta 62: 481

Müller WR, Szymanski K, Knop JV, Nikolić S, Trinajstić N (1990) On the Enumeration and Generation of Polyhex Hydrocarbons. J Comput Chem 11: 223

Nikolić S, Trinajstić N (1990) Compact Molecular Codes for Annulenes, Aza–annulenes, Annulenoannulenes, Aza–annulenoannulenes, Cyclazines and Aza–cyclazines. Croat Chem Acta 63: 155

Nikolić S, Trinajstić N, Knop JV, Müller WR, Szymanski K (1990) On the Concept of the Weighted Spanning Tree of Dualist. J Math Chem 4: 357

Nikolić S, Trinajstić N, Knop JV, Müller WR, Szymanski K (1991) On the Classification and Enumeration of Planar Polyhex Hydrocarbons. J Mol Struct (Theochem) 231: 219

Ohkami N (1990) Graph–Theoretical Analysis of the Sextet Polynomial — Proof of the Correspondence Between the Sextet Patterns and Kekulé Patterns. J Math Chem 5: 23

Ohkami N, Motoyama A, Yamaguchi T, Hosoya H, Gutman I (1981) Graph–Theoretical Analysis of the Clar's Aromatic Sextet — Mathematical Properties of the Set of the Kekulé Patterns and the Sextet Polynomial for Polycyclic Aromatic Hydrocarbons. Tetrahedron 37: 1113

Osawa E (1970) 超芳香族 Kagaku 25: 854

Papulov YuG, Rosenfeld VR, Kemenova TG (1990) Молекулярные графы. Тверской государственный университет, Tver

Perry D (1932) The Number of Structural Isomers of Certain Homologs of Methane and Methanol. J Am Chem Soc 54: 2918

Peter R, Jenny W (1965) Ein zyclischer Kohlenwasserstoff der Benzo(c)phenanthrenreihe. Chimia 19: 45

Peter R, Jenny W (1966) Höhere, kondensierte Ringsysteme — 1. Mitteilung — Untersuchungen in der [10]–Coronaphenreihe — Synthese von Di–[benzo(c)–phenanthren–3,10–dimethylen]. Helv Chim Acta 49: 2123

Plavšić D, Nikolić S, Trinajstić N (1992) The Conjugated–Circuit Model — Application to Non–Alternant Hydrocarbons and a Comparison with Some Other Theoretical Models of Aromaticity. J Mol Struct (Theochem) 277: 213

Polansky OE, Rouvray DH (1976) Graph–Theoretical Treatment of Aromatic Hydro–carbons I — The Formal Graph–Theoretical Description. Match 2: 63

Polansky OE, Rouvray DH (1977) Graph–Theoretical Treatment of Aromatic Hydro–carbons III — Corona–Condensed Systems. Match 3: 97

Pólya G, Read RC (1987) Combinatorial Enumeration of Groups, Graphs, and Chemical Compounds. Springer–Verlag, New York

Ramaraj R, Balasubramanian K (1985) Computer Generation of Matching Polynomials of Chemical Graphs and Lattices. J Comput Chem 6: 122

Randić M (1976a) Conjugated Circuits and Resonance Energies of Benzenoid Hydrocarbons. Chem Phys Letters 38: 68

Randić M (1976b) Enumeration of the Kekulé Structures in Conjugated Hydrocarbons. J Chem Soc Faraday Trans 2 72: 232

Randić M (1977) Aromaticity and Conjugation. J Am Chem Soc 99: 444

Randić M (1986) A Statistical Approach to Resonance Energies of Large Molecules. Chem Phys Letters 128: 193

Randić M, Gimarc BM, Nikolić S, Trinajstić N (1989) On the Aromatic Stability of Helicenic Systems. Gazz Chim Ital 119: 1

Randić M, Henderson LL, Stout R, Trinajstić N (1988) Conjugation and Aromaticity of Macrocyclic Systems. Int J Quant Chem, Quant Chem Symposium 22: 127

Randić M, Nikolić S, Trinajstić N (1987) On the Aromatic Stability of a Conjugated C_{60} Cluster. Croat Chem Acta 60: 595

Randić M, Nikolić S, Trinajstić N (1988) Enumeration of Kekulé Structures for Helicenic Systems. Croat Chem Acta 61: 821

Randić M, Trinajstić N, Henderson LL, Stout RP (1993) A Note on the Enumeration of Kekulé Structures in a Class of Coronoids. J Mol Struct (Theochem) 285: 121

Ranshoff JEB, Staab HA (1985) En Route to Hexaaza–Kekulene. Tetrahedron Letters : 6179

Read RC (1978) On General Dissections of a Polygon. Aequationes Mathematicae 18: 370

Redelmeier DH (1981) Counting Polyominoes – Yet Another Attack. Discrete Math 36: 191

Rouvray DH (1974) Isomer Enumeration Methods. Chem Soc Reviews 3: 355

Rouvray DH (1991) The Origin of Chemical Graph Theory [in] Chemical Graph Theory – Introduction and Fundamentals (Bonchev D, Rouvray DH, Edit). Abacus Press/Gordon and Breach, New York; Mathematical Chemistry – Volume 1: 1

Sachs H (1984) Perfect Matchings in Hexagonal Systems. Combinatorica 4: 800

Schiff H (1875a) 4. October 1875. Ber Dtsch Chem Ges 8: 1350

Schiff H (1875b) Zur Statistik chemischer Verbindungen. Ber Dtsch Chem Ges 8: 1542

Schmalz TG, Seitz WA, Klein DJ, Hite GE (1986) C_{60} Carbon Cages. Chem Phys Letters 130: 203

Schmalz TG, Seitz WA, Klein DJ, Hite GE (1988) Elemental Carbon Cages. J Am Chem Soc 110: 1113

Schmidt–Radde RH, Vollhardt KPC (1992) Total Synthesis of Angular [4]Phenylene and [5]Phenylene. J Am Chem Soc 114: 9713

Schulman JM, Peck RC, Disch RL (1989) Ab Initio Heats of Formation of Medium–Sized Hydrocarbons – 11 – The Benzenoid Aromatics. J Am Chem Soc 111: 5675

Schweitzer D, Hausser KH, Vogler H, Diederich F, Staab HA (1982) Electronic Properties of Kekulene. Mol Phys 46: 1141

Scott LT, Hashemi MM, Bratcher MS (1992) Corannulene Bowl–to–Bowl Inversion Is Rapid at Room Temperature. J Am Chem Soc 114: 1920

Scott LT, Hashemi MM, Meyer DT, Warren HB (1991) Corannulene – A Convenient New Synthesis. J Am Chem Soc 113: 7082

Sheng RQ (1987) Identification of the Kekulé Structures of a Hexagonal System. Chem Phys Letters 142: 196

Sheng RQ (1989) A Necessary and Sufficient Condition for the Existence of Kekulé Structures in Benzenoid Systems. Match 24: 207

Sheng RQ (1990) Rapid Ways to Recognize Kekuléan Benzenoid Systems [in] Advances in the Theory of Benzenoid Hydrocarbons (Gutman I, Cyvin SJ, Edit). Springer–Verlag, Berlin; Topics in Current Chemistry 153: 211

Sheng RQ (1991) The Number of Kekulé Structures of Unbranched Catacondensed Benzenoid Systems and Primitive Coronoid Systems. Match 26: 247

Sheng RQ, Cyvin SJ, Gutman I (1989) Remarks on Kekuléan and Non–Kekuléan Benzenoid Hydrocarbons. J Mol Struct (Theochem) 187: 285

Sheng RQ, Cyvin SJ, Gutman I (1991) "Essentially Disconnected Benzenoids" are Essentially Disconnected. Match 26: 191

Smith FT (1961) Capacitive Energy and the Ionization of Aromatic Hydrocarbons. J Chem Phys 34: 793

Staab HA, Binnig F (1964) Synthese und Eigenschaften von Hexa–m–phenylen. Tetrahedron Letters : 319

Staab HA, Binnig F (1967) Zur Konjugation in makrocyclischen Bindungssystemen, VII – Synthese und Eigenschaften von Hexa–m–phenylen und Octa–m–phenylen. Chem Ber 100: 293

Staab HA, Bräunling H (1965) Zur Konjugation in makrocyclischen Bindungssystemen II – Synthese und Eigenschaften des 3,6';3',6";3",6–Triphenanthrylens. Tetrahedron Letters : 45

Staab HA, Bräunling H, Schneider K (1968) Zur Konjugation in makrocyclischen Bindungssystemen, X – Über 3.6';3'.6";3".6–Triphenanthrylen und verwandte Verbindungen. Chem Ber 101: 879

Staab HA, Diederich F (1983) Cycloarenes, a New Class of Aromatic Compounds, I – Synthesis of Kekulene. Chem Ber 116: 3487

Staab HA, Diederich F, Čaplar V (1983) Cycloarenes, a New Class of Aromatic Compounds, III – Studies towards the Synthesis of Cyclo[$d.e.d.e.e.d.e.d.e.e$]decakis-benzene. Liebigs Ann Chem : 2262

Staab HA, Diederich F, Krieger C, Schweitzer D (1983) Cycloarenes, a New Class of Aromatic Compounds, II – Molecular Structure and Spectroscopic Properties of Kekulene. Chem Ber 116: 3504

Staab HA, Sauer M (1984) Cycloarene, eine neue Klasse aromatischer Verbindungen, IV – Versuche zur Synthese des Cyclo[d.e.e.d.e.e.d.e.e]nonakisbenzens und des Cyclo–[d.e.d.e.d.e.d.e.d.e]decakisbenzens. Liebigs Ann Chem : 742

Stojmenović I, Tošić R, Doroslovački R (1986) Generating and Counting Hexagonal Systems [in] Proceedings of the Sixth Yugoslav Seminar on Graph Theory, Dubrovnik 1985 (Tošić R, Acketa D, Petrović V, Edit). Novi Sad: 189

Temperley HNV, Fisher ME (1961) Dimer Problem in Statistical Mechanics – An Exact Result. Phil Mag (Ser 8) 6: 1061

Tomescu I, Balaban AT (1989) Decomposition Theorems for Calculating the Number of Kekulé Structures in Coronoids Fused via Perinaphthyl Units. Match 24: 289

Tošić R, Mašulović D, Brunvoll J, Cyvin BN, Cyvin SJ (1993) Enumeration of Some Classes of Polyhex Hydrocarbons. J Mol Struct (Theochem) 282: 295

Trinajstić N (1990) On the Classification of Polyhex Hydrocarbons. J Math Chem 5: 171

Trinajstić N (1992a) On the Classification of Polyhexes. J Math Chem 9: 373

Trinajstić N (1992b) Chemical Graph Theory – Second Edition. CRC Press, Boca Raton

Trinajstić N, Jeričević Ž, Knop JV, Müller WR, Szymanski K (1983) Computer Generation of Isomeric Structures. Pure & Appl Chem 55: 379

Trinajstić N, Klein DJ, Randić M (1986) On Some Solved and Unsolved Problems of Chemical Graph Theory. Int J Quant Chem, Quant Chem Symposium 20: 699

Trinajstić N, Nikolić S, Knop JV, Müller WR, Szymanski K (1991) Computational Chemical Graph Theory – Characterization, Enumeration and Generation of Chemical Structures by Computer Methods. Ellis Horwood, Chichester/New York

Trinajstić N, Schmalz TG, Živković TP, Nikolić S, Hite GE, Klein DJ, Seitz WA (1991) [N]Phenylenes – A Theoretical Study. New J Chem 15: 27

Vögtle F (1983) Concluding Remarks [in] Cyclophanes II (Vögtle F, Edit). Springer–Verlag, Berlin; Topics in Current Chemistry 115: 157

Vögtle F, Staab HA (1968) Zur Konjugation in makrocyclischen Bindungssystemen, XII – Versuche zur Darstellung des Cyclo[d.e.d.e.d.e.d.e.d.e.d.e]dodecakisbenzens – Eine neue Synthese von 1.2;7.8–Dibenzo–anthracenen. Chem Ber 101: 2709

Vogler H (1985) Theoretical Study of Geometries and 1H–Chemical Shifts of Cycloarenes. J Mol Struct (Theochem) 122: 333

Vogler H (1986) Structures and 1H–Chemical Shifts of Conjugation Deficient Hydrocarbons. Int J Quant Chem 30: 97

Wilcox CF Jr (1968) Stability of Molecules Containing (4n)–Rings. Tetrahedron Letters : 795

Wilcox CF Jr, Gutman I, Trinajstić N (1975) Graph Theory and Molecular Orbitals – XI – Aromatic Substitution. Tetrahedron 31: 147

Wilcox CF Jr, Lahti PM, Rocca JR, Halpern MB, Meinwald J (1978) Strained Polycyclic Aromatic Molecules — I — Synthesis of 9—Bromodinaphth[1,2—a:2',1'—j]Anthracene. Tetrahedron Letters : 1893

Wynberg H, Nieuwpoort WC, Jonkman HT (1973) Flexible Aromatic Rings. Tetrahedron Letters : 4623

Zhang FJ, Chen RS (1987) 苯类碳氢化合物拓扑性质的某些研究 J Nature 10: 163

Zhang FJ, Chen RS (1989) A Theorem Concerning Perfect Matchings in Hexagonal Systems. Acta Math Appl Sinica 5: 1

Zhang FJ, Chen RS (1991) When Each Hexagon of a Hexagonal System Covers It. Discrete Appl Math 30: 63

Zhang FJ, Chen RS, Guo XF (1985) Perfect Matchings in Hexagonal Systems. Graphs and Combinatorics 1: 383

Zhang FJ, Cyvin SJ, Cyvin BN (1990). "Crowns", and Aromatic Sextets in Primitive Coronoid Hydrocarbons. Monatsh Chem 121: 421

Zhang FJ, Guo XF (1988) The Necessary and Sufficient Conditions for Benzenoid Systems with Small Number of Hexagons to have Kekulé Patterns. Match 23: 229

Zhang FJ, Guo XF (1991) Directed Tree Structure of the Set of Kekulé Patterns of Generalized Polyhex Graphs. Discrete Appl Math 32: 295

Zhang FJ, Guo XF, Chen RS (1990) The Existence of Kekulé Structures in a Benzenoid System [in] Advances in the Theory of Benzenoid Hydrocarbons (Gutman I, Cyvin SJ, Edit). Springer—Verlag, Berlin; Topics in Current Chemistry 153: 181

Zhang FJ, Guo XF, Cyvin SJ, Cyvin BN (1992) The Enumeration of Catafusenes Rooted at a Polygon Core. Chem Phys Letters 190: 104

Zhang FJ, Lin YX (1992) An Algorithm for Recognizing Kekulé Structures in Benzenoid and Coronoid Systems. J Mol Struct (Theochem) 257: 167

Zhang FJ, Lin YX (1993) A Note on the Enumeration of Equivalence Classes in Spectro—scopy. J Mol Struct (Theochem) 282: 291

Zhang FJ, Liu YT, Chen RS (1991) A Note on Properties of Normal Benzenoids and Coronoids. Match 26: 243

Zhang FJ, Zheng ML (1992) Generalized Hexagonal Systems with Each Hexagon Being Resonant. Discrete Appl Math 36: 67

SUBJECT INDEX

addition of hexagon 165
addition mode 235
addition unit 168
adjacency matrix 34
algebraic structure count 15
alkane 4
alternating cycle 235
annelation 165
annulene 14
annulenoannulene 14
annulenoid Kekulé structure 32
antiaromatic conjugated circuit 13
antikekulene 14
aromatic conjugated circuit 13
aromatic sextet 238
aromaticity 13
ASC (Algebraic Structure Count) 15
associated benzenoid 62
aufbau unit 168
available for addition 170

Balaban picture 114
basic single coronoid 165
bay 165
benzenoid 19
biphenylenoid 51
boundary vertex 56
branched 20
buckminsterfullerene 17
building–up 165

catacondensed 20
cardinality 89
characteristic polynomial 41
chemical formula 84
chemical graph 1
circular benzenoid 110
circular single coronoid 110
circumextremal benzenoid 134
circumextremal single coronoid 134
circumscribing 119
Clar structure 40
cluster 52
color excess 20
combinatorics 211
concealed non–Kekuléan 20
conjugated circuit 236
connectivity 97
constant–isomer benzenoid series 119
corannulene 15
core coronoid 125
corofusene 45
corohelicene 44
corona hole 19
corona–condensation 235

coronafusene 45
coronaphene 1
coronenic coronoid 228
coronoid 19
corrected structure count 15
cove 165
crude total 212
CSC (Corrected Structure Count) 15
cut segment 236
cyclacene 16
cycloarene 9
cyclo–decakisbenzene 1
cyclo–dodecakisbenzene 1
cyclo–nonakisbenzene 1

degenerate single coronoid 24
Dias parameter 56
Dias periodic table 85
dot diagram 89
dualist 43

e–cut 236
e–cut segment 236
edge 55
edge cut 236
effective unit 25
elementary aufbau unit 168
elementary cut (e–cut) 236
elementary cut segment 236
embeddable 43
enantiomorphic 47
essentially disconnected 20
even–carbon formula 143
excised internal structure 119
excising 119
extra long step 173
extremal benzenoid 64
extremal coronoid 64
extremal single coronoid 99
extreme benzenoid 139
extreme single coronoid 139
extreme–left benzenoid 139

Fibonacci number 31
figure–eight 48
fissure 165
five–contact addition 165
fixed bond 20
fjord 165
fluoranthenoid 51
fluorenoid 51
formula index 159
four–contact addition 165
fractal benzenoid 3
free edge 165

Editorial Policy

This series aims to report new developments in chemical research and teaching - quickly, informally and at a high level. The type of material considered for publication includes:

1. Preliminary drafts of original papers and monographs

2. Lectures on a new field, or presenting a new angle on a classical field

3. Seminar work-outs

4. Reports of meetings, provided they are

    a) of exceptional interest and

    b) devoted to a single topic.

Texts which are out of print but still in demand may also be considered if they fall within these categories.

The timeliness of a manuscript is more important than its form, which may be unfinished or tentative. Thus, in some instances, proofs may be merely outlined and results presented which have been or will later be published elsewhere. If possible, a subject index should be included. Publication of Lecture Notes is intended as a service to the international chemical community, in that a commercial publisher, Springer-Verlag, can offer a wider distribution to documents which would otherwise have a restricted readership. Once published and copyrighted, they can be documented in the scientific literature.

Manuscripts

Manuscripts should comprise not less than 100 and preferably not more than 500 pages. They are reproduced by a photographic process and therefore must be typed with extreme care. Symbols not on the typewriter should be inserted by hand in indelible black ink. Corrections to the typescript should be made by pasting the amended text over the old one, or by obliterating errors with white correcting fluid. Authors receive 50 free copies and are free to use the material in other publications. The typescript is reduced slightly in size during reproduction; best results will not be obtained unless the text on any one page is kept within the overall limit of 18 x 26.5 cm (7 x 10$^1/_2$ inches). The publishers will be pleased to supply on request special stationary with the typing area outlined.

Manuscripts should be sent to one of the editors or directly to Springer-Verlag, Heidelberg.

Lecture Notes in Chemistry

For information about Vols. 1–22
please contact your bookseller or Springer-Verlag